T0273825

Quantitative Genetics

Quantitative genetics is the study of continuously varying traits, which make up the majority of biological attributes of evolutionary and commercial interest. This book provides a much-needed up-to-date, in-depth, yet accessible text for this field. In lucid language, the author guides readers through the main concepts of population and quantitative genetics and their applications. Written to be approachable even to those without a strong mathematical background, applied examples, a glossary of key terms, and problems and solutions support students in grasping important theoretical developments and their relevance to real-world biology. This is an engaging, must-have textbook for advanced undergraduate and postgraduate students. Given its applied focus, it also equips researchers in genetics, genomics, evolutionary biology, animal and plant breeding and conservation genetics with the understanding and tools for genetic improvement, comprehension of the genetic basis of human diseases and conservation of biological resources.

Armando Caballero is Professor of Genetics at the University of Vigo, Spain, with research interests in quantitative and population genetics, conservation genetics and evolution. He has served as Associate Editor for the journals *Evolution*, *American Naturalist*, *Journal of Evolutionary Biology*, *Genetics Selection Evolution* and *Heredity*.

Quantitative Genetics

ARMANDO CABALLERO

University of Vigo, Galicia, Spain

CAMBRIDGE
UNIVERSITY PRESS

University Printing House, Cambridge CB2 8BS, United Kingdom

One Liberty Plaza, 20th Floor, New York, NY 10006, USA

477 Williamstown Road, Port Melbourne, VIC 3207, Australia

314–321, 3rd Floor, Plot 3, Splendor Forum, Jasola District Centre, New Delhi – 110025, India

79 Anson Road, #06–04/06, Singapore 079906

Cambridge University Press is part of the University of Cambridge.

It furthers the University's mission by disseminating knowledge in the pursuit of
education, learning, and research at the highest international levels of excellence.

www.cambridge.org
Information on this title: www.cambridge.org/9781108481410
DOI: 10.1017/9781108630542

Originally published in Spanish as *Genética Cuantitativa* (2017) by Editorial Síntesis, Madrid, Spain.

© Armando Caballero Rúa and Editorial Síntesis 2017

First published in English by Cambridge University Press in 2020, translated and updated from the original
Spanish edition by Armando Caballero

© Armando Caballero Rúa 2020

First published 2020

Printed in the United Kingdom by TJ International Ltd. Padstow Cornwall

A catalogue record for this publication is available from the British Library.

Library of Congress Cataloging-in-Publication Data
Names: Caballero, Armando, 1960– author.
Title: Quantitative genetics / Armando Caballero, University of Vigo, Spain.
Other titles: Genética cuantitativa. English
Description: Cambridge, United Kingdom ; New York : Cambridge University Press, 2020. | Originally
published in Spanish as Genética Cuantitativa (2017) by Editorial Síntesis, Madrid, Spain. | Includes
bibliographical references and index.
Identifiers: LCCN 2019039768 (print) | LCCN 2019039769 (ebook) | ISBN 9781108630542 (epub) | ISBN
9781108481410 (hardback) | ISBN 9781108722353 (paperback)
Subjects: LCSH: Quantitative genetics.
Classification: LCC QH452.7 (ebook) | LCC QH452.7 .C3313 2020 (print) | DDC 572.8/6–dc23
LC record available at https://lccn.loc.gov/2019039768

ISBN 978-1-108-48141-0 Hardback
ISBN 978-1-108-72235-3 Paperback

To Bill Hill

Contents

Preface

I want to thank Cambridge University for allowing me to produce the English version of my Spanish book and for the continuous help and support. For this version, I have updated some content and references, correcting a few minor errors present in the Spanish version. I want to thank Professor Carlos López-Fanjul for having read the complete draft, making additional comments to improve the book, and to Professor Peter Keightley for reading Chapters 7 and 8 and making useful suggestions and corrections. I also want to thank Raquel Sampedro for help in managing the text and Miguel Toro, Beatriz Villanueva, Antonio Carvajal and Humberto Quesada for helpful discussions on specific parts of the book. Any errors in the content are, however, my sole responsibility, and I would appreciate any comments in this regard. Finally, I want to express my gratitude to my family for their understanding and support during the preparation of the manuscript.

Preface to the Spanish Version

Quantitative genetics is the branch of genetics dedicated to the study of quantitative traits, which are all those biological attributes with continuous variation. The vast majority of traits of evolutionary interest and commercial value in animals and domestic plants are quantitative, so that quantitative genetics contributes to the understanding of adaptation and evolution of living beings and provides the necessary tools for the genetic improvement and conservation of biological resources. A large number of human diseases are also of a quantitative nature, so that the study of quantitative genetics is fundamental for their understanding.

This book aims to build the foundations on which the subject is based, trying to maintain a balance between explanations of the most basic concepts and the most modern methods of analysis. It is therefore intended to be not only an introduction for students, particularly postgraduates, but also a reference book for researchers. Understanding this subject requires an elementary knowledge of mathematics and statistics, and, whenever possible, I have attempted to explain, or at least indicate, the origin of the equations that reflect the different concepts. Either the original sources of the main contributions or useful reviews or books have been cited so that the reader can get more insight into each subject. Each of the 11 chapters includes some problems and questions, solved at the end of the book, and there is a glossary of the most important terms.

I want to express my gratitude to several colleagues who have contributed with data or comments or by amending errors. First of all, and especially, I have to thank Professor Carlos López-Fanjul for having read the complete draft and for making numerous corrections, comments and suggestions that have substantially improved the content and its presentation. I owe him, as well as Professor Bill Hill of the University of Edinburgh, a great deal of my knowledge in this field, as well as teaching me how to carry out high-quality research. I also want to thank Professors Aurora García-Dorado and Miguel Toro for their help with various chapters. Other colleagues have contributed comments and discussions for specific topics or have contributed data, photos or figures, among whom are my colleagues from the University of Vigo, Humberto Quesada, Paloma Morán, Emilio Rolán-Alvarez, Juan Galindo and Daniel Estévez-Barcia, and from other institutions, Beatriz Villanueva, Jesús Fernández, Almudena Fernández and Andrés Legarra. I also want to express my gratitude to the coordinator of the series of which this book is a part, César (Mario) Benito, for reading the complete draft, making corrections to the text and suggesting improvements. Any errors that remain in the book are, however, my responsibility. Finally, I have to thank my wife, Esther, and my children, Alberto and Laura, for their understanding and support during the long work sessions.

1 Continuous Variation

Concepts to Study

- Quantitative traits
- Meristic and threshold traits
- Genotypic and phenotypic values
- Additivity, dominance and epistasis
- Major gene
- Pleiotropy
- Fitness and its main components
- Infinitesimal model

Objectives for Learning

- To understand the definition of quantitative trait, the reason for the different names by which they are known and their types depending on whether they are expressed with continuous or discrete observable variation
- To distinguish the concepts of phenotypic and genotypic value and the types of intra-and interlocus gene action
- To understand the concept of fitness and its main components
- To know the definitions of major gene and pleiotropy
- To understand the infinitesimal model and the partition of phenotypic variation in genetic and environmental components

1.1 Quantitative Traits

Some heritable characteristics are qualitative, with an expression clearly identifiable in discrete classes. Such is the case of attributes like some differences in colour, shape or structure, by which individuals of a population or species can be classified. The analysis of this type of simple character was what allowed Mendel to describe the bases of inheritance and many other geneticists, later, to understand the relation between this and the chromosomal behaviour during reproduction, as well as the interactions between genes. However, most of the traits that we find in nature present a continuous variation. Even some of the seemingly discrete attributes, such as colour, may show gradual variation if analysed in detail. These types of characters with gradual variation are called quantitative traits and, sometimes, metric or continuous traits. Among them it is possible to find many with purely continuous variation, with a priori infinite possibilities of expression, whose analysis is based on measurement, such as body dimensions or weight, but also those with discrete variation whose characterization is carried out by counting, the so-called meristic traits, such as the morphological structures that vary in number, or any discrete characteristic that implies numerical variation in its expression, such as the number of offspring of the same birth or the number of matings that an individual carries out throughout its life. Even some characters whose expression is displayed in only a few possible categories, the so-called threshold traits, such as susceptibility or resistance to certain substances, death or survival or the circumstance of suffering, or not, a disease, often imply an underlying continuous variation.

In fact, there are an indefinitely large number of heritable biological characteristics that can present a continuous variation, observable as such or underlying, and this becomes evident when verifying that practically all variable characters that have been studied in laboratory organisms or domestic species have a certain hereditary component, which can be exploited by artificial selection. The potential of genetic change is evident in the enormous morphological differences generated, for example, between dog breeds, where the weight of the largest breeds is of the order of 200 times greater than that of the smallest ones, or the tremendous improvement obtained in the productive traits of domestic animals and plants, for example, the duplication of milk production in cattle in the last 50 years, it being possible to attribute at least half of this increase to genetic changes (Hill, 2014).

There are two fundamental premises that characterize this variation. First is the genetic control by a large enough number (say greater than 5) of generally unknown genes – the reason why the quantitative traits are often called polygenic or multifactorial traits. Second is that its expression is influenced, to a greater or lesser extent, by environmental factors intrinsic or extrinsic to the individual. As we will see in the different chapters of this book, these are precisely the two complications that characterize the analysis of continuous variation: (1) the ignorance of all or a large part of its genetic bases and (2) the ignorance of the relative influence of inheritance and environment in its phenotypic expression.

The importance of quantitative traits is fundamental. First, from an evolutionary point of view, all the attributes of a reproductive nature that are the direct object of natural selection, such as viability, fertility or mating success, are quantitative. Second, in the practical aspects of human consumption, the large majority of productive characteristics of plants and domestic animals, such as milk production, egg laying, quality and quantity of meat, plant biomass and a long endless list, are also quantitative. A large part of the development of quantitative genetics has occurred thanks to the continuous interest to improve animal and plant production through artificial selection and crossing methods. Third, with respect to human longevity and well-being, many of the most common diseases in our species, such as cancer, psychiatric disorders, autoimmune disorders and excess cholesterol or blood pressure, among many others, are also quantitative traits, which in the medical jargon are also called complex traits. A large part of the current genetic research focuses precisely on finding out the genetic basis of the aforementioned characters and on delimiting the relative importance of the environment in their expression. This has allowed us to find thousands of genes involved in hundreds of traits, both in wild species and those of economic interest, as well as in our own (MacArthur et al., 2017).

Quantitative genetics focuses mainly on the analysis of continuous variation and its applications to the study of evolution, animal and plant breeding and medicine. However, its imbrication with population genetics is total, since the analysis of quantitative traits is always carried out in the context of a population. This latter can be defined, in this context, as a set of individuals that constitute a reproductive unit, that is, those that are connected by a spatial and temporal relationship and share the same gene pool, the genes of which they are carriers. Population genetics includes the study of the heritable variation in general, qualitative or quantitative, and the forces of change of the gene frequencies that act on it. Such study is essential to understand biological evolution because this is, fundamentally, the result of the spatial-temporal change of the genetic composition of the populations. Since, as we have already indicated, most traits of evolutionary importance are quantitative, a fundamental objective of population genetics and, therefore, of quantitative genetics, is the understanding of the evolutionary process by means of the study of the genetic (mutation, chromosome segregation, recombination, etc.), ecological and demographic (population census size, migration, geographic distribution of populations, etc.) and adaptive processes (natural selection) that act on populations. The basic questions refer to which are the forces that maintain the variability in the populations, what role they play with respect to each other and how the integration of the phenotypes in the environments is accomplished, that is, how does the adaptation of the organisms occur to the environment in which they live. The application during millennia of artificial selection on a great number of quantitative traits, and the analysis and development of mathematical methods to reveal the consequences of this type of selection, has allowed us to understand, in turn, a good part of the way in which natural selection operates. The instruments used in quantitative genetics encompass the statistical characterization of populations for different traits, the performance of controlled experiments and the formulation of theoretical models. The latter are the primary tool, allowing a simplified description of the observed phenomena, which facilitates the prediction of natural processes and the response to artificial selection of domestic plants and animals.

1.2 Basic Concepts and Definitions

The genome of an individual is constituted by a great variety of elements, with the traditional concept of the gene being a functional unit located in a unique position of the genome. For practical purposes we will use the term *locus* to refer to a genomic element located in a fixed position of the genome, which may have different variants (alleles) which may or may not have an impact on the expression of the trait in the individual. Although the term *gene* refers to a functional unit, in this book we will use locus and gene interchangeably in most cases. For a particular locus, the genotype of a diploid individual can be homozygous, if it carries two copies of the same allele, or heterozygous, if it carries different alleles. As we will see in the next chapter, the genetic description of the populations can be done in terms of the allele or gene frequencies when the genotypes are distinguishable. For most quantitative characteristics, however, this is not possible, and the analysis will be limited to the calculation of general trait parameters in the population, such as means and variances. Sometimes there are loci with such a large effect on a trait that it allows us to distinguish their genotypic classes. These loci are called major genes, unlike the rest, which are called minor genes.

For a quantitative trait, the effects of the alleles of the different genes of which an individual is a carrier can, in a simplified form, be added to constitute the genotypic value of the individual, in which case it is said that there is additive gene action or additivity, but the relationships between alleles intra- or interlocus can be more complex. The partial or total prevalence of the effect of one allele on another at a given locus constitutes the concept of dominance, analogous to that applied for qualitative characteristics (the dominant allele is the one that prevails over the recessive allele). Likewise, the lack of additivity between the effects of different loci is called, generically, epistasis or epistatic interaction. The expression of the trait for the individual, its phenotypic value, is the result of the combination of its genotypic value and the effect of the environmental factors that surround it, as well as a possible interaction between both that we will discuss in later chapters. Finally, when a given locus has an effect on more than one quantitative trait, it is said that there is pleiotropy. All these concepts will be explained in greater depth in later chapters.

Any trait with a continuous genetic basis is an object of study by quantitative genetics. In the evolutionary context, however, the most relevant characters are those related to the reproduction of individuals. Darwin's (1859) Theory of Natural Selection is fundamentally based on the competition of individuals for resources, which determines their survival and/or differential reproduction, which constitutes their fitness. This trait can be defined as the contribution of the individual with offspring to the next or future generations, and it is the subject of direct natural selection. As we will see in later chapters, the indirect impact of natural selection on any other quantitative trait depends on the relationship between this latter trait and fitness. For natural selection to act, it is enough that there are inherited differences in fitness between the different individuals. Given the abstraction of its definition, fitness is difficult to evaluate, and in practice, it is specified in more empirically accessible traits, grouped under the name of the main components of fitness that are, among others of lesser importance, viability or survival, fecundity and mating success. A substantial part of quantitative genetics is dedicated to the study of the genetic variation for these traits and their implication in the context of evolution, animal and plant breeding and conservation.

1.3 Historical Perspective

1.3.1 Beginnings of Quantitative Genetics: Heritable Variation and Evolution

The beginnings of quantitative genetics, like those of genetics in general, are linked to the development of the theory of evolution and the elucidation of the bases of inheritance. The evolutionary theory consists essentially of two complementary parts. One of them proposes the mechanism of natural selection whose action can result in adaptive change. The Theory of Natural Selection by Charles Darwin (1859) offers an adaptation mechanism that represents the response of the individual to the constant challenge posed by the variation of the environment, in which individuals are passive subjects selected by it. The other part of the evolutionary theory specifies the hereditary principle, whose purpose is to guarantee the relative permanence of the acquired change. In this respect, Darwin held to the opinions of his time, accepting, as was general then, the mixed inheritance of the characters of both parents, which explained the similarity between the parents and their progeny. However, this type of inheritance implied a continuous loss of variation with the passing of generations, which provoked important criticisms. Gregor Mendel's experiments, based on the study of qualitative pea characters, and his idea of particle inheritance, published in 1865, although almost totally unknown until 1900, could explain the resemblance between parents and offspring without implying a constant loss of genetic variation. At the end of the nineteenth century, William Bateson distinguished two types of variability: continuous and discontinuous. The first one referred to the small differences between the organisms that constituted, for Darwin, the raw material on which natural selection would act. The second was, however, the one that Bateson considered essential to explain the observable variability. This notion was later developed by Hugo de Vries, who introduced at the beginning of the twentieth century the term *mutation* to designate the hereditary changes he considered as a norm: those with large and discontinuous effects. The legacy of Bateson and de Vries was embraced by other researchers, and the new science, baptized by Bateson as genetics, was built, in a sense, as the strongest opponent of Darwin's theory, since it presented mutation as the engine of evolutionary change.

During the period between the publication of Mendel's experiments and his re-discovery, Francis Galton, Darwin's cousin, also tried to elucidate the principles of hereditary transmission. However, Galton's (1889) approach was different from Mendel's. On the one hand, he was interested in the transmission of those characters that presented a continuous variation instead of focusing on those that showed clearly differentiable alternative forms. On the other, his study focused on the measurement of the average similarity between individuals with a certain degree of kinship in order to achieve a predictive model. Galton pointed out the fact that by representing the average height of the offspring against the average height of the parents, a linear relationship was obtained (Figure 1.1). However, the slope of the line indicated that children's height deviated less from the average population than that of their parents, that is, parents who were shorter than average had children with a stature somewhat greater than their own, and those who had a height higher than the average had shorter children. Galton called this circumstance 'regression towards mediocrity' and interpreted that this would prevent any selective progress, concluding that evolution should be based on variants of great effect rather than on the result of the action of selection on continuous variation. Karl Pearson would later prove that the regression to mediocrity deduced by Galton did not necessarily imply a problem for evolution. The linear regression technique, widely used in statistical analysis in all fields, comes from these Galton studies and was formally developed by Pearson along with the correlation technique. Likewise, other statistical

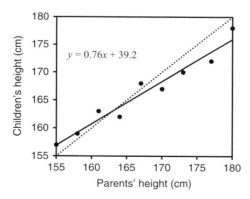

$y = 0.76x + 39.2$

Figure 1.1 Type of representation similar to those made by Galton, comparing the average height of children with that of their parents. The regression line presents a slope of 0.76, lower than that which would be expected with a perfect resemblance relationship between parents' and children's heights (dotted line).

techniques, such as the analysis of variance and various hypothesis testing methods, introduced by Ronald A. Fisher (1918), also have their origin in the analysis of quantitative traits. Finally, the path analysis of Sewall Wright (1921), applied to the study of inbreeding, has had some applications in the social and ecological sciences. Therefore, it can be said that quantitative genetics and its applications, such as animal genetic improvement (Gianola and Rosa, 2015), have contributed in an essential way to the development of many of the statistical techniques of universal use.

Biometry began with Galton and also the conflict between biometric and Mendelian conceptions of inheritance. For the Mendelians, with Bateson in the lead, the object of the science of inheritance was to develop a model of the process of transmission from parents to offspring of those factors that determined observable characteristics of individuals. For the biometricians, led by Pearson and Weldon, the objective, on the contrary, was to measure the phenotypic resemblance between the individuals of a population, since they doubted that the Mendelian laws could be applied to the traits with continuous variation. These two visions of inheritance also reflected a different conception of evolution. For the Mendelians or mutationists, this would be produced by rapid changes of great magnitude, while for the biometricians, it would occur continuously and gradually, more in accordance with the Darwinian model. The Mendelians argued that the quantitative characteristics reflected, basically, environmental differences, creating the debate of inheritance versus environment, or nature–nurture. A methodological interpretation that led on this and other occasions to error is to think that the observable is the general norm. Given that the effects of the major genes are the only ones observed empirically, it is easy to assume that only these exist and to ignore the contribution of others of smaller effect. Something very similar occurred with the denial until the 1960s of the additive gene action. When studying major genes, generally recessive, it was assumed that this was the general form of gene action. That is, again, the experimental need was taken as a general rule. Udny Yule (1902) was the first to seek the connection between both visions of inheritance, studying the application of Mendelian laws to panmictic populations, although the general formulation was proposed in 1908 independently by Godfrey Hardy in England and Wilhelm Weinberg in Germany, arriving at the Hardy–Weinberg principle that we will study in the next chapter.

Wilhelm Johannsen (1903) managed to take a big step towards the reconciliation of the two positions, showing the intervention of the medium in the expression of quantitative traits in his theory of pure lines. With his experiments he showed that the continuous variation observed for the trait weight in beans was produced by the combined influence of the genes and the environment. The bean is, like the pea used by Mendel, an autogamous legume species. Starting from 19 seeds that differed in weight, Johannsen established 19 lines by continuous self-fertilization that also differed in the average weight of the seeds, between 34 and 64 cg. He also observed a certain variation in weight between the individuals of each of the lines around their average. He showed first that any individual of a line with heavy or light beans gave rise to descendants which maintained the average parental weight. Johannsen also found that if the same line was selected for several generations to increase or decrease seed weight, no response was obtained, which indicated that there was no genetic variation within that line. His conclusion was that, since it is a selfer and each line was produced by self-fertilization from a single initial seed, each line would be formed by individuals with the same homozygous genetic constitution (pure lines) and that, therefore, the great variation between the average weights of the lines was due to the different genetic constitution of each of them, while the smaller variation in weight observed among the individuals of a given line came from environmental sources. Johannsen coined the terms genotype, to denote the genetically identical individuals of each line, and phenotype, for the observable value of the trait in each individual that would be the result of genetic and environmental effects.

Another important advance in the reconciliation between Mendelian and biometric hypotheses came from the hand of George Shull in 1908, studying characters of corn, and, above all, Herman Nilsson-Ehle, working with cereals, who attributed the hereditary determination of a trait to the segregation of several genes of similar and cumulative effects, giving support to the multifactorial hypothesis of quantitative traits or theory of the polymeric factors. Through crosses of varieties of wheat with flowers of different colours, Nilsson-Ehle (1909) found that several genes contributed to the variation of colour tones of the flower, from white to intense red, and that the individual effects of the genes were small and summable. Subsequent studies by other researchers (mainly R. A. Emerson and E. M. East with corn) confirmed the multifactorial hypothesis of quantitative traits.

1.3.2 The Development of the Central Body of Quantitative and Population Genetics

Both quantitative genetics and theoretical population genetics were developed gradually from the early twentieth century and reached maturity in the early 1930s with the publication of three essential works: *The Genetic Theory of Natural Selection*, by Ronald A. Fisher (1930), *Evolution in Mendelian Populations*, by Sewall Wright (1931), and *The Causes of Evolution*, by John B. S. Haldane (1932). Fisher mathematically demonstrated that natural selection, by acting on the genetic variability of populations, could perfectly explain the evolutionary change. Fisher's legacy to the theoretical body of quantitative genetics is essential, in particular the proposal of the Fundamental Theorem of Natural Selection, which sets the theoretical basis of the consequences of the action of natural selection on fitness, and the development of the so-called infinitesimal model, which proposes an interpretation of the nature of continuous variation in discrete Mendelian terms. Using the techniques of analysis of variance, also devised by him, the phenotypic variance of a certain character can be broken down into a series of components attributable to different genotypic and environmental causes. Thus, the genotypic variance can be ascribed to an additive component, due to the average effect of genes, and others due to the effects of dominance and epistasis, on which we will get insight in later chapters. The additive genetic component can be estimated with relative ease and is of great importance because it is the determinant of the immediate response to selection and family resemblance. The proportion of the phenotypic variance explained by this additive component constitutes the concept of heritability.

Among the different authors who carried out notable extensions to Fisher's theoretical central body is Jay Lush, who is considered the father of animal breeding. His book *Animal Breeding Plans* (Lush, 1945) gathers numerous applications of quantitative genetics to the genetic improvement of animals, among which stands out the well-known 'breeder's equation', by which selection response is predicted in terms of the heritability of the trait and the selection differential applied, which is the selection pressure exerted on the population.

The development of selection indices, through which individual and family information of a trait can be combined to obtain a greater response, is also due to him (Lush, 1947). Michael Lerner also made important contributions to quantitative theoretical and practical genetics collected in his book *Population Genetics and Animal Improvement* (Lerner, 1950), and more evolutionary aspects, such as the development of the theory of genetic homeostasis (Lerner, 1954), which explains the greater plasticity of heterozygotes in the face of environmental variation. In the decades of the 1960s and 1970s the contributions of Alan Robertson stood out, with the theory of limits to selection under the infinitesimal model and the Second Theorem of Natural Selection, which explains how natural selection can result in adaptation through the trait-fitness additive covariance. Together with William (Bill) Hill he also made important contributions to animal breeding, as well as to population and evolutionary genetics, highlighting his predictions about the behaviour under selection of genes physically linked in the chromosome. Hill also extended Robertson's selection limits theory by introducing the impact of mutation. Many of these applications have been collected in the different editions of Douglas Falconer's famous book, *An Introduction to Quantitative Genetics* (Falconer and Mackay, 1996), and a lucid historical summary of the most important contributions of quantitative genetics to animal breeding can be seen in Hill's (2014) review.

Haldane made important contributions to the theoretical body of population genetics, with immediate application in quantitative genetics, establishing expressions on the fate of advantageous mutations in populations and deriving the conditions of polymorphisms in some situations. His most interesting contribution is the deduction of the balance that is reached between the appearance of deleterious mutations, those that reduce fitness, and their elimination by selection. Natural selection acts as a purging factor of the deleterious variability that is constantly generated by mutation, and the populations carry a 'mutation load', a concept due to Hermann Muller, which is simply equal to the diploid mutation rate. Thus, a high mutation rate could lead to negative implications for the population, particularly in asexual species (Muller, 1932). In these latter, the accumulation of deleterious mutations is very fast due to the lack of genetic recombination, which would allow their better elimination by natural selection. Thus, each time a mutation is fixed in an asexual population, the genome with fewer mutations will carry one more, without the possibility of going back to its previous state, a phenomenon known as Muller's Ratchet. The classic studies of Terumi Mukai in the 1960s and 1970s indicated that the deleterious mutation rate for viability in *Drosophila* was very high and mutational effects were small but high enough to be harmful, corroborating Muller's vision. The high mutation load could only be overcome by a greater effectiveness of selection with a synergistic effect of mutations, that is, when the combined effect of two or more mutations is greater than the sum of the effects of each of them separately.

Wright's contribution to population genetics is probably the most extensive and is summarized in the four volumes of his book *Evolution and the Genetics of Populations* published between 1968 and 1978. Wright emphasized the importance of epistatic interactions to produce advantageous combinations of genes on which selection could act and, especially, developed most of the theoretical body of work on the possible impact of chance (genetic drift) on evolution. We owe him the concept of effective population size, which allows us to quantify the effects of inbreeding and genetic drift and which was developed later by James F. Crow and Motoo Kimura. Wright discussed intensively with Fisher and

Haldane about the relative role of evolutionary forces in populations. For Fisher and Haldane, the most advantageous scenario from the evolutionary point of view was that of a large population rich in variability, where selection could act on individual genes of predominantly additive gene action. For Wright, on the contrary, the most favourable situation, gathered in his Shifting Balance Theory, would be that of a population subdivided into small isolated sub-populations in which genetic drift could expose to selection novel combinations of interactive genes, where dominant and epistatic gene actions could play an important role. One of the best introductions to this and other debates on evolutionary issues is found in Crow's (1986) book.

Although with the works of Fisher, Haldane and Wright the mechanisms that act on the genetic variability present in populations and, especially, the mode of action of natural selection on genetic variability were established in the 1930s, its excessively mathematical view meant that a few more years were needed before the neo-Darwinian theory of evolution was popularized thanks to the biological foundation provided by other fields of biology. The so-called modern synthesis represented the integration of the different biological disciplines, previously very separated from each other, in a global context, that of evolutionary biology. Modern population and quantitative genetics represent the contribution of genetics to the synthesis. This contribution conferred to Darwinian evolutionism the capacity to elaborate mathematical models that allow us to treat microevolutionary processes in a general way, thus becoming the core of neo-Darwinism.

The first half of the century of neo-Darwinism was dominated by the pre-eminence given to the selective force as an agent of evolutionary change, with a panselectionist view proposed by the experimentalists and headed by T. Dobzhansky. At the end of the 1960s, an anti-selection reaction occurred, whose most representative aspect is the formulation of the so-called Neutralism, formally developed by Kimura (1983) in his book *The Neutral Theory of Molecular Evolution*. This is not an opposition to neo-Darwinism but an orthodox version of it, in which it is desired to objectively establish the evolutionary importance of the forces that modify the composition of gene pools. Its essential contribution lies in not considering natural selection as a proven fact, but in establishing the null hypothesis against which the selection alternative can be verified in each specific case. A large part of molecular variation behaves as practically neutral, and selection can act very subtly when moulding such variation. The neutral theory was refined with the qualification of *quasi-neutral*, defended by Tomoko Ohta, one of Kimura's collaborators, and the study of the impact of the hitchhiking of deleterious or advantageous mutations on neutral variation, particularly in genomic regions of low recombination, is one of the most active areas of current research (Charlesworth and Charlesworth, 2010).

Kimura's contribution to quantitative and theoretical population genetics is not restricted to the neutral theory, and can be compared in quality and magnitude with that of Wright, Fisher and Haldane. In addition to developing most of the theoretical framework of the study of molecular variation, he made important advances in aspects related to linkage, population structure and sexual reproduction. Most of the theoretical framework of population genetics is condensed in his book with James Crow, *An Introduction to Population Genetics Theory* (Crow and Kimura, 1970), which still has full validity in many aspects.

Many of the fields of study of quantitative genetics developed in the second half of the twentieth century focused on prediction models of genetic variability and evolutionary change, the elucidation of the nature of quantitative traits and the study of the consequences of inbreeding, selection, population structure and crossing (Barton and Turelli, 1989; Lynch and Walsh, 1998; Walsh and Lynch, 2018). In the area of animal breeding, research focused on the development of more efficient selection methods, making use of information from relatives for the prediction of genetic values and the estimation of fixed effects (Henderson, 1984). In the last 50 years the development of molecular genetic techniques has been strengthened in the study of genetic variability at the level of DNA itself. The discovery of restriction enzymes in the late 1960s and, above all, the polymerase chain reaction technique, in the late 1980s, triggered the development of protocols to allow molecular analysis to be carried out from a few loci to the sequencing of complete genomes. This has exponentially increased the battery of genetic markers available for the elucidation of kinships, with applications in animal and plant breeding, such as genomic selection (Hayes et al., 2013), and in the mapping of genes, with repercussions on animal and plant improvement (Blasco and Toro, 2014), as well as in medicine (Visscher et al., 2017).

1.4 The Infinitesimal Model

As we have previously emphasized, the defining characteristics of quantitative traits are their polygenic nature and the modulation of gene effects by environmental influences, precisely the reasons that explain the continuous variation presented by many of them. To illustrate this, let us consider the possibility that a quantitative trait is controlled by a single locus with two alleles, A and a, so that the presence of allele a determines the addition of 1/2 unit of the trait, while allele A does not add anything to its value. From the offspring of a heterozygous individual Aa, reproduced by self-fertilization, we will find the distribution of genotypic values represented in Figure 1.2a. There will be, therefore, three genotypic classes of individuals, AA, Aa and aa, whose genotypic values are 0, 1/2 and 1 units, respectively.

Suppose now that there are two loci involved in the determination of the trait and, again, an individual heterozygous for the two loci self-fertilizes. If the effect of one of the alleles of each locus is now 1/4, to maintain the same total range of variation as in the case of one locus, and the other is null, and if the individual genotypic values are obtained as the sum of the effects of the alleles of the two loci, we have that there are five genotypic classes in the population (Figure 1.2b). In general, the frequencies of the classes are obtained with the development of the binomial $(0.5 + 0.5)^n$, where n is the number of loci. When more loci are considered, the number of genotypic classes increases at a rate of $2n + 1$ (Figures 1.2c and 1.2d). With 32 loci it can be observed that the genotypic classes begin to be so numerous that the unit of measurement of the trait can become incapable of distinguishing them. If, in addition, the loci had different effects, the number of genotypic classes would increase substantially. Note also that genotypic classes of more extreme values have a non-negligible probability when the number of loci is small, but when this increases the probability of finding individuals from the extreme classes is negligible and huge populations would be needed to detect their presence. For example, with 32 loci, the probability of appearance of an individual of the most extreme class (with 64 alleles that produce no effect, that is, with genotypic value 0, or with 64 alleles that add 1/64 effect, that is, with genotypic value 1) is only 5.42×10^{-20}.

If to the possibility that there are many genotypic classes we add the effect of the environment to configure the phenotypic classes, we can find that the phenotypic variation is

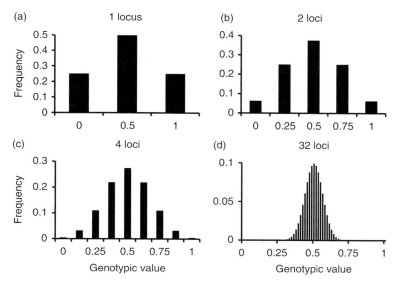

Figure 1.2 Distribution of frequencies of individuals from a population classified by their genotypic value for a quantitative trait controlled by 1, 2, 4 or 32 loci. In each case, the population comes from the self-fertilization of a single individual heterozygous for the loci involved. For each locus, one allele has no effect on the trait, and the other has effect $1/(2n)$, where n is the number of loci, and the effects of the different loci are added to obtain the genotypic value.

purely continuous, with an indefinitely large number of classes that could be approximated by a normal distribution, as can be deduced from Figure 1.2. The vast majority of quantitative traits, following the law of large numbers, have a phenotypic distribution in the form of a Gaussian bell, either in the original units or by making a change of scale, such as taking logarithm or arcsine transformations (see Falconer and Mackay, 1996, chapter 7).

The basic model of variation, proposed by Fisher, is the infinitesimal model, where the genotypic value of an individual (G) is determined by the joint effect of many loci (theoretically an indefinitely large number), with independent segregation, whose effects are small (in theory infinitesimal) and additively cumulative. An environmental deviation (E) is added to the genotypic value of the individual to determine the phenotype (P) of the individual, that is,

$$P = G + E. \tag{1.1}$$

The characteristics of the infinitesimal model are described in Figure 1.3. The distribution of phenotypic values is determined by a normal distribution, whose phenotypic mean $\overline{P}$ is equal to the genotypic mean $\overline{G}$. The reason for this equality is that it is assumed that environmental deviations can increase or decrease the phenotype of the individual with equal probability and similar magnitude, that is, $\overline{E} = 0$. This assumption is fundamental when it comes to carrying out the partition of the genetic variance in its components, as we will see in Chapter 3. The genotypic values are then distributed as a normal distribution with mean $\overline{G}$ and genotypic variance V_G. Environmental deviations are also distributed as a normal distribution

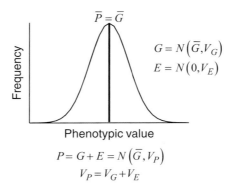

Figure 1.3 Characteristics of Fisher's infinitesimal model. The phenotypic value (P) is the result of the sum of the genotypic value (G) and the environmental deviation (E), which follows the normal distribution, so that the phenotypic variance V_P is the sum of the genotypic V_G and environmental V_E variances.

with mean 0 and environmental variance V_E. The phenotypic values are distributed, likewise, with normal distribution and phenotypic variance

$$V_P = V_G + V_E. \tag{1.2}$$

One of the fundamental objectives of quantitative genetics is to quantify the proportion of the phenotypic variation due to its genetic and environmental components, since the resemblance between relatives, the relative influence of environmental factors and the response to natural and artificial selection, will depend on this partition, as we will see in other chapters. On some occasions, the variance of the trait is scaled to the square of the mean (squared coefficient of variation, $CV_P^2 = V_P/\overline{P}^2$) to avoid scale effects and to be able to compare the magnitude of the variation between different quantitative traits.

The characteristics of the infinitesimal model are always violated in practice, since the number of genes is not infinite and their effects are not equal and infinitesimal. In fact, the experimental results indicate that most gene effects are of small magnitude with a lower proportion of loci with effects of great magnitude. In addition, quantitative trait loci are frequently subject to interactions, and their frequencies show dependence. However, the model serves as a predictive approach in a large number of scenarios. Later versions of the model include, in addition, dominance, the possibility of frequency dependence between loci (the so-called linkage disequilibrium) and interaction between effects (epistasis), circumstances that we will study in later chapters.

We have previously commented that quantitative traits can also be expressed discretely instead of continuously. Figure 1.4 illustrates two examples of such traits in *Drosophila melanogaster*. Panel (a) presents the number of sternopleural bristles in a large sample of individuals from a laboratory population. Note that the distribution is approximately normal with a slight asymmetry. Panel (b) shows the distribution of the number of pupae produced per female, a measure of fecundity, in the same population. This trait, like other main components of fitness, usually shows asymmetry towards low values due to the presence of deleterious genes of substantial effect that, although rare, can be carried by some individuals, positioning them very far from the average.

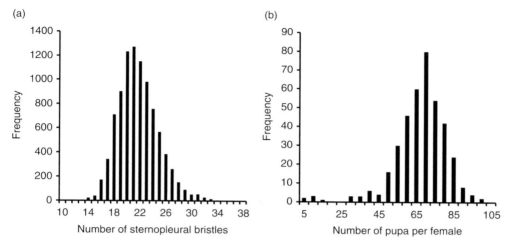

Figure 1.4 Distribution of two quantitative traits with discrete expression in a population of *Drosophila melanogaster* maintained in the laboratory with a large census size. (a) Number of sternopleural bristles (sum of both side plates). Vilas' (2014) data, corresponding to 9202 individuals (design of 1200 full-sib families with 10 individuals per family). The mean bristle number was $\bar{P} = 21.74$, and the variance $V_P = 9.80$. (b) Productivity per female measured as the number of pupae produced 11 days after mating with a male. Vilas' (2014) data corresponding to 388 females of the population. The mean pupae number was $\bar{P} = 64.35$, and the variance $V_P = 188.42$.

The most extreme case of quantitative traits with discrete expression are threshold traits, where only two or three phenotypic classes occur. A typical example is the susceptibility to a disease: individuals suffer it or not. The idea is that an underlying trait, which is called propensity, or liability in the context of human diseases, is a continuous trait determined perhaps by several or many loci and environmental factors, and that if a certain 'threshold' value of the liability is exceeded, the trait changes of phenotypic expression (Figure 1.5). This model can be applied to a large number of human diseases whose polygenic nature has been clearly demonstrated. The proportion or percentage of affected individuals is called, in general, incidence and, in the context of human diseases, prevalence, where incidence is the new number of cases in a given period of time.

Problems

1.1 In the cross between two pure lines, a heterozygous hybrid was obtained for 20 biallelic loci that affect a quantitative trait. For these loci, one allele has no effect on the trait and the other increases it by one unit. (a) How many genotypic classes would be found in the offspring by self-fertilization of the hybrid? (b) How often would descendants be found with a heterozygous genotype at all 20 loci? (c) With what probability would descendants with a phenotypic value equal to 10 be found?

1.2 The following table shows the number of sternopleural bristles in 50 individuals of *Drosophila melanogaster*. Knowing that the genetic variance of the trait is $V_G = 4$, deduce

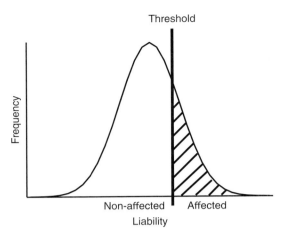

Figure 1.5 Model of threshold trait applicable, for example, to suffering a certain disease or not. There is an underlying continuous trait (liability in the context of human diseases) determined by the combined effect of a group of genes and environmental effects. Once a certain threshold of the underlying trait has been overcome, the disease is manifested.

the value of the environmental variance (V_E) and the phenotypic (CV_P) and genotypic (CV_G) coefficients of variation.

24	24	21	21	21	20	19	19	18	24
23	25	22	24	22	22	18	20	22	28
23	27	23	20	23	21	26	27	27	28
26	26	24	26	27	26	20	22	21	18
18	21	17	21	18	20	23	22	34	22

Self-Assessment Questions

1 Threshold traits are those determined by two or three loci.
2 Quantitative traits are affected by many loci and, therefore, are also called polygenic traits.
3 Additivity, or additive gene action, implies that the phenotypic value of the heterozygote is intermediate between that of the homozygotes.
4 Major genes are those that affect qualitative traits, while minor genes affect quantitative ones.
5 Threshold traits are a type of meristic trait.
6 When a locus has an effect on two or more traits, we speak of pleiotropy.
7 In the infinitesimal model, the phenotypic mean is not expected to be equal to the genotypic mean.
8 In the infinitesimal model, the loci have generally small effects, but there may be large-effect genes.
9 The main components of fitness generally present asymmetric phenotypic distributions.
10 The incidence in a threshold trait is the proportion of individuals affected by that trait.

2 Forces of Change in the Allele Frequencies

Concepts to Study

- Allele, gamete and genotype frequencies
- Hardy–Weinberg equilibrium
- Expected heterozygosity or gene diversity and allelic diversity
- Gametic or linkage disequilibrium
- Genetic drift
- The ideal population of Wright–Fisher
- Equilibrium between mutation and back-mutation
- Migration models
- Selection and dominance coefficients
- Types of within-locus gene action
- Stable and unstable equilibria
- Antagonistic pleiotropy and marginal overdominance

Objectives for Learning

- To learn how to calculate allele frequencies from genotype frequencies
- To know the conditions for Hardy–Weinberg equilibrium, its implications and the definition of expected heterozygosity and allelic diversity
- To know how to calculate linkage disequilibrium
- To understand the process of genetic drift
- To learn the basic characteristics of the Wright–Fisher idealized population
- To know how to calculate the changes in allele frequency by mutation and how an equilibrium between mutation and back-mutation is reached
- To understand the homogenizing effect of migration and the different population models used for its description
- To know the general model of fitness and the concepts of selection coefficient and dominance coefficient
- To understand how the changes in allele frequency for deleterious or beneficial alleles take place with different types of within-locus gene action
- To learn how to distinguish stable and unstable models of allele frequencies
- To comprehend the concepts of antagonistic pleiotropy and marginal overdominance
- To understand the impact of selection on the test for Hardy–Weinberg equilibrium

2.1 Allele, Gamete and Genotype Frequencies

The genetic description of a population can be done at three different levels, the locus, the gamete or the individual genotype, by specifying the different variants in each case

Table 2.1 *Illustration of the calculation of genotype frequencies and allele or gene frequencies*

Genotype	*AA*	*Aa*	*aa*	Total
No. individuals	40	50	10	100
Genotype frequency	0.4	0.5	0.1	1

Allele	*A*	*a*		
Number	130	70		200
Allele frequency	$p = 0.65$	$q = 0.35$		1

(allele, gamete or genotype) and their respective frequencies. A population of a diploid species is composed of individuals (genotypes) that reproduce by the union of their gametes to form zygotes that will give rise to the individuals of the next generation, hence the interest of a genotypic and gametic description. But genotypes and gametes are sets of alleles, two for each locus in the first case and one in the second, hence the interest of the allelic description.

Consider the simplest case: a biallelic locus A with alleles *A* and *a*, therefore, genotypes *AA*, *Aa* and *aa*, and suppose that in a population formed by 100 individuals, the number of those corresponding to each genotype is 40, 50 and 10, respectively (Table 2.1). Genotype frequencies are normally given in relative values, so that their sum is unity. From the genotype frequencies we obtain the allele frequencies, p that of *A* and $q = 1 - p$ that of *a*, which are also the gamete frequencies because the gametes carry a single copy of each of the chromosomes of the individual that produce them. Since homozygotes for one locus carry two copies of the relevant allele, and heterozygotes only one, allele frequencies can be obtained quickly as the sum of the frequency of the homozygotes carrying the allele in question plus half the frequency of the heterozygotes, that is, $p = 0.4 + 0.5/2 = 0.65$ and $q = 0.1 + 0.5/2 = 0.35$.

Suppose now that we consider two loci A and B, with alleles *A* and *a*, and *B* and *b*, respectively, there being nine possible genotypes formed by the combination of the three genotypes for each locus: *AABB, AaBB, . . . aabb*. The constitution of the four gametes that can exist will be *AB, Ab, aB* and *ab*, whose gametic frequencies we will denominate $P_{AB}, P_{Ab}, P_{aB}, P_{ab}$, respectively, and whose values will depend on the allele frequencies for each locus and the possible physical and/or genetic association between the two loci in question.

2.2 Hardy–Weinberg Equilibrium

In order to establish the genetic constitution of a filial population based on the genetic description of the parental population, it is necessary to specify the way in which the gametes produced by the parents are united in pairs to form the offspring. The simplest case is one in which gametic paring is random, called panmixia. If there were no differentiation in allele frequencies between sexes and we consider male and female gametes that carry the allele *A* with frequency p or the allele *a* with frequency q, random pairing generates the descendants shown in Figure 2.1, whose expected genotype frequencies are obtained developing the binomial expression $(p + q)^2$, that is,

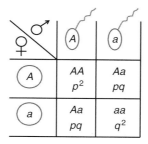

Figure 2.1 Calculation of the expected genotype frequencies from the allele or gametic frequencies in a scenario under panmixia. The frequency of the gamete carrying allele A is p and that of allele a is q in both sexes.

$$\begin{array}{lll} \text{Genotypes:} & AA \quad Aa \quad aa & \\ \text{Genotype frequencies:} & p^2 \quad 2pq \quad q^2. & \end{array} \qquad (2.1)$$

This is known as the Hardy–Weinberg principle or equilibrium, since it was independently deduced in 1908 by the English mathematician G. H. Hardy and the German physician W. Weinberg. Although very elementary, this principle is basic in population and quantitative genetics, since it greatly simplifies most of the theoretical developments formulated for the processes that affect populations. Note that random pairing is only required for the locus (or loci) under study. Although gametes do not unite in a strictly random manner, for most of the genome the assumption is usually valid. The possible discrepancy between the expected and observed genotype frequencies can provide information about the forces of change acting on the locus, as we will see later. In the example described in the previous section, the observed genotype frequencies were 0.4, 0.5 and 0.1 for the genotypes AA, Aa and aa, respectively, and the corresponding allele frequencies were $p = 0.65$ and $q = 0.35$. Therefore, the expected frequencies of the three genotypes in Hardy–Weinberg equilibrium would be $p^2 = 0.4225$, $2pq = 0.455$ and $q^2 = 0.1225$, which are similar to those observed, although to determine this it would be necessary to carry out a statistical test that will be illustrated in a problem solved at the end of the chapter.

The expected genotype frequencies in Hardy–Weinberg equilibrium for a locus are illustrated in Figure 2.2 as a function of allele frequencies. Note that the maximum frequency of heterozygotes is 0.5, and that when one allele is rare, most of its copies are found in heterozygous individuals. For example, if allele A were at a frequency of $p = 0.01$, the expected frequency of AA homozygotes could be only 0.0001, one in 10,000 individuals, while heterozygotes Aa would constitute approximately 0.2% of the population, or 2 out of 1000 individuals. This observation will be essential for some of the arguments that we will develop in later sections.

The Hardy–Weinberg principle can refer not only to a locus but to other elements, such as a chromosomal organization (for example, an inversion or a translocation). The extension to multiallelic loci is immediate. Suppose that locus A has n alleles, $A_1, A_2, \dots A_n$, with frequencies $p_1, p_2, \dots p_n$, respectively. The frequency of the homozygous

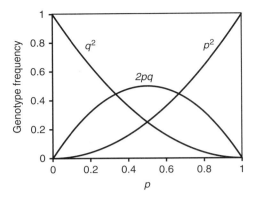

Figure 2.2 Genotype frequencies expected with Hardy–Weinberg equilibrium for a locus with alleles A and a and different values of their allele frequencies p and q, respectively.

genotype A_iA_i is p_i^2, and that of heterozygote A_iA_j is $2p_ip_j$, being n homozygotes and $n(n-1)/2$ heterozygotes, and the expected global frequency of heterozygotes is

$$H = 1 - \sum_{i=1}^{n} p_i^2, \qquad (2.2)$$

which is called expected heterozygosity or gene diversity, and is a measure of the genetic variation present in the population for the locus in question. Another measure of genetic diversity in the case of multiallelic loci is the so-called allelic diversity or allelic richness, which is defined as the average number of alleles per locus that segregate in the population.

The composition of genotype frequencies determined by the Hardy–Weinberg principle could remain unchanged throughout generations if there are no forces of change acting on the locus considered. For this, a series of conditions must be met, in addition to the aforementioned random pairing of gametes, which are the following:

1 The population census size must be infinite, so that changes in frequencies in successive generations would not be expected as a result of chance.

2 The Mendelian segregation of the locus should be correct, so that the heterozygotes produce an equal number of gametes carrying alleles A and a.

3 For a species with separate sexes, there should be equal allele frequencies in both of them.

4 No mutation or immigration of individuals from other populations should occur in the population.

5 The individuals carrying the three genotypes must have the same probability of survival or viability, and equal fecundity, and the fertilizing capacity of the two gametes must also be the same. In other words, there must be absence of any form of selection that benefits or disfavours any of the alleles or genotypes.

If the above conditions are met and maintained throughout generations, the allelic and genotype frequencies will also remain unchanged, hence the denomination of Hardy–Weinberg equilibrium. However, if a manipulation or pure chance modifies the frequencies in a generation, a new equilibrium will be established with the new frequencies. That is to say, the Hardy–Weinberg equilibrium is not properly a balance in the physical sense of the term, since it does not have a tendency to recover the equilibrium prior to the change. For example, suppose that the allele frequencies in a given generation are $p = q = 1/2$, and the genotype frequencies are 1/4, 1/2 and 1/4 for the genotypes AA, Aa and aa, respectively, that is, those of equilibrium. Suppose that, for some reason, in a given generation, the individuals aa disappear, with which the allele frequencies will have changed to $p' = 2/3$ and $q' = 1/3$ and the genotypic ones will be 1/3, 2/3 and 0, respectively, that differ from those of equilibrium. If the survivors mate at random, the allele frequencies will remain at their new values and the genotypic ones will correspond to a new equilibrium, $p'^2 = 4/9$, $2p'q' = 4/9$ and $q'^2 = 1/9$. On the contrary, if the cause for which the homozygotes disappear is permanent, for example because the allele a is lethal in homozygosis (it produces the mortality of the individual), and it is repeated generation after generation, the disequilibrium will also be permanent. As we see, in the absence of permanent sources of disequilibrium, the Hardy–Weinberg equilibrium is recovered in a single generation of panmixia and this is one of the reasons why it is so common to find loci in equilibrium, or close to it, in nature. Note, however, that if we refer to loci located on the X chromosome, the equilibrium is not reached immediately, but will be achieved progressively, being necessary that the difference in allele frequencies between males and females disappears.

The causes that can produce disequilibria are multiple, some of which will affect specific loci and others to the entire genome. An example of the first case is that of lethality that we have just described. Another is the non-random pairing for the locus, which can be of different types. For example, positive assortative mating occurs when individuals of the same genotype tend to mate with each other more frequently than expected by chance, whereas negative assortative mating (also called disassortative) implies the opposite. In the first case there will be excess and, in the second, defect of homozygotes with respect to the expected values with Hardy–Weinberg equilibrium. An example of a situation that generates Hardy–Weinberg disequilibrium for the whole genome is that of species with inbred matings. For example, in many plants there is partial or total self-fertilization and a generalized excess of homozygotes will be found in the whole genome.

2.3 Gametic or Linkage Disequilibrium

When we consider two or more loci, gametic frequencies can provide important information about the physical and/or genetic relationship between them. The gametic, gametic phase or linkage disequilibrium is the possible association in frequencies between two or more loci, although generally only two are considered. Consider two loci, A (with alleles A and a and allele frequencies p_A and p_a) and B (with alleles B and b and allele frequencies p_B and p_b) and the four possible gametes carrying the AB, Ab, aB and ab alleles, with corresponding gametic frequencies P_{AB}, P_{Ab}, P_{aB} and P_{ab}. If there is no association between the allele frequencies of loci A and B, that is, if their segregation is independent, it is said that there is gametic equilibrium and the frequency of each gametic type is equal to the product of the frequencies of the alleles of each locus, for example $P_{AB} = p_A p_B$. When this equality is not met, it is said that there is gametic or linkage disequilibrium (D), which is quantified as the difference between the real frequency and that expected in the equilibrium,

$$D = P_{AB} - p_A p_B \tag{2.3}$$

(Lewontin and Kojima, 1960). For example, suppose the allele frequencies are $p_A = 0.6$; $p_a = 0.4$; $p_B = 0.7$; $p_b = 0.3$; and the gametic ones are $P_{AB} = 0.4$; $P_{Ab} = 0.2$; $P_{aB} = 0.3$; and $P_{ab} = 0.1$. We find then that $D = P_{AB} - p_A p_B = -0.02$, indicating disequilibrium (although a statistical test must be applied to confirm or reject it as we will see in a problem solved at the end of the chapter). If the same calculation is repeated for the remaining gametes, we obtain $D = P_{ab} - p_a p_b = -0.02$, $D = P_{Ab} - p_A p_b = 0.02$ and $D = P_{aB} - p_a p_B = 0.02$, indicating that, at disequilibrium, gametes Ab and aB are in excess and gametes AB and ab in defecit with respect to their expected equilibrium values. From the four formulas above it can be deduced that the disequilibrium can also be expressed in terms of the four gametic types as

$$D = P_{AB}P_{ab} - P_{Ab}P_{aB}, \tag{2.4}$$

which also results in $D = -0.02$. Equation (2.4) can be interpreted as the difference in frequency between the double heterozygote in coupling phase (AB/ab), which would be formed by the union of gametes AB and ab ($P_{AB} \times P_{ab}$) and the double heterozygote in repulsion phase (Ab/aB) due to the union of gametes Ab and aB ($P_{Ab} \times P_{aB}$).

There are numerous reasons why there may be gametic phase disequilibrium, including chance, the mixing of populations with disparate genetic constitutions, and natural selection favouring certain gametic combinations over others. Consider an example of the second scenario. Suppose that two populations, historically separated, meet at a given moment and cross to each other, something that may have been frequent in human evolutionary history. Imagine that, for the two-loci system considered, all individuals are homozygotes, $AABB$ those of the first population, and $aabb$ those of the second. Let us also suppose that the two populations have the same census size, sexes are equally represented, and that when individuals of the two populations meet, they mate in panmixia. At the time of the encounter, there will be only individuals with genotype $AABB$ or $aabb$, both at frequency $1/2$, allele frequencies $p_A = p_a = p_B = p_b = 1/2$ and gametic frequencies $P_{AB} = P_{ab} = 1/2$ and $P_{Ab} = P_{aB} = 0$. It is obvious that each of the loci will be in Hardy–Weinberg disequilibrium since the observed heterozygous frequency is zero versus its expected value of $1/2$. It is also evident that the set formed by the two loci will be in gametic disequilibrium, $D = 1/4$.

The panmictic pairing would generate $AABB$, $AaBb$ and $aabb$ genotypes with frequencies $1/4$, $1/2$ and $1/4$, respectively, confirming that the Hardy–Weinberg equilibrium is reached after a single generation for autosomal loci. Note that the double heterozygotes are in coupling phase, AB/ab, since they come from the union of AB and ab gametes. In this first generation after the encounter, the gametic frequencies of the parents will depend on whether the A and B loci are located or not in the same chromosome and, if they are in the same chromosome, whether recombination occurs, or not, between both loci in the double heterozygotes.

Suppose, first, the extreme situation in which the two loci are in the same chromosome, so intimately close that recombination never occurs between them (Table 2.2-I). In this case, all individuals will always be heterozygotes AB/ab, the gametic frequencies will be $P_{AB} = P_{ab} = 1/2$ and $P_{Ab} = P_{aB} = 0$, and the disequilibrium will remain at its value $D = 1/4$.

Table 2.2 *Illustration of the calculation of gametic frequencies in a scenario without recombination (I) or with free recombination (II) between loci A and B*

Parental genotype and frequency	Gametes produced			
	AB	Ab	aB	ab
I. Scenario without recombination ($c = 0$)				
1/4 $AABB$	1	0	0	0
1/2 $AaBb$	1/2	0	0	1/2
1/4 $aabb$	0	0	0	1
Total	1/2	0	0	1/2
II. Scenario with free recombination ($c = 0.5$)				
1/4 $AABB$	1	0	0	0
1/2 $AaBb$	1/4	1/4	1/4	1/4
1/4 $aabb$	0	0	0	1
Total	3/8	1/8	1/8	3/8

As can be seen, if there is no recombination between the two loci, that is, if there is total linkage, the disequilibrium will remain unchanged, hence the traditional denomination of linkage disequilibrium.

Now suppose the opposite extreme in which the two loci are in different chromosomes, or in the same but so separated that there is always crossing over between them (Table 2.2-II). In this case, the double heterozygotes will generate the four types of gametes with equal probability, the recombinant gametes (Ab and aB) will appear with global frequency $c = 1/2$, and the disequilibrium will be reduced by a half, $D = P_{AB} P_{ab} - P_{Ab} P_{aB} = (3/8 \times 3/8) - (1/8 \times 1/8) = 1/8$. If we consider a new generation of panmixia we will again obtain a reduction by a half of the previous disequilibrium, and so on. That is, in the best case, when the loci in question are not linked, the disequilibrium will be reduced by a factor of 1/2 per generation. Note, therefore, that two loci do not have to be linked to exhibit disequilibrium. However, the more extreme the linkage, the longer it will take to reach equilibrium.

In general, the disequilibrium in generation t will be reduced with respect to that of the previous generation by a fraction $(1 - c)$, where c is the frequency of recombinants, that is, $D_t = D_{t-1}(1 - c)$. If there is no recombination ($c = 0$), the disequilibrium remains unchanged. If, on the contrary, the loci are unlinked ($c = 1/2$) the disequilibrium is reduced by a half in each generation. If it is expressed in terms of the disequilibrium of the initial generation,

$$D_t = D_0(1 - c)^t. \tag{2.5}$$

Solving for t, we obtain $t = [\ln(D_t/D_0)]/[\ln(1 - c)] \approx \ln(D_0/D_t)/c$ from which the time necessary for the disequilibrium to be reduced by a certain proportion can be calculated. For example, if two loci are at a genetic distance of 1 cM, which would be equivalent to 1% recombination ($c = 0.01$), the time necessary for a given disequilibrium to be reduced by a half would be $t \approx [\ln(2)]/0.01 \approx 70$ generations. If we assume that the generation interval is 30 years, applicable for example to humans, the time required would be about 2000 years, which

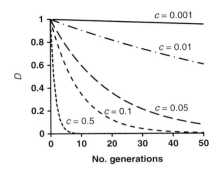

Figure 2.3 Evolution of gametic disequilibrium (D) (relative to the initial one) for two loci located at a genetic distance such that recombination frequency is c.

illustrates the long time required for gametic disequilibrium to disappear if the loci in question are closely linked. Figure 2.3 illustrates the time needed to reduce the initial disequilibrium between two loci located at different genetic distances. Even if the loci are not linked, the disequilibrium takes about 10 generations to disappear.

The value of the disequilibrium expressed by equations (2.3) or (2.4) depends on the allele frequencies of the two loci involved and, as we saw in the example, even with total absence of recombinant gametes, their value is less than unity. For this reason, a disequilibrium value (D') relative to its maximum possible value (D_{max}) is often used given the allele frequencies (Lewontin, 1964),

$$D' = \frac{D}{D_{max}}. \tag{2.6}$$

The value of D_{max} is equal to the smallest of $p_A p_b$ or $p_a p_B$ if D is positive, or the smallest of $p_A p_B$ or $p_a p_b$ if D is negative. In the example proposed at the beginning of the section, where $p_A = 0.6, p_B = 0.7$ and $D = -0.02$, the maximum possible disequilibrium value would be $D_{max} = p_a p_b = 0.12$ and $D' = -0.17$, that is, 17% of the maximum possible.

Recall that, at the beginning of this section, we indicated that gametic disequilibrium is synonymous of lack of independence of the allele frequencies of a pair of loci. Therefore, another way of presenting the standardized disequilibrium is as the correlation (r) between the allelic frequencies of the two loci,

$$r^2 = \frac{D^2}{p_A p_a p_B p_b} \tag{2.7}$$

(Hill and Robertson, 1968). The statistical significance of the disequilibrium is carried out with a χ^2 test of contingency or independence (with one degree of freedom in the case of biallelic loci), resulting that the value of χ^2 is equal to the product of r^2 and the sample size.

It is called perfect linkage disequilibrium to that corresponding to a total association between the alleles of the two loci, that is, there are only gametes AB and ab, in which case, $|D'| = r^2 = 1$. And it is called complete linkage disequilibrium when referring to two loci whose allele frequencies are different but there are only three gametic types, in which case $|D'| = 1$ but r^2 can take different values. For example, if $p_A = 0.2$; $p_a = 0.8$; $p_B = 0.3$; $p_b = 0.7$; and the gametic frequencies are $P_{AB} = 0.2$; $P_{Ab} = 0$; $P_{aB} = 0.1$; and $P_{ab} = 0.7$, so that the gamete Ab is not present in the sample, we would have that, using equations (2.3), (2.6) and (2.7), $D = P_{AB} - p_A\,p_B = 0.14$, $D' = 1$ and $r^2 = 0.58$.

There are numerous applications of linkage disequilibrium analysis, such as the estimation of the effective population size (Chapter 5), the detection of selection footprint and the mapping of genes with effect on quantitative traits (Chapter 11). In all cases, the analysis is based on the study of the observed disequilibrium between identifiable genetic markers that, at least in the latter case, must necessarily be of known genomic position. The idea is that genes physically close on the chromosome are more likely to be in strong linkage disequilibrium. The effective population size is a parameter that quantifies the magnitude of genetic drift, a concept that we will study in the next section of this chapter and that refers to the genetic change occurring by chance in populations of reduced size. Genetic drift generates linkage disequilibrium and, therefore, the degree of this disequilibrium provides a measure of the intensity of the drift and, hence, of the magnitude of the effective population size. If selection acts on a given locus, producing an increase or decrease in its allele frequency, linkage will drag the nearby loci, so that drastic changes in the frequency of genetic markers in a specific region of the genome will give a clue, called selection footprint, on the possibility that there are one or more selected loci in that region. Following the same principle, a genetic marker of known position associated with a high value of a quantitative trait can be used for genetic mapping, providing an indication of the proximity of a locus that controls the trait. Thus, for example, the analysis of hundreds of quantitative traits related to human characters and diseases has revealed the association of these traits with thousands of SNP markers, allowing the approximate localization of the genes that affect these traits (Visscher et al., 2017).

2.4 Forces of Change in the Allele Frequencies

Real populations are subject to different natural or artificially induced processes that modify their genetic composition. These forces of change can be a dispersive process, such as genetic drift, the random change in allele frequencies due to chance, which is predictable in magnitude but not in direction, or systematic processes, predictable both in magnitude and direction, such as mutation, migration and selection. In this section we will study each of these processes separately, using to a large extent the theoretical developments due to Fisher (1930), Haldane (1932) and Wright (1931, 1969). In later chapters we will combine the different factors, in a progressive way, to understand the complex conjunction of forces that can act on the populations and are responsible for evolution and the genetic improvement of animals and plants.

2.5 Genetic Drift: The Ideal Population of Wright–Fisher

Genetic drift is the random change of the allele frequencies as a consequence of the sampling of gametes that takes place in reproduction. Let us suppose a population with a census size of $N = 10$ parents in which the allele frequencies of the alleles A and a are $p_0 = 0.6$ and $q_0 = 0.4$

initially. If the 10 individuals mate at random to produce the same number of descendants, it can be assumed that the reproduction implies a random sampling of 20 gametes of the large number produced by the parents. In this process it would be expected to sample 60% of gametes (i.e. 12) with allele A (which would mean that in the next generation its expected frequency is $p_1 = 0.6$) and 40% (i.e. 8) with allele a. However, sampling can give other results, for example, suppose that it gives 14 alleles A and 6 alleles a, that is, p_1 has increased to 0.7 and p_2 has been reduced to 0.3. This change could also have occurred in the opposite direction (a reduction of p_1 and an increase of p_2). In the third generation it will be expected, therefore, that 70% of the gametes carry allele A and 30% carry allele a. But again, the result that occurs may be other, again involving an increase or decrease in the frequencies over their expected value. It could occur, in its most extreme case, that only gametes with allele A were sampled, in which case $p = 1$ and $q = 0$, all descendants being homozygotes of the AA genotype. In this case it is said that the allele A has been fixed in the population and allele a has been lost. This situation is one of no return, which will be attained sooner or later, unless there are other processes of change in the allele frequencies, and the frequencies reached will be unchanged. Whatever the size of the population, genetic drift leads, inexorably, to the fixation of a given allele and the loss of the others (if there were more than two) and this process is faster the smaller the population size.

Figure 2.4 (upper panels) shows the process of genetic drift considering multiple genes with the same initial allele frequency ($q_0 = 0.5$) in populations of two different sizes (this could also refer to multiple populations and a single gene). Note the dispersion of allelic

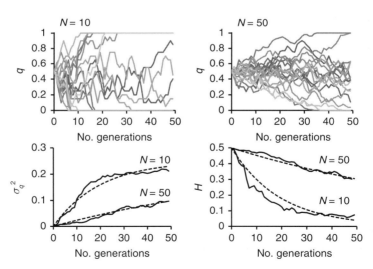

Figure 2.4 Changes in allele frequencies (q) by genetic drift of 20 biallelic loci with initial frequency $q_0 = 0.5$ in a population with census size N. The lower left graph shows the variance of the allele frequencies observed in each generation (solid lines) for each of the population sizes and the expected variance (dashed lines), obtained with equation (2.9). The lower right graph shows the corresponding results for the observed heterozygosity, calculated with equation (2.2), and its expected value, obtained with equation (2.11).

frequencies from the first generation, which qualifies genetic drift as a dispersive process. Eventually, after a number of generations, one of the alleles is fixed at each locus. In the example, we would expect that each of the alleles will become fixed 50% of the time.

The process of genetic drift can be quantified, but for this it is necessary to assume very simplified conditions. The simplest way to do this quantification is to assume that the population fulfils some basic characteristics, although in practice they may or may not hold, which allows us to develop predictive equations that can then be extended to more complex situations. The so-called ideal population of Wright–Fisher assumes the following characteristics:

1 A population of census size N, constant throughout generations, belonging to a diploid species without separate sexes.

2 Reproduction occurs by random pairing of the gametes produced by the parents, including random self-fertilization.

3 The predictions refer to an autosomal locus that, for convenience, has two alleles, A and a, with initial frequencies p_0 and q_0, respectively. Therefore, the locus is in Hardy–Weinberg equilibrium with initial genotype frequencies p_0^2, $2p_0q_0$ and q_0^2, for the AA, Aa and aa genotypes, respectively.

4 There are no other forces of change of the allele frequencies (mutation, migration and selection).

In the ideal population, therefore, all individuals have exactly the same probability of contributing offspring to the next generation and mating between gametes is random. The simulations presented in Figure 2.4 correspond to the characteristics of the ideal population. It is considered a set of loci whose initial allele frequencies are equal ($p_0 = q_0 = 0.5$) for all of them. This could also refer to the situation in which a single locus is evaluated but in different equivalent populations, starting with the same allele frequencies in all of them. The random sampling of gametes of generation 0 to form the individuals of generation 1 is a binomial process with a mean expected frequency for the different loci equal to $\bar{q}_1 = q_0$. The magnitude of the dispersion in allele frequencies for the different loci (or the different populations) is given by the expected variance of the frequencies reached by the different loci in that generation 1, which is equal to

$$\sigma_{q_1}^2 = \frac{p_0q_0}{2N},$$
(2.8)

that is, the probability of a binomial distribution with $2N$ samples. Note that the expected variance of the allele frequencies and, therefore, the genetic drift, is greater at a lower population size, since the populations with the lowest size will be subject to more erratic changes in frequencies. The prediction of the expected variance of allele frequencies in any generation can be obtained through the general expression

$$\sigma_{q_t}^2 = p_0q_0\left[1 - \left(1 - \frac{1}{2N}\right)^t\right],$$
(2.9)

being able to verify that, for $t = 1$, equation (2.8) is recovered.

The dispersive process leads to the fixation and loss of alleles, which implies an increase in the frequency of homozygotes at the expense of heterozygotes. Noting that, by definition, the variance of the frequencies of the different loci (or populations for the same locus) is equal to the mean of the squares of the frequencies minus the square of the mean, that is, $\sigma^2_{q_t} = \overline{q^2_t} - \overline{q}^2_t$ and that the average expected frequency is equal to the initial frequency, $\overline{q}_t = q_0$, we have that the expected frequency of homozygotes aa in generation t is $\overline{q^2_t} = q_0^2 + \sigma^2_{q_t}$. That is, the frequency of homozygotes AA and aa increases by an amount $\sigma^2_{q_t}$ with respect to the initial frequency due to the drift process, consequently reducing that of the heterozygotes (H_t). Therefore, the expected frequencies of the three genotypes for the set of loci (or populations) are

$$AA : p_0^2 + \sigma^2_{q_t}, \tag{2.10}$$

$$Aa : 2p_0 q_0 - 2\sigma^2_{q_t} = H_t, \tag{2.11}$$

$$aa : q_0^2 + \sigma^2_{q_t} . \tag{2.12}$$

Over generations, the bracketed term in equation (2.9) tends to one and $\sigma^2_{q_{t=\infty}} = p_0 q_0$, so the expected genotype frequencies are p_0, 0 and q_0, respectively, for the AA, Aa and aa genotypes. That is, it is expected that in a p_0 proportion of the loci (or populations) under study, the allele A will be fixed, in a proportion q_0 allele a will be fixed, and the heterozygous individuals will disappear completely. Genetic drift, therefore, implies the reduction in the heterozygosity or gene diversity expressed in equation (2.2) at the beginning of this chapter.

2.6 Change in Allele Frequencies by Mutation

Mutation is the ultimate source of variation and, therefore, the process generating new allelic variants on which selection acts. In Chapter 7 we will study the nature of mutation from the point of view of quantitative traits, the methods for estimating the mutation rate and mutational effects, and their implications on the mean and variance of quantitative traits. In this chapter we will limit ourselves to establishing its impact as a force for systematic change of allele frequencies in the absence of other evolutionary factors. Consider the model of a locus A, with alleles A and a and frequencies p and q, respectively. Suppose that allele A, considered the wild or common allele in the population, mutates to form a with a probability u per generation, the mutation rate (Figure 2.5). In practice it is likely that different mutations that affect a given locus will produce different alleles, however it can be assumed that the different mutant allelic variants are encompassed in a single allelic form. Although less likely, the opposite process, the reversal of the mutated allele, a, to its wild version, may also occur, the back-mutation rate being v.

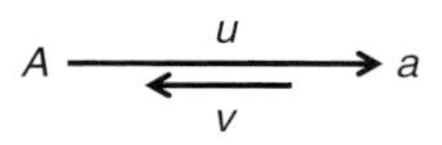

Figure 2.5 Mutation and back-mutation process between two allelic forms, A and a, of a locus, where u is the mutation rate and v is the back-mutation rate.

In this situation, the change in frequency of the A allele produced by mutation and back-mutation from one generation to the next is

$$p_t = p_{t-1}(1-u) + q_{t-1}v. \tag{2.13}$$

Since there are two mutational processes that occur in opposite directions, it ends up reaching a situation of equilibrium in which $p_{t-1} = p_t = \hat{p}$. Substituting in (2.13) we obtain $\hat{p} = \hat{p}(1-u) + (1-\hat{p})v$, from where we arrive at the equilibrium frequencies,

$$\hat{p} = \frac{v}{u+v}, \qquad \hat{q} = \frac{u}{u+v}. \tag{2.14}$$

The mutation rate per locus is usually assumed to be of the order of $u \approx 10^{-5}$–10^{-6}, whereas the back-mutation rate towards the wild allele is usually an order of magnitude smaller, $v \approx 10^{-6}$–10^{-7}. Therefore, the equilibrium frequencies applying the expressions (2.14) would be $\hat{p} \approx 0.1$ and $\hat{q} \approx 0.9$, that would be reached very slowly given the low magnitude of the change in frequencies generated by mutation in each generation. Obviously, in the previous argument we have only taken into account the mutation process, ignoring all the others. In practice, if the wild allele, A, is a functional allele, and the mutant allele, a, is not, selection will prevent the latter from increasing significantly in frequency, although it will also depend on other factors that we will discuss later. In general, changes in allele frequency generated by mutation are extremely slow. The importance of mutation does not reside in these changes in frequency, which are generally negligible, but in the generation of new allelic forms in the whole genome and population. For example, if we consider a population size of $N = 1000$ individuals and assume that the rate of appearance of new allelic variants per locus and generation is $u \approx 10^{-5}$, it is expected that $2Nu = 0.02$ individuals will be born in the population, that is, two out of every 100, with a mutation in the locus in question. If we consider the whole genome and assume that there are 20,000 loci subject to the same mutation rate, the number of new mutations that would appear in the whole population would be 400 each generation.

2.7 Change in Allele Frequencies Due to Migration

Another source of systematic change in allele frequencies that can occur in a population is that generated by immigrant individuals from another population with different frequencies. The simplest model for its study is the so-called island-continent model (Figure 2.6a), where a large population (continent) and a much smaller one (island) are considered, there being a gene flow due to migration between both at a rate m per generation. Given the large relative size of the continent with respect to the latter, the changes by migration only significantly affect the island. If the frequency of allele A in the continent is P, which is assumed invariable, and in the island it is p_{t-1} in a given generation, the frequency of the allele in the island in the next generation will be

$$p_t = p_{t-1}(1-m) + Pm. \tag{2.15}$$

Subtracting P in each side of equation (2.15) and operating, we have $(p_t - P) = (p_{t-1} - P)(1-m)$, from which it follows that the difference in allele frequencies

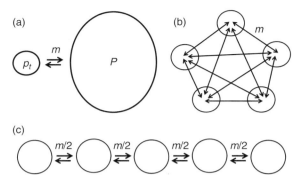

Figure 2.6 Population models for the study of migration. (a) Island-continent model. (b) Island model. (c) Stepping-stone model. In all cases, *m* is the fraction of genes that migrate from each island in each generation.

between the island (p) and the continent (P) for the allele in question is reduced each generation by the effect of migration in a fraction $(1 - m)$. Expressed with respect to the initial difference,

$$(p_t - P) = (p_0 - P)(1 - m)^t. \tag{2.16}$$

Another model widely used to analyse migration is the island model (Figure 2.6b), which assumes a number of populations of the same size and panmictic mating, where migration occurs randomly between any of them, so that, each generation, a fraction *m* of genes leaves any population and is incorporated into any other, including the source, with equal probability. The change in allele frequencies by generation in this model is also expressed by equation (2.16), so that *p* is the frequency in a given population and *P* is the average of the frequencies of all populations.

The result of the process is that the allele frequency of the island (both in one model or in the other) will gradually tend to approach the frequency on the continent (or the average of the islands) until the difference disappears. That is, migration produces a homogenization in allele frequencies. The impact of migration on allele frequencies is considerably greater than that of mutation, simply because migration rates (m) are usually much higher than mutation rates (u). There are other models of biological interest, such as the stepping-stone model (Figure 2.6c), where populations are placed one after another, as would occur in species that live in linear habitats, such as rivers or coasts, and where migration occurs between contiguous populations. In this case, the homogenizing effect of migration is less than that of the island model for the same migration rate.

2.8 Change in Allele Frequencies by Natural Selection

2.8.1 General Model of Fitness for a Locus

As discussed in Chapter 1, natural selection can be thought of as the survival and/or differential reproduction of any unit capable of replication, such as a gene, a genotype or an

Table 2.3 *Example of the calculation of the fitness of the three genotypes for a locus with respect to viability and the general formulation where* s *is the selection coefficient against the homozygote* aa *and* h *is the dominance coefficient*

	AA	*Aa*	*aa*
No. zygotes	100	100	100
No. survivals	80	60	20
Proportion	0.8	0.6	0.2
Fitness (w)	1	0.75	0.25
General formulation	1	$1 - sh$	$1 - s$
Selection coefficient			$s = 0.75$
Dominance coefficient			$h = 0.333$

individual. Natural selection acts exclusively on the trait fitness (Darwin, 1859), which can be defined as the global capacity of these replicative units (usually genotypes or individuals) to contribute offspring to the next generation. Evolution is, then, the spatial-temporal genetic change due mainly to natural selection, which acts directly on the genes that affect fitness and, indirectly, on those that determine, by pleiotropy, the different facets of adaptation. The generic concept of fitness can be concretized, in practice, in a series of reproductive traits that, together, constitute fitness itself. These traits are called main components of fitness. Natural selection, therefore, can act at different levels, corresponding to the different components. Thus, selection can act on the differential production of gametes or zygotes in a component of fitness that is often called fecundity. Selection on the differential survival of zygotes or juvenile individuals up to reproductive age acts on the main component called viability or survival. Selection at the level of mating between individuals (preferential mating, vigour of the male, choice of the female, etc.), called sexual selection, acts on the component called mating success. Selection can even occur at the gametic level (meiotic drive), although its consequences tend to be less relevant than in the previous cases.

In order to analyse the changes in frequency by selection, consider, for example, a viability model. Suppose that, in a given generation of a population, after reproduction, 100 zygotes are produced from each of the three genotypes of locus A (Table 2.3), but the differential mortality associated with each of them implies a different number of survivors that reach the reproductive age. The proportion of surviving zygotes in relation to the initial ones is then a measure of fitness, which is generally scaled to unity, a value normally assigned to the genotype with the highest survival, and is usually represented by w. In the example, the fitness of the heterozygote is equal to three quarters of that of the homozygote *AA*, and that of the homozygote *aa* is only a quarter of it. Allele *A* is therefore the advantageous allele, and allele *a* is the deleterious one. The calculations would be similar if they were carried out in terms of the differential fecundity of the genotypes or other reproductive traits implying differentiable contributions with descendants to the next generation, although with some nuances, since some reproductive traits depend on the overall fitness of the pairs that mate instead of the individual genotypes.

In the general model, the selection coefficient (s) is defined as the reduction in fitness of the most disadvantageous genotype (*aa*) in relation to the most advantageous one (*AA*), and the dominance coefficient (h) as the factor that modulates the effect on the heterozygote, that

Table 2.4 *Types of intralocus gene action depending on the value of the dominance coefficient*

h value	Gene action
$0 \leq h < 0.5$	Dominance of allele A and recessivity of allele a
$h = 0.5$	Additivity
$0.5 < h \leq 1$	Dominance of allele a and recessivity of allele A
$h < 0$	Overdominance
$h > 1$	Underdominance

is, it indicates the type of gene action with respect to fitness in the locus considered. The selection coefficient can take values between 0, which indicates absence of selection or neutrality, that is, the fitness of all genotypes is the same, and 1, which indicates lethality of the a allele in homozygosis. The dominance coefficient can take any value, implying different intralocus gene actions (Table 2.4). If $h = 0.5$ the value of the heterozygote is intermediate between that of the two homozygotes and this is referred to as additivity or additive gene action. Deviations from 0.5 to 0 or 1 imply dominance or recessivity of a given allele, incomplete if $0.5 < h < 1$ or $0 < h < 0.5$, or complete if $h = 1$ or $h = 0$, respectively. Finally, h could take values lower than zero, which would imply that the heterozygote's fitness is greater than 1 and, therefore, greater than that of any of the homozygotes, which is called overdominance. In contrast, if $h > 1$ the value of the heterozygote is lower than that of any of the homozygotes, which is referred to as underdominance. In these situations, the fitness of the different genotypes will be scaled to that with the highest fitness, which will be unity. In the example of Table 2.3 the advantageous allele (A) is dominant over the recessive deleterious allele (a).

Now suppose that the genotype frequencies in the population hold the Hardy–Weinberg principle, that is,

	AA	Aa	aa
Frequency	p^2	$2pq$	q^2
Fitness (w)	1	$1 - sh$	$1 - s$

The average fitness of the population ($\overline{w}$) is obtained by multiplying the genotype frequencies by the corresponding fitness values, that is,

$$\overline{w} = p^2 + 2pq(1 - sh) + q^2(1 - s) = 1 - 2pqsh - q^2s = 1 - sq(q + 2ph). \quad (2.17)$$

2.8.2 Change of Frequency of a Lethal Recessive Allele

Let us first illustrate the change in allele frequencies produced by the selection against a deleterious allele and suppose the most extreme case in which that allele is a recessive lethal gene, that is, $s = 1$ and $h = 0$, since the aa genotype is not viable and the allele is only expressed in double dose. In a fecundity model it would be equivalent to a genotype which is sterile. The calculations of the genotypic frequencies after a generation of selection are shown in Table 2.5.

Table 2.5 *Changes in the genotype frequencies for a lethal recessive allele* (s = 1, h = 0) *in the initial generation*

Genotype	AA	Aa	aa
Zygote frequency	p_0^2	$2p_0q_0$	q_0^2
Fitness (w)	1	1	0
Frequency after selection	$\frac{p_0^2}{p_0^2+2p_0q_0}$	$\frac{2p_0q_0}{p_0^2+2p_0q_0}$	$\frac{0}{p_0^2+2p_0q_0}$

From the frequencies of the surviving genotypes after selection in generation 0, the frequency of allele a in generation 1 (q_1) is calculated, as indicated at the beginning of the chapter, as the homozygote aa frequency (which is zero in this case) plus half the frequency of heterozygotes,

$$q_1 = \frac{p_0 q_0}{p_0^2 + 2p_0 q_0} = \frac{q_0}{p_0 + 2q_0} = \frac{q_0}{1 + q_0},$$

which indicates that selection has caused a reduction in the frequency of the allele a in generation 1 with respect to 0. In the second generation we will have, similarly, that $q_2 = q_1/(1 + q_1)$ and, substituting the value of q_1, we obtain $q_2 = q_0/(1 + 2q_0)$. Repeating the procedure in successive generations we arrive at the general expression

$$q_t = \frac{q_0}{1 + t q_0}. \tag{2.18}$$

From equation (2.18) it can be deduced that the time necessary for the frequency of a lethal recessive allele to be reduced by the effect of natural selection to a given value q_t will be $t = (1/q_t) - (1/q_0)$. For example, if the initial frequency of the recessive lethal is $q_0 = 0.01$, the number of generations needed for the frequency to be reduced, say to a half ($q_t = 0.005$), will be $t = 1/q_0 = 100$ generations. This long period of time illustrates a result of great practical interest. Deleterious alleles are usually found at low frequencies in the populations due to selection. We saw in Section 2.2 that when one allele is at a low frequency, if there is Hardy–Weinberg equilibrium, the vast majority of copies of that allele are found in hetero-zygotes. However, for this model of lethal recessive, the fitness of the heterozygote is equal to that of the wild homozygote (AA). Consequently, the low presence of recessive homozygotes (aa) prevents the effective action of natural selection on all alleles a and, therefore, the frequency of the deleterious allele, even if lethal, will be reduced very slowly throughout generations. This reduction is illustrated in Figure 2.7, where the decline in frequency by selection of a recessive lethal gene is presented for different initial frequencies. If we take into account that, as we will see in Chapter 7, deleterious alleles appear continuously by mutation, the result will be that, in practice, the elimination of recessive deleterious alleles, even if they are lethal, is practically impossible, and there will always be alleles of this type present in populations 'hidden' in heterozygotes. Equation (2.18) shows the ineffectiveness of discoura-ging the homozygotes for undesirable recessive alleles from reproducing in order to reduce their frequency, since it is equivalent to the artificial imposition of lethality. This phenomenon explains the presence of a large number of recessive deleterious alleles segregating in natural

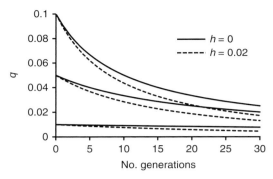

Figure 2.7 Allele frequency reduction throughout generations of a lethal allele ($s = 1$), partially ($h = 0.02$) or completely recessive ($h = 0$), with different initial frequencies in an infinite size population.

populations, which are responsible for some important consequences of the combined effect of selection and inbreeding on quantitative traits, and which we will study in Chapter 8.

If, on the contrary, the lethal alleles are not completely recessive, that is to say, they have a certain deleterious effect in the heterozygote, selection will be more efficient, since it will act against all a alleles, and the allele frequency will be reduced more quickly. Figure 2.7 compares the frequency decline of completely recessive lethals with others whose fitness in the heterozygote is 2% lower than that of the wild homozygote. As can be seen, this minimal deleterious effect in the heterozygote is responsible for natural selection being more efficient in the elimination of the allele.

The fact that deleterious alleles escape the action of natural selection when protected in heterozygotes partially disappears in the case of sex-linked genes. Half of the male offspring of females heterozygous for lethal alleles will be unviable, carrying the allele in hemizygosis, so that the frequency of the lethal will be reduced by half in each generation. This explains why the X chromosomes carry a number of lethal genes (and deleterious genes in general) much lower than the autosomal chromosomes.

2.8.3 Change in the Allele Frequency of a Favourable Allele

So far we have talked about changes in the allele frequency of deleterious alleles due to natural selection. Let us now consider the process from the point of view of beneficial alleles, following the general model of Table 2.6. Suppose that in a given population there is one, or a few, copies of the advantageous allele A, with most individuals being homozygotes aa. It is expected that, due to the effect of selection (and excluding overdominance and under-dominance gene action models), the advantageous allele will increase progressively in frequency until it is fixed, substituting allele a, previously predominant. The changes in frequency that will occur in this general model are presented in Table 2.6.

From Table 2.6 we can calculate the frequency of allele A in the next generation (p'), as $p' = [p^2 + pq(1 - sh)]/\overline{w}$, and the change in allele frequency from one generation to the next (Δp) is

Table 2.6 *Changes in the genotype frequencies for a locus after one generation of selection*

Genotype	AA	Aa	aa
Zygote frequency	p^2	$2pq$	q^2
Fitness (w)	1	$1 - sh$	$1 - s$
Frequency after selection	$\frac{p^2}{\overline{w}}$	$\frac{2pq(1-sh)}{\overline{w}}$	$\frac{q^2(1-s)}{\overline{w}}$
Average fitness (equation (2.17))		$\overline{w} = 1 - sq(q + 2ph)$	

Table 2.7 *Change in allele frequency (Δp) of a favourable allele A under different types of gene action*

	h	Δp
A is dominant	0	$spq^2/\overline{w}$
Additive gene action	1/2	$spq/(2\overline{w})$
A is recessive	1	$sqp^2/\overline{w}$

Note: h is the dominance coefficient, and $\overline{w}$ is the average fitness of the population.

$$\Delta p = p' - p = \frac{p(1 - qsh)}{\overline{w}} - p = \frac{spq[q + h(p - q)]}{\overline{w}}. \tag{2.19}$$

The values of Δp for different types of gene action are presented in Table 2.7.

Note that if the favourable allele is recessive ($h = 1$) and its initial frequency (p) is low, as we had assumed, the change in allele frequency will be very slow while the frequency is low, that is, during the first generations, since Δp is a function of p^2. The explanation is the same as that given to explain the changes in frequencies of deleterious alleles since, obviously, $\Delta q = -\Delta p$. When the advantageous allele is rare in the population, most of its copies are in the heterozygotes, where the selection is not effective given its recessive status. This change is illustrated in Figure 2.8. In contrast, if the allele is dominant ($h = 0$), since the change in frequency per generation is a function of q^2, the initial increase in frequency (p) will be rapid but when p becomes large and, therefore, q small, the change will be very slow then. For a gene with additive gene action, the changes will be rapid with intermediate frequencies of the allele, since Δp is a function of pq.

2.8.4 Overdominance and Underdominance

In the models considered so far, there is unequivocally an advantageous allele and, therefore, the most favourable genotype is one of the two homozygotes, while the fitness of the heterozygote is somewhere between the fitnesses of the two homozygotes. This leads, in a population of infinite size, to the fixation of the favourable allele and the elimination of

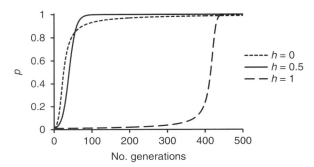

Figure 2.8 Change in allele frequency by selection of a favourable allele (A, with frequency p) initially at low frequency with the course of generations in an infinite size population. The dominance coefficient (h) indicates the type of locus gene action: $h = 0$ (allele A is dominant); $h = 1/2$ (the gene action is additive); $h = 1$ (allele A is recessive).

the deleterious one, reaching in theory a final state characterized by the maximum population average fitness, $\overline{w} = 1$. There are two models of gene action, presented in Table 2.4, in which the situation is different because in none of them is there an unequivocally advantageous allele. In the case of overdominance ($h < 0$), the heterozygote presents greater fitness than any of the homozygotes, so it is more convenient to express the fitnesses relative to that of the heterozygote, as indicated in Table 2.8, so that the homozygotes AA and aa have a selective disadvantage s_A and s_a, respectively. Since the most favourable genotype is that carrying both alleles, natural selection does not lead to the fixation of any of them, but to an equilibrium frequency that is obtained when $s_A p = s_a q$, from where the equilibrium frequencies of Table 2.8 are obtained. For given selection coefficients, the fitness of the population ($\overline{w}$) turns out to be the maximum possible when these frequencies are reached, that is, $\overline{w} = 1 - \hat{p}^2 s_A - \hat{q}^2 s_a$ is the maximum possible fitness at Hardy–Weinberg equilibrium. Therefore, under these conditions, selection will tend to reach and maintain these equilibrium frequencies in a stable way and the permanent conservation of the polymorphism at that locus in the population considered. For this reason, this type of selection is called balancing selection.

For the case of underdominance (the heterozygote has a fitness lower than that of any of the homozygotes), it is convenient to use the same model but changing the signs of the selection coefficients (Table 2.8). This leads to the conclusion that the underdominant case also maintains the same equilibrium frequencies, with the difference that with these frequencies the average fitness of the population, $\overline{w} = 1 + \hat{p}^2 s_A + \hat{q}^2 s_a$, is the minimum possible and, therefore, the equilibrium is unstable. That is, any change of the frequencies will lead to the fixation of allele A (if $p > \hat{p}_A$) or allele a, otherwise. An illustration of the changes in allele frequency with the overdominance and underdominance models is presented in Figure 2.9 for different initial values of allele A frequency. In the overdominant case (panel a), the frequencies will tend to change until reaching the value of equilibrium given by $\hat{p}_A = s_a/(s_A + s_a) = 0.2/(0.1 + 0.2) = 0.666$, that corresponds to the maximum average fitness of the population, and the locus will be maintained at equilibrium in a stable manner. In the underdominant case (panel b), selection leads to the same equilibrium frequencies ($\hat{p}_A =$

Table 2.8 *Models of overdominance and underdominance, where the equilibrium frequencies and the average fitness of the population in each case are indicated*

Genotype	AA	Aa	aa
Genotype frequency	p^2	$2pq$	q^2
Fitness (w) with overdominance	$1 - s_A$	1	$1 - s_a$
Fitness (w) with underdominance	$1 + s_A$	1	$1 + s_a$

	Population mean fitness ($\overline{w}$)	Frequencies at equilibrium	
Overdominance	$1 - p^2 s_A - q^2 s_a$	$\hat{p} = \frac{s_a}{s_A + s_a}$	$\hat{q} = \frac{s_A}{s_A + s_a}$
Underdominance	$1 + p^2 s_A + q^2 s_a$		

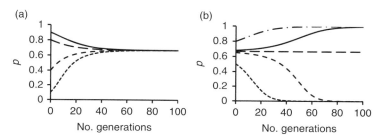

Figure 2.9 Change in allele frequency by selection of an allele with (a) an overdominance or (b) an underdominance model in an infinite size population, following the model in Table 2.8 with selection coefficients $s_A = 0.1$ and $s_a = 0.2$.

0.666) that determine, in this case, that the average fitness of the population is minimal. Therefore, this balance is unstable, so that any modification of these frequencies would produce the fixation of one allele or the other.

Overdominance could be expected to be a frequent phenomenon, since it could be thought that a heterozygous individual with two different alleles would be fitter than a homozygous individual that only has one. For example, it is known that the different allelic variants of an enzyme usually have maximum activity at different pH or temperature, so having two variants can provide greater versatility against environmental changes. However, overdominance has only been demonstrated on very few occasions. Most of the known cases are due to what is called marginal overdominance or pleiotropic overdominance, which is the result of the effect of the gene on two traits, in the so-called antagonistic pleiotropy. Pleiotropy is a frequent phenomenon, as occurs with genes that encode enzymes, since many of them are involved in various metabolic pathways with different phenotypic effects on the individual. An example of antagonistic pleiotropy is presented in Table 2.9, with reference to a gene with effect on mice body size. This trait affects, among other possible, two main components of fitness, fecundity and viability. In the case of fecundity, the allele that increases body size would increase fecundity, because it is known that larger mothers have a larger uterus and, therefore, produce a greater number of offspring. In parallel, a larger body size can be a disadvantage, since larger mice will have more difficulty in hiding to escape predators. Therefore, body size will have a positive correlation with fecundity and negative with viability. The total fitness, which can be obtained from the multiplicative combination of both components, will produce the same result as overdominance for a single trait, although this does not occur for any of the two components but for their combination in the margin of the table, hence the name of marginal overdominance.

Another example of apparent overdominance, or pseudo-overdominance (Charlesworth and Willis, 2009), occurs by the combined action of two or more recessive deleterious loci closely linked or located in regions where recombination is restricted without the need for any of them to be overdominant. For example, a known case is that of some inversions that maintain equilibrium frequencies in natural populations. Since recombination is limited in the inverted region, it is possible that deleterious (even lethal) recessive alleles are accumulated in that region in repulsion phase. That is, while for some loci the wild allele will be found in the inverted region, the corresponding deleterious recessive allele will be in the standard one, and vice versa for other loci. This would be a disadvantage for homokaryotypes (deleterious homozygotes for some loci) for both standard and inverted regions, but an advantage for the heterokaryotype, which would be heterozygous for those loci and, therefore, with maximal fitness.

A known example of marginal overdominance by antagonistic pleiotropy is that of sickle cell haemoglobin in endemic regions of malaria. The mutant allele produces sickle cell anaemia in homozygosis, but provides resistance to malaria in homozygosis and heterozygosis, being the sickle red blood cells inhospitable to the protozoan *Plasmodium falciparum*, transmitted by the *Anopheles* mosquito and responsible for malaria. Of the combined effect of anaemia and malaria, the heterozygous genotype is the most favourable, maintaining equilibrium frequencies of the sickle cell allele relatively high (of the order of 10%) for a very deleterious allele in regions of endemic malaria.

There are several examples of underdominance, among which we can mention the case of the Rh blood group, and that of inversions and translocations, since heterokaryotypes (heterozygotes for inversion or translocation) produce unviable gametes when there is crossing over within the inverted region or for some types of chromosomal segregation in the case of translocations. In many species where inversions and translocations are frequent, parallel systems have evolved to avoid the deleterious effects mentioned. For example, in the genus *Drosophila* the disadvantage of heterokaryotypes is avoided because there is no recombination in males and because unviable recombinant chromosomes are incorporated into the polar

Table 2.9 *Illustrative example of marginal overdominance by antagonistic pleiotropy of a locus, where it is assumed that allele A is favourable dominant for fecundity and unfavourable recessive for viability*

Genotype	AA	Aa	aa
Gene effect on fecundity (f)	1	1	$1 - s$
Gene effect on viability (v)	$1 - s'$	1	1
Global effect on fitness ($w = f \times v$)	$1 - s'$	1	$1 - s$

Note: The overall fitness, obtained in the margin of the table as the product of both components, presents overdominance.

bodies, which do not produce gametes, in females. In other species where these mechanisms do not exist, the inversions and translocations that appear by mutation will generally be eliminated, since their frequency will be less than that of equilibrium (see Figure 2.9b). Its fixation can only be achieved if the frequency exceeds the equilibrium threshold, a phenomenon that is only possible due to genetic drift. This explains why the fixation of inversions and translocations has taken place, preferably, in species with low population sizes, such as in primates.

2.8.5 Natural Selection and Hardy–Weinberg Disequilibrium

As discussed above, one of the possible causes of discrepancy between the observed genotype frequencies and those expected at Hardy–Weinberg equilibrium is the action of natural selection on the locus considered. However, on some occasions, the inference of the action of selection from the disequilibrium can lead to errors. The reason is that the allele frequencies used to obtain the expected genotype frequencies at equilibrium are generally those of the adult individuals present once the selection has taken place, whereas they should be the frequencies prior to selection. To illustrate this, consider the example in Table 2.10. Suppose that, before selection (for example in the zygote or juvenile state), the genotype frequencies for a locus with genotypes AA, Aa and aa are 1/4, 1/2 and 1/4, respectively, and the fitnesses that determine survival from this state to the adult are 1, 0.8 and 0.4, respectively. From the frequencies of surviving adults, the allele frequencies $p = 0.6$ and $q = 0.4$ can be obtained and, from them, the expected ones with Hardy–Weinberg equilibrium (0.36, 0.48 and 0.16, respectively). Note that the observed frequencies (0.333, 0.533, 0.133) imply an excess of heterozygotes and a defect of homozygotes with respect to the expectations, which could lead us to think, erroneously, that the gene action of the locus is overdominance, or that the reason for the disequilibrium is other, for example, negative assortative mating, which would also generate a similar result. The problem arises from using the allele frequencies after selection. If these frequencies had been known before selection ($p = q = 0.5$), the expected values would

Table 2.10 *Illustrative example that selection on viability generates Hardy–Weinberg disequilibrium but that the comparison between expected and observed genotype frequencies can result in an erroneous inference on the type of gene action of the selected locus*

Genotype	AA	Aa	aa
Genotype frequency before selection	0.25	0.5	0.25
Fitness (viability)	1	0.8	0.4
Frequency in the surviving adults	0.25/0.75 = 0.333	0.4/0.75 = 0.533	0.1/0.75 = 0.133
Allele frequency of the survivors: $p = 0.333 + (0.533/2) = 0.6$ $q = 1 - p = 0.4$			
Expected frequency with Hardy–Weinberg equilibrium	p^2 0.36	$2pq$ 0.48	q^2 0.16
Difference between observed and expected frequencies	$0.333 - 0.36 = -0.027$	$0.533 - 0.48 = 0.053$	$0.133 - 0.16 = -0.027$

be 0.25, 0.5 and 0.25, so that the actual fitnesses could have been correctly deduced as follows. The proportion of survivors is, for the three genotypes: $0.333/0.25 = 1.322$; $0.533/0.5 = 1.066$; and $0.133/0.25 = 0.532$. Dividing the three values between the highest one we obtain 1, 0.8 and 0.4, respectively, recovering the correct fitnesses.

Finally, note also that if selection acts at the fecundity level of the parents and the number of individuals of each genotype is calculated before reproduction, no Hardy–Weinberg deviations due to selection can be detected.

2.8.6 Recreation of the Evolutionary Change in the Laboratory

The change in allele frequencies by natural selection is a difficult phenomenon to observe since the evolutionary changes are generally slow, except those triggered by human intervention such as the increase of antibiotic or insecticide resistance genes or climate change (Shaw and Etterson, 2012). However, it has been possible to recreate evolutionary changes by natural selection in the laboratory using microorganisms, such as bacteria and viruses. The advantages of these are their easy propagation, rapid reproduction, the possibility of handling large populations in small spaces, and their asexuality that allows easy cloning (Elena and Lenski, 2003). One of these experiments is described next in which evolution was evaluated by selection of *Escherichia coli* strains for 20,000 generations (Cooper and Lenski, 2000).

A sample of the starting population (ancestral population) remained frozen indefinitely, while others evolved in a minimal medium supplemented with glucose, transferring the cultures every day after a dilution that maintained census sizes of the order of millions of bacteria. The number of cell generations per day was 6.6, which implies that this experiment was carried out in about 3000 days or about 10 years. At certain periods of time, the reproductive capacity of the evolved populations was compared with the ancestral one. As illustrated in Figure 2.10, in each of these analyses the ancestral and evolved population were

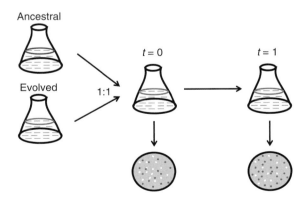

Figure 2.10 Design for estimating the relative fitness of an evolved bacterial population with respect to the ancestral population from which it comes. After mixing the two populations in equal parts, the differential population growth between both populations was evaluated when they competed in the experimental environment. This growth was established from the number of colonies in a plate on days $t = 0$ and 1, distinguishing the colonies of the two types by their colour (red and white). (Adapted with permission from Elena and Lenski, 2003.)

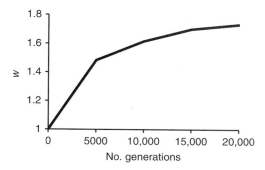

Figure 2.11 Evolution of the average fitness (w) of *Escherichia coli* populations throughout generations, estimated as the ratio of the bacterial growth rate of the evolved populations with respect to the growth rate of the ancestral population. (Data taken from Cooper and Lenski, 2000.)

mixed in equal parts and diluted in the experimental culture, evaluating the number of colonies in both populations on the plate, distinguishable by the colour associated with a different genetic marker in each of them. After a day of competition between the two populations, another evaluation was carried out. The relative fitness of the evolved population with respect to the ancestral one was measured as the ratio between the observed growth rate from $t = 1$–2 in the evolved population and in the ancestral population.

    The evolution of the relative fitness of the evolved populations (average of 12 popula-tions) with respect to the ancestral population is illustrated in Figure 2.11. Note that this figure does not show the increase in frequency of a single advantageous gene, but the overall increase in fitness due to changes in frequency of several, or perhaps many, of them, present at the beginning of the experiment at low frequencies or which appeared by mutation throughout it. During the first 5000 generations, the increase in fitness was quick, the average fitness of the evolved populations (that is, their growth rate) reaching a value 50% higher than that of the ancestral population. However, the rate of increase in fitness was substantially reduced in subsequent generations, due to the fixation of advantageous genes as well as the antagonistic pleiotropic effects of genes involved in different metabolic functions (Cooper and Lenski, 2000).

Problems

2.1 In a population of gazelles, 200 individuals were analysed for the α-amylase gene, finding three allelic variants (A_1, A_2 and A_3) and the observed genotype frequencies indicated in the following table. (a) What is the value of the expected heterozygosity in the popula-tion? (b) Is the population in Hardy–Weinberg equilibrium for that locus?

Genotype	A_1A_1	A_1A_2	A_1A_3	A_2A_2	A_2A_3	A_3A_3	Total
Number	23	61	28	39	41	8	200

2.2 In an analysis of a sample of 200 individuals, haplotypes were found for two SNPs with alleles A and a, and B and b, respectively, as shown in the following table. (a) Calculate the linkage disequilibrium (D), its value relative to the maximum possible (D') and the value of the squared correlation between the allele frequencies of the loci, r^2. (b) Is the observed disequilibrium significant? (c) If the recombination frequency between the two SNPs is 0.05, how many generations would it take for the disequilibrium to be reduced by a half?

Haplotype	AB	Ab	aB	ab	Total
Number	29	37	88	46	200

2.3 A large population of *Drosophila melanogaster* has been subdivided into 64 lines of constant census size $N = 40$ that are maintained with panmixia for 50 generations. A biallelic genetic marker is found in generation $t = 0$ with allele frequency $q_0 = 0.4$ for the minority allele. Calculate for generation $t = 50$: (a) the expected frequency of the allele in the lines. (b) The expected variance of the allele frequencies between lines. (c) The expected heterozygosity for the marker.

2.4 The frequencies of an allele in four island populations between which migration is assumed to occur at a rate $m = 0.1$ per generation are 0.3, 0.2, 0.4 and 0.1, respectively. (a) What will be the allele frequencies on each island after 10 generations? And after 50? (b) If an allele frequency of 0.275 is estimated for the first island in generation 10, what would be the most likely estimate of the migration rate?

2.5 Individuals from a population of snails were sampled with three colours of the shell and, after marking them for identification, they were released in the field. The number of individuals of each phenotype was 80 black, 240 brown and 180 white, a character determined by a biallelic locus with *AA*, *Aa* and *aa* genotypes, respectively. After five days, the individuals were re-sampled, finding 72 blacks, 130 browns and 33 whites. (a) What are the fitnesses of each genotype and the selection and dominance coefficients of the locus? (b) What is the average fitness of the population and what would be the expected change in allele frequency after one generation of selection?

2.6 The genotype frequencies for a biallelic locus in the juvenile stage of a population of 325 worms were: 35 of genotype *AA*, 144 of genotype *Aa* and 146 of genotype *aa*. In the adult stage the numbers became: 22, 130 and 75, respectively. (a) What are the relative fitnesses of the three genotypes? What kind of gene action does the locus have? (b) What is the average fitness of the population? Is this the maximum possible fitness? Why?

Self-Assessment Questions

1 Allele and gametic frequencies coincide in the case of a single locus.
2 Gene diversity is the observed frequency of heterozygotes.
3 Positive assortative mating implies a heterozygous defect.
4 Linkage disequilibrium occurs only between linked genes.
5 In an ideal population, an allele with an initial frequency of 0.2 is expected to be fixed with a probability of 80%.

6 If the mutation rate of a locus is $u = 10^{-6}$ per generation, in a diploid population of size $N = 10^6$ a mutation is expected at that locus each generation.

7 The change in allele frequency of a beneficial recessive allele will be slow when the allele is at low frequency.

8 Overdominance maintains allele frequencies equal to 1/2.

9 The Rh blood group is an example of underdominance.

10 An excess of heterozygotes necessarily indicates overdominant gene action of the locus for fitness.

3 Components of Phenotypic Values and Variances

Concepts to Study

- Genotypic values, additive values and dominance deviations
- Additive, dominance and epistatic variance
- Linkage disequilibrium variance
- Heritability and genetic correlation
- Phenotypic plasticity
- Genotype–environment interaction and correlation
- Repeatability

Objectives for Learning

- To understand the partition of the phenotypic value and variance into their genotypic and environmental components and the genotypic component into its additive, dominance and epistatic ones
- To learn the concept of heritability and its importance in the determination of the resemblance between relatives and the genetic improvement of plants and animals
- To understand the importance of phenotypic plasticity and the interaction between genotype and environment in evolution, as well as its implications in the fields of animal and plant breeding and conservation
- To learn how to undertake the partition of environmental variance in its known or controlled and unknown or uncontrolled components

3.1 Decomposition of the Genotypic Value for a Locus

As already indicated in Chapter 1, the phenotypic value (P) of an individual for a quantitative trait, deviated from the population mean, is decomposed into the genotypic value (G), determined by the genetic endowment of the individual, and the environmental deviation (E), that is, $P = G + E$. The genotypic value for a locus, in turn, can be decomposed into two elements, the additive value (A), also called genic value and breeding value, the latter in the context of animal breeding, and the dominance deviation (D), that is,

$$G = A + D. \tag{3.1}$$

This decomposition is of great practical use, since, as we will see later, the additive value is the key component in determining the resemblance between parents and offspring. For this reason, the additive value is also essential for the prediction of the response to selection, which is one of the main goals of evolutionary genetics and plant and animal breeding.

To understand the partition of the genotypic value into components, it is necessary to refer to the model of a single locus even though, as we have already indicated, it is not usual to know the frequency and individual effects of the genes that contribute to genetic variation for a quantitative character. Let us start by explaining a very simple model, presented in Table 3.1, which allows us to understand the concept. Consider a given locus A that affects human height, for which there are two alleles, A_1 and A_2, present in a large size population. Let us suppose that the genotypic value of the genotype A_2A_2 is an increase of 10 mm in the height of its carriers relative to the genotype A_1A_1, which would be, therefore, zero. Let us also suppose that the heterozygotes A_1A_2 have an intermediate value between the two homozygotes, that is, 5 mm, so that the locus gene action is additive. This case corresponds to that presented in the left part of Table 3.1. It is evident that genes do not confer genotypic value until they are united in pairs to form genotypes, but it is useful to attribute them a value prior to the union. The additive value of an individual is then defined as the sum of the effects of the alleles carried by that individual. In the example of Table 3.1 the effects of the A_1 and A_2 alleles are, respectively, 5 mm and 0, and the genotypic and additive values are the same in this case. The dominance deviations, which are simply the difference between the genotypic and the additive values, are, in this case, nil. In the right part of the table a somewhat different model is assumed, in which the heterozygote has a genotypic value closer to that of the homozygote A_2A_2, that is, there is partial dominance of the A_2 allele over the A_1. The additive values remain the same but now there is a dominance deviation in the heterozygote of 3 mm. The fact of associating a value to each allele of the locus has a great utility, since it is the genes, and not the genotypes, that are transmitted to the offspring. In practice, as we will see later, the additive value of an individual can be estimated from the average genotypic value of their progeny or by using information about the resemblance between relatives of different types.

The model shown in Table 3.1 is illustrative for understanding the concept of additive value and dominance deviation, but it entails two important simplifications: (1) that the frequencies of the two alleles in the population are equal, and (2) that the mean of the trait for the locus in the population is zero. In fact, the additive and dominance values are defined as deviations from the population mean and also depend on the allele frequencies, so they can vary from one population to another. The general decomposition is presented below.

Consider the genotypic values of the dominance model in Table 3.1 and suppose that the frequencies of the three genotypes in the population are in Hardy–Weinberg equilibrium, with p and q being the frequencies of alleles A_1 and A_2, respectively (Table 3.2). The table shows the algebraic expressions of the general model on the left and the solutions corresponding to the numerical example on the right, where the allele frequencies are $p = 0.6$ and $q = 0.4$.

Table 3.1 *Simplified decomposition of the genotypic value for a locus*

		Additive model			Dominance model	
Genotype	A_1A_1	A_1A_2	A_2A_2	A_1A_1	A_1A_2	A_2A_2
Genotypic value (G)	0	5	10	0	8	10
Additive value (A)	0	$5 + 0$	$5 + 5$	0	$5 + 0$	$5 + 5$
Dominance deviation (D)	0	0	0	0	3	0

Table 3.2 *General decomposition of the genotypic value for a locus*

	General model			Numerical example		
Genotype	A_1A_1	A_1A_2	A_2A_2	A_1A_1	A_1A_2	A_2A_2
Genotype frequency	p^2	$2pq$	q^2	0.36	0.48	0.16
Genotypic value (G)	0	ah	a	0	8	10
G value deviated from M	$0 - M$	$ah - M$	$a - M$	−5.44	2.56	4.56
Additive value (A)	$2\alpha_1 = -2q\alpha$	$\alpha_1 + \alpha_2 = (p - q)\alpha$	$2\alpha_2 = 2p\alpha$	−4.48	1.12	6.72
Dominance deviation (D)	$-2dq^2$	$2dpq$	$-2dp^2$	−0.96	1.44	−2.16
Dominance effect		$d = a(h - 1/2)$			3	
Genotypic mean		$M = aq + 2dpq$			5.44	
Average effect of allelic substitution		$\alpha = ah - 2dq$			5.60	
Average effect of allele A_1		$\alpha_1 = -q\alpha$			−2.24	
Average effect of allele A_2		$\alpha_2 = p\alpha$			3.36	

Note: The genotypic mean M corresponds to a scale whose origin is the homozygote A_1A_1 value.

Note that the model of genotypic values is analogous to that described in Chapter 2 for fitness (Table 2.3), the generic value a being equivalent to the selection coefficient s, although a different scale is used. The effect in the heterozygote is modulated by the term h or dominance coefficient, described in Table 2.4 of Chapter 2, which indicates the degree of deviation with respect to additivity. Dominance can also be expressed as the difference d (the dominance effect) between the genotypic value of the heterozygote and the average of the values of the two homozygotes, that is,

$$d = ah - (a/2) = a(h - \tfrac{1}{2}). \tag{3.2}$$

Recall that a value of $h = 0.5$ implies that the value of the heterozygote is intermediate between those of the homozygotes, that is, that the gene action is additive ($d = 0$). Values of $0 \le h \le 0.5$ or $0.5 < h < 1$ correspond to dominant gene action ($d \ne 0$), A_1 being the dominant allele and A_2 the recessive, or vice versa. Finally, $h < -1$ indicates underdominance (the heterozygote confers a lower value than any of the homozygotes) and $h > 1$ overdominance (the heterozygote confers a greater value than any of the homozygotes). In the numerical example of Table 3.2, $a = 10$, $h = 0.8$ and $d = 8 - [0 + 10]/2 = 3$.

By means of the sum of the products of genotypic values and frequencies, the contribution of the locus to the population mean is calculated as

$$M = 2ahpq + aq^2 = 2ahq - 2dq^2 = aq + 2dpq, \tag{3.3}$$

which implies, in the example, an average deviation of 5.44 mm with respect to the height of the A_1A_1 individuals.

Let us now calculate the average effect associated with each allele, which, as we have said, will depend on the allele frequencies and the population mean, M. The average effects can be obtained in several ways. The simplest is to calculate them as the average of the trait in the individuals carrying at least one copy of the allele in question, while the other comes at random from the population. For the A_1 allele, the mean of the individuals with genotype A_{1-} (where the low line indicates that the other allele can be A_1 with probability p or A_2 with probability q) would be $0p + ahq = ahq$. Likewise, the mean of the individuals A_{2-} would be $ahp + aq$. These values, deviated from the population mean (M; equation (3.3)) are the average effects of the alleles,

$$\alpha_1 = ahq - M = -q[ah - 2dq] = -q\alpha \tag{3.4}$$

$$\alpha_2 = ahp + aq - M = p[ah - 2dq] = p\alpha, \tag{3.5}$$

for alleles A_1 and A_2, respectively. The calculations would be done in the same way if there were more alleles in the locus. The common term

$$\alpha = ah - 2dq \tag{3.6}$$

is also called the average effect of allelic substitution and can be obtained simply as $\alpha_2 - \alpha_1$ with biallelic loci.

With the definitions of the average effects of the alleles represented in equations (3.4) and (3.5), the additive values (A) can be obtained, defined as the sum of the effects of the genes carried by each genotype, that is, $2\alpha_1$, $\alpha_1 + \alpha_2$ and $2\alpha_2$ for genotypes A_1A_1, A_1A_2 and A_2A_2, respectively.

The dominance deviations (D) can be obtained simply as the difference between the genotypic values deviated from the population mean and the additive values. For example, for the A_1A_1 genotype, the dominance deviation is obtained as $(0 - M) + 2q\alpha = -[aq + 2dpq] + 2q\alpha = -2dq^2$, and analogously for the other genotypes. The resulting terms are shown in Table 3.2.

Note that, since additive values and dominance deviations are defined as deviations from the population mean, M, their means, $\overline{A}$ and $\overline{D}$, are zero. This can be checked by adding the products of the A or D values by the genotype frequencies of each genotype in Table 3.2. Note also that in the absence of dominance ($d = 0$), that is, with additive gene action, the genotypic values are perfectly known from the average effects of the genes that compose them, that is, $G = A$. If there is dominance, that knowledge still exists but it is no longer perfect and the additive values depend on both a and d.

Originally, Fisher (1918, 1930) defined the average effects of alleles from the linear regression of the genotypic values, weighted by their respective frequencies, on gene dose, this being the number of alleles of a given type in the different genotypes, for example, zero, one and two A_2 alleles for genotypes A_1A_1, A_1A_2 and A_2A_2, respectively. The regression coefficient obtained is α (Figure 3.1), that is, α is the slope of the line that allows the prediction of the genotypic value to be obtained from the number of alleles. The dominance deviations are then established as the differences between the genotypic values and their linear prediction, in particular as those values that minimize the average quadratic difference between G and A. Therefore, from a statistical point of view, the dominance deviations can be considered as a factor of mismatch to the linear prediction or, what is the same, as the interaction between the allelic effects for the determination of the genotypic value. The regression of the values of the character on the number of allelic copies per individual is, as will be seen below, a general method with multiple applications, such as the identification and mapping of genes with effect on quantitative traits using molecular markers.

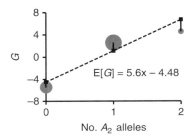

Figure 3.1 Regression of the genotypic value (G, deviated from the population mean) on gene dose (number of A_2 alleles in the genotype) for the numerical example of Table 3.2. The circular areas are proportional to the frequency of each genotype. The slope of the regression line is the average effect of an allelic substitution ($\alpha = 5.6$). The squares on the line represent the additive values (A), and the vertical bars between them and the centre of the circular areas indicate the dominance deviations (D).

As we discussed earlier, the additive value of an individual, precisely because of its additive nature, can easily be estimated as twice the deviation of the average trait value of its offspring from the population mean when the other parent is taken at random. Suppose that an individual of genotype A_1A_1, for example a male, is crossed with multiple females taken at random from the population and the character is evaluated in a descendant of each of these crosses, which is called a progeny test. The male A_1A_1 will provide the A_1 allele to all its offspring, the maternal allele being A_1 with probability p or A_2 with probability q, and the average of its offspring will be $ahq - M = -q\alpha = \alpha_1$, that is, the average effect of the A_1 allele. Since the male only contributes one gamete to each descendant, the additive value of the male is double, $2\alpha_1$, as would be expected. If the individual were a heterozygote, the argument is the same except that half of its progeny carries the A_1 allele and the other half the A_2 allele, with its additive value being $\alpha_1 + \alpha_2$. Therefore, the additive value of an individual can be estimated through the value of the trait in its progeny and it is precisely this method that has been commonly used to evaluate the genetic values of the available parents to select the most suitable in plant and animal breeding.

In the same way that the additive value of an individual can be estimated from the genotypic value of their offspring, it can also be said that the expected genotypic value of an individual is equal to the average of the additive values of their parents,

$$E[G_{\text{offspring}}] = \frac{A_{\text{father}} + A_{\text{mother}}}{2}, \tag{3.7}$$

since the average value of the gametes transmitted by the father and the mother to a descendant is precisely its additive value. This additive property is responsible for the resemblance between relatives, and the variation of the additive values of the individuals of a population, as we will see later, is the main factor for the genetic improvement of plants and animals.

3.2 Decomposition of the Genotypic Variance for a Locus

As discussed in Chapter 1, the genetic description of a population for a quantitative trait is based primarily on the mean and the variance of the character in the population. The

Table 3.3 *Decomposition of the genotypic variance for a locus*

	General model	Numerical example
Additive variance (V_A)	$2\alpha^2 pq$	15.05
Dominance variance (V_D)	$(2dpq)^2$	2.07
Genotypic variance (V_G)	$2\alpha^2 pq + (2dpq)^2$	17.12

phenotypic variance is constituted by the genotypic and environmental components, and just as the genotypic value can be decomposed into additive and dominance components, the genotypic variance (V_G) is decomposed into additive variance (V_A), the variance of the additive values, and dominance variance (V_D), the variance of dominance deviations,

$$V_G = V_A + V_D. \tag{3.8}$$

We will start by obtaining these components with the expressions in Table 3.2 referring to a locus. Since the additive and dominance values are deviated from the population mean, their variance is obtained simply by squaring the values and weighing them by the genotype frequencies, that is,

$$V_A = 4\alpha^2 p^2 q^2 + 2pq\alpha^2(p-q)^2 + 4\alpha^2 p^2 q^2 = 2\alpha^2 pq \tag{3.9}$$

$$V_D = 4d^2 p^2 q^4 + 8d^2 p^3 q^3 + 4d^2 p^4 q^2 = (2dpq)^2. \tag{3.10}$$

Note that $2pq$, the common term in the two expressions, is the expected heterozygosity or expected frequency of heterozygotes with Hardy–Weinberg equilibrium. The numerical values corresponding to the example of Table 3.2 are presented in Table 3.3.

The decomposition of the genotypic value in the sum of its components implies that the additive and dominance values are independent, since it is an orthogonal partition, so the covariance between both components is zero. We can verify this by calculating the corresponding covariance cov(A, D) as the sum of $A \times D$ products weighted by their frequencies, cov(A, D) = $4p^2 q^3 \alpha d - 4p^2 q^2 (p-q)\alpha d + 4p^3 q^2 \alpha d = 0$.

The relative magnitude of the additive and dominance variances depends on the type of gene action and the allele frequencies. Figure 3.2 shows the values of α and of the components of additive (V_A) and dominance (V_D) variance for the numerical example of Tables 3.2 and 3.3, considering different frequencies of the A_1 allele, that is, p. Note that the magnitude of V_D is much smaller than that of V_A for most allele frequencies, since while V_A is proportional to heterozygosity, V_D is proportional to the square of heterozygosity. If there is complete dominance ($h = 0$ or 1) and the frequency of the recessive allele is low, V_D may be greater than V_A. However, even with full dominance, $V_A > V_D$ if the frequencies are intermediate ($1/3 > p > 2/3$). For example, if $p = q$, we obtain that $\alpha = a/2$, and $V_A = a^2/8$, twice V_D. This is why most quantitative traits tend to present an additive variance substantially greater than the dominance variance, as will be seen in more detail.

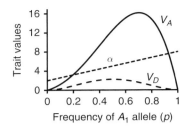

Figure 3.2 Values of the average effect of allelic substitution, α, and of the components of additive (V_A) and dominance (V_D) variance for the numerical example of Tables 3.2 and 3.3 (where $a = 10$ and $h = 0.8$), depending on the allele A_1 frequency (p). The values of α in Table 3.2 and of V_A and V_D in Table 3.3 correspond to a value of $p = 0.6$.

So far we have considered models of gene action with additivity or dominance, where the maximum value of the character is given for one of the homozygous genotypes. In Chapter 2 we studied the cases of overdominance and underdominance for fitness (advantage and disadvantage of the heterozygote, respectively). With these models, a balance in gene frequencies can be reached (stable in the case of overdominance and unstable in the case of underdominance). Recall that, in the case of overdominance, where the genotypic values of the genotypes A_1A_1, A_1A_2 and A_2A_2 were $1 - s_A$, 1, $1 - s_a$, respectively (see Table 2.8 of Chapter 2), the equilibrium frequency of the A allele was $\hat{p} = s_a/(s_A + s_a)$, in which case the average fitness of the population is the maximal one. If we use the equivalent model in the terms presented in this chapter (Table 3.2), the genotypic values will be 0, ah ($h > 1$) and a. The corresponding equilibrium frequency is $\hat{q} = [(a/2) + d]/2d$ and, substituting the value of the dominance effect d from equation (3.2), $d = a(h - \frac{1}{2})$, we obtain $\hat{q} = h/(2h - 1)$. Note that, with the equilibrium frequencies, the additive variance is zero, which can be checked by substituting the corresponding values in the equation of α in Table 3.2, $\alpha = ah - 2d\hat{q} = 0$, with which $V_A = 2\alpha^2 pq = 0$. As we will see in Chapter 10, because the mean fitness in the equilibrium is maximal, the change in the mean by natural selection must be null, which implies absence of additive variance.

3.3 Decomposition of Genotypic Values and Variances for More than One Locus

3.3.1 Gametic Disequilibrium Variance

The decomposition of the genotypic value described above refers to the case of a single locus. If we consider two or more loci with effect on the quantitative trait under study, the additive and dominance variances are simply the sum of the corresponding contributions from each locus. However, the variance of the multilocus genotypic values may be different from the sum of the contributions of each locus when the frequencies and/or effects at different loci are not independent. The first case refers to the possible existence of variance generated by gametic phase or linkage disequilibrium. As already explained in Chapter 2, gametic disequilibrium is the association in gene frequencies between the alleles of two or more loci and can be generated by various reasons, among which selection, genetic drift and

Table 3.4 *Example of effects and frequencies for two loci*

Genotype of locus A	A_1A_1	A_1A_2	A_2A_2
Genotypic value	0	8	10
Frequency	0.36	0.48	0.16
Genotype of locus B	B_1B_1	B_1B_2	B_2B_2
Genotypic value	0	6	12
Frequency	0.25	0.5	0.25

Loci A and B		A_1A_1	A_1A_2	A_2A_2
Multilocus genotypic value	B_1B_1	0	8	10
	B_1B_2	6	14	16
	B_2B_2	12	20	22

population mixing stand out. Gametic disequilibrium generates a component of variance that can increase or reduce the total genotypic variance, depending on the sign of the disequilibrium in relation to the effects of the alleles. The variance of disequilibrium is in fact a covariance, since the variance of the sum of the genotypic values of two loci, A and B, is $V(G_{loc.A} + G_{loc.B}) = V_{G(loc.A)} + V_{G(loc.B)} + 2\text{cov}(G_{loc.A}, G_{loc.B})$, and this last term can be positive or negative depending on linkage disequilibrium. To understand this variance component let us return to a numerical example and suppose that in addition to the A locus presented in Table 3.2 there is another locus controlling the character (locus B), also biallelic, with the genotypic effects and frequencies shown in Table 3.4. The new locus considered is additive and segregates in the population with equal frequencies of each allele ($p = q = 0.5$) in Hardy–Weinberg equilibrium.

If there were no correlation in frequencies between the alleles of the two loci, that is, with gametic equilibrium ($D = 0$ with equation (2.4) of Chapter 2; do not confuse this term with the dominance deviation, which is traditionally expressed with the same symbol), the four gametic types (A_1B_1, A_1B_2, A_2B_1 and A_2B_2) would have frequencies equal to 0.3, 0.3, 0.2 and 0.2, respectively (Table 3.5-I), and the genotype frequencies resulting from the random union of these gametes would be those indicated in the table. With these genotype frequencies and the corresponding multilocus genotypic values of Table 3.4, the contribution of both loci to the mean of the population is 11.44 and to the genotypic variance $V_G = 35.13$.

Suppose now that there is a correlation between alleles A_1 and B_1, so that gametes A_1B_1 and A_2B_2 are more frequent than expected at equilibrium (and the alternative ones, A_1B_2 and A_2B_1 are, therefore, less frequent), being the gametic frequencies of A_1B_1, A_1B_2, A_2B_1 and A_2B_2 equal to 0.4, 0.2, 0.1 and 0.3, respectively (Table 3.5-II), corresponding to a disequilibrium coefficient $D = 0.1$). The allele frequencies of the two loci do not change, and the same occurs with the genotype frequencies of each locus separately (see the marginal frequencies in Table 3.5-II), as well as with the mean (11.44). However, the multilocus genotype frequencies resulting from the union of gametes change with respect to the case of equilibrium, becoming the genotypes with more extreme values ($A_1A_1B_1B_1$ and $A_2A_2B_2B_2$, with effects 0 and 22, respectively) more common in the population. This induces a greater genotypic variance (48.57) than that obtained with independent segregation of the two loci (35.13).

Table 3.5 *Example of gametic phase disequilibrium genotypic variances*

I Gametic equilibrium		A_1B_1	A_1B_2	A_2B_1	A_2B_2	D
Gametic frequency		0.3	0.3	0.2	0.2	0
			A_1A_1	A_1A_2	A_2A_2	
Genotype frequency	B_1B_1		0.09	0.12	0.04	0.25
Mean = 11.44	B_1B_2		0.18	0.24	0.08	0.50
$V_G = 35.13$	B_2B_2		0.09	0.12	0.04	0.25
			0.36	0.48	0.16	1
II Positive gametic disequilibrium		A_1B_1	A_1B_2	A_2B_1	A_2B_2	D
Gametic frequency		0.4	0.2	0.1	0.3	0.1
			A_1A_1	A_1A_2	A_2A_2	
Genotype frequency	B_1B_1		0.16	0.08	0.01	0.25
Mean = 11.44	B_1B_2		0.16	0.28	0.06	0.50
$V_G = 48.57$	B_2B_2		0.04	0.12	0.09	0.25
			0.36	0.48	0.16	1
III Negative gametic disequilibrium		A_1B_1	A_1B_2	A_2B_1	A_2B_2	D
Gametic frequency		0.1	0.3	0.4	0.2	−0.1
			A_1A_1	A_1A_2	A_2A_2	
Genotype frequency	B_1B_1		0.04	0.12	0.09	0.25
Mean = 11.44	B_1B_2		0.16	0.28	0.06	0.50
$V_G = 21.69$	B_2B_2		0.16	0.08	0.01	0.25
			0.36	0.48	0.16	1

Note: D is the gametic disequilibrium corresponding to the gametic frequencies and is obtained with equation (2.4) of Chapter 2.

If, on the contrary, the association occurs between A_1 and B_2 alleles, so that the gametes A_1B_2 and A_2B_1 increase in frequency with respect to the equilibrium case (Table 3.5-III, disequilibrium coefficient of $D = -0.1$), the individuals with more extreme genotypes will become less common, which implies that the variance of disequilibrium will reduce the genotypic variance (to 21.69). The disequilibrium variance is transient and is reduced as the gametic disequilibrium is diminished by chromosome segregation or recombination, as presented in equation (2.5) of Chapter 2, but can be maintained for many generations if the loci are tightly linked. The component of the additive genetic variance that is unaltered by changes in the gametic phase disequilibrium, that is, that obtained by the sum of the additive variances for the different loci considered separately, is called additive genic variance or simply genic variance (Walsh and Lynch, 2019, p. 504).

There is another type of gametic disequilibrium variance, different from linkage disequilibrium, which arises when crosses do not occur randomly but there is assortative mating between genotypes. For example, when individuals of genotype $A_1A_1B_1B_1$ tend to mate more frequently with individuals of the same genotype than with those of any other. In this case, a covariance is generated due to the correlation between the alleles carried by the uniting gametes. This source of disequilibrium, however, is much more labile than the previous one since it disappears as soon as individuals mate randomly. It would be permanent, however, if the type of assortative mating persists.

3.3.2 Epistatic Variance

Regardless of the existence of gametic disequilibrium variance, the multilocus genotypic variance can deviate from the sum of the contributions of each locus if the combined genotypic effects are not the simple sum of the effects of each locus separately, which is known as epistasis or epistatic interaction between loci. That is, the global multilocus genotypic value (G) would be the sum of the additive components (ΣA) and the dominance deviations (ΣD) of the different loci, plus the sum of all the possible interactions between these values, which can be collected in a global term, I:

$$G = \Sigma A + \Sigma D + I. \tag{3.11}$$

Interactions between the effects of loci can occur between their additive values (I_{AA}, for two loci, I_{AAA} for three loci, etc.), between dominance values (I_{DD}, I_{DDD}, ...) or between additive and dominance values (I_{AD}, I_{AAD}, ...), that is to say,

$$I = I_{AA} + I_{AD} + I_{DD} + I_{AAA} + I_{AAD} + \ldots \tag{3.12}$$

Generally, only interactions of pairs of loci are considered, since higher order interactions have proportionally less importance the greater the number of loci involved in the interaction, as the probability that all of them are segregating in the population is reduced.

As for the other components of the genotypic value, the epistatic interactions constitute a source of genotypic variation, the epistatic variance (V_I), so that, in a multilocus model, the global decomposition of the genotypic variance is

$$V_G = V_A + V_D + V_I. \tag{3.13}$$

The epistatic variance can, in turn, be decomposed according to the factors that give rise to interaction, that is, the additive by additive (V_{AA}), additive by dominance (V_{AD}), or dominance by dominance (V_{DD}) variance, for interactions between two loci, V_{AAA}, V_{AAD} and so on, for three loci, and so on:

$$V_I = V_{AA} + V_{AD} + V_{DD} + V_{AAA} + V_{AAD} + \ldots \tag{3.14}$$

As for the other components of the genotypic variance, the magnitudes of the epistatic variance depend on the model of gene action and the genotype frequencies.

Obviously, multiple epistatic models can be considered (see the analysis of different models by López-Fanjul et al., 2000). As an illustration, we will study one of the simplest and possibly most realistic models, which consists in assuming that the double homozygote $A_2A_2B_2B_2$ for two biallelic epistatic loci presents an effect different from the sum of the effects of the two loci separately. The algebraic model and a numerical example are shown in Table 3.6. The epistatic factor (k) can be greater than 1, in which case we speak of synergistic epistasis (the value of the double homozygote is greater than the sum of the values of the two homozygotes separately), less than 1 (antagonistic epistasis), or equal to 1 when there is no epistasis. In the numerical example we assume the effects and frequencies

Table 3.6 *Example of an epistatic model*

General model					Numerical example		
	A_1A_1 (p^2)	A_1A_2 ($2pq$)	A_2A_2 (q^2)		A_1A_1 (0.36)	A_1A_2 (0.48)	A_2A_2 (0.16)
B_1B_1 (r^2)	0 (p^2r^2)	ah ($2pqr^2$)	a (q^2r^2)	B_1B_1 (0.25)	0 (0.09)	8 (0.12)	10 (0.04)
B_1B_2 ($2rs$)	$a'h'$ ($2p^2rs$)	$ah + a'h'$ ($4pqrs$)	$a + a'h'$ ($2q^2rs$)	B_1B_2 (0.5)	6 (0.18)	14 (0.24)	16 (0.08)
B_2B_2 (s^2)	a' (p^2s^2)	$ah + a'$ ($2pqs^2$)	$k(a + a')$ (q^2s^2)	B_2B_2 (0.25)	12 (0.09)	20 (0.12)	44 (0.04)

Epistatic factor: k 2

Value in excess   22

for $A_2A_2B_2B_2$: $c = (k-1)(a+a')$

Locus A	Locus B	Locus A	Locus B
a; h	a'; h'	10; 0.8	12; 0.5
$d = a(h - 1/2)$	$d' = a'(h' - 1/2)$	3	0
$d_e = d - (c/2)s^2$	$d_e' = d' - (c/2)q^2$	0.25	−1.76
$\alpha_e = ah - 2d_e q$	$\alpha_e' = a'h' - 2d_e's$	7.8	7.76

Epistatic effects

$i(\alpha_e \times \alpha_e') = cqs$	4.4
$i(\alpha_e \times d_e') = (c/2)q$	4.4
$i(d_e \times \alpha_e') = (c/2)s$	5.5
$i(d_e \times d_e') = (c/4)$	5.5

Global components

$M = (aq + 2dpq) + (a's + 2d'rs) + cq^2s^2$	12.32
$V_A = 2\alpha_e^2 pq + 2\alpha_e'^2 rs$	59.31
$V_D = (2d_e pq)^2 + (2d_e'rs)^2$	0.79
$V_{AA} = 4i(\alpha_e \times \alpha_e')^2 pqrs = 4c^2q^2s^2 pqrs$	4.65
$V_{AD} = 8i(\alpha_e \times d_e')^2 pqr^2s^2 + 8i(d_e \times \alpha_e')^2 p^2q^2 rs =$ $= 2c^2q^2s^2(pqr^2 + p^2rs) = 2c^2pq^2rs^2(qr + ps)$	5.81
$V_{DD} = 16\, i(d_e \times d_e')^2 p^2q^2r^2s^2 = c^2p^2q^2r^2s^2$	1.74
$V_I = V_{AA} + V_{AD} + V_{DD} = c^2q^2(1 - q^2)s^2(1 - s^2)$	12.20
$V_G = V_A + V_D + V_I$	72.30

of Tables 3.4 and 3.5-I including synergistic epistasis with an epistatic factor value of $k = 2$, that is, the excess of the double homozygote value in relation to its expected additive value is $c = (k - 1)(a + a') = a + a'$. The marginal dominance effects for each locus (d_e and d_e') can be obtained from the partial derivatives of the mean with respect to the allele frequencies (see Kojima, 1959) and are a function of the degree of dominance at each locus (equation (3.2)) and the magnitude of c (see Table 3.6).

Note that although there was no dominance in either of the two loci ($h = h' = 0.5$; $d = d' = 0$), epistasis generates marginal dominance effects, that is, d_e and d_e' would be different from zero. The average effects of an allelic substitution (α_e and α_e') are obtained from the derivative of the genotypic mean with respect to the allele frequencies, and are expressed simply by means of equation (3.6) applied to the dominance effects d_e and d_e'. The epistatic effects of additive by additive $[i(\alpha_e \times \alpha_e')]$, additive by dominance $[i(\alpha_e \times d_e')$ and $i(d_e \times \alpha_e')]$, and dominance by dominance $[i(d_e \times d_e')]$ interactions are obtained from the second, third and fourth derivatives, respectively, of the genotypic mean with respect to the allele frequencies and are only a function of c and of the allele frequencies. Note that epistatic effects only exist if there is epistasis ($k \neq 1$) and do not depend, in this model, on the type of intralocus gene action, that is, they do not depend on the degree of dominance (h and h'). Finally, the components of the epistatic variance are obtained from the corresponding average, dominance and epistatic effects. The resulting expressions and their numerical example can be seen in Table 3.6. Note that, in this example, the total additive variance (59.31) is larger than the sum of the variances for each locus in the absence of epistasis ($2\alpha^2 pq + 2\alpha'^2 rs = 15.05 + 18 = 33.05$). Therefore, epistasis generates in this case an increase of the additive variance. Dominance variance is also changed by epistasis.

The interaction between locus effects is a phenomenon increasingly found due to the development of refined genomic analyses that allow for estimating the effects of individual loci (Phillips, 2008). However, the epistatic variance, when it has been estimated adequately (normally it is difficult and it is usually confounded with the dominance variance in a set that is called non-additive variance), is generally small, even smaller than the dominance variance. It is necessary to note the difference between the so-called functional epistasis, which refers to the molecular interactions between particular genotypes, and statistical epistasis, that is, the statistical deviation from a multilocus additive model (Fisher, 1918), as defined by equation (3.11). Thus, a low value of epistatic variance does not necessarily imply that functional epistasis is rare. The low values of epistatic variances commonly observed are a consequence of the very definition of intralocus (dominance) and interlocus (epistasis) interaction, where the quadratic deviations of the dominance and epistatic values with respect to the additive effects become minimal. Except for high epistatic effects and specific combinations of allele frequencies, the relative magnitude of the epistatic variance is usually small. Figure 3.3 illustrates the components

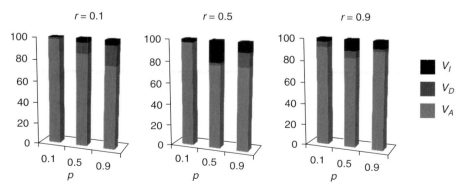

Figure 3.3 Magnitude of the additive variance (V_A, light grey bar), dominance (V_D, dark grey bar) and epistatic (V_I, black bar) corresponding to the numerical example of Table 3.6 considering three values of allele A_1 (p) and allele B_1 (r) frequencies.

of additive, dominance and epistatic variance corresponding to the numerical example of Table 3.6 considering three values of allele A_1 (p) and allele B_1 (r) frequencies. Note that, except for intermediate frequencies, the epistatic variance is smaller than the dominance variance and both are much smaller than the additive variance.

3.4 Concepts of Heritability and Genetic Correlation

The magnitude of the genetic variation for a quantitative trait and the covariation between two traits is usually expressed with a dimensionless measure, the heritability or the correlation, respectively. In this section we will analyse these concepts and their importance.

3.4.1 Degree of Genetic Determination

As explained in Chapter 1, one of the main objectives in the study of quantitative genetic variation is the determination of the proportion of this variation that is attributable to the genetic endowment of individuals over that which is due to the environment. The index that quantifies this proportion is called heritability.

Heritability in the broad sense, also called degree of genetic determination (H^2), is defined as the ratio between the genotypic (V_G) and phenotypic (V_P) variances,

$$H^2 = \frac{V_G}{V_P}. \tag{3.15}$$

The exponent refers to the fact that the index is a ratio of variances and, therefore, of quadratic values. The value of the heritability ranges between 0 and 1, extreme values that imply that all the phenotypic variation is of environmental or genetic origin, respectively. It is a common mistake among beginners to assume that a heritability equal to zero means that the relevant character is not inherited. This is incorrect, since it is one thing that a character has no genetic basis and another that it is not genetically variable. Heritability does not only depend on the nature of the trait studied but also on the genotypic constitution of the population under study. To understand the concept, let us think of two examples of extreme heritability.

Suppose that from a plant with vegetative reproduction multiple cuttings are taken that are planted in the open field. After some time, the plants will grow and may present certain differences in relation to various quantitative traits such as, for example, the number of leaves per plant. The phenotypic variation observable for this character and population will respond only to environmental variations that have affected each plant differently, since these are genetically identical (they are clones of the initial plant). The sources of environmental variation taking place during the development of the plant can be the degree of insolation, the water supplied, the concentration of nutrients in each area of the field, and so on. Therefore, all phenotypic variance is of environmental type, $V_P = V_E$, the genotypic variance $V_G = 0$, so that $H^2 = 0$. Obviously, the character number of leaves per plant is heritable and will have a genetic basis constituted by the expression of many genes. The value of $H^2 = 0$ simply indicates that the variation observed for the character in this particular population is due only to environmental causes.

Suppose, now, that cuttings of different plants are grown in a greenhouse where the environmental conditions are kept as uniform as possible, controlling humidity, lighting, soil nutrient supply, and so on, so that the environmental conditions are as similar as possible for all plants. This is precisely what is sought in most experimental studies, since the source of environmental variability is usually the noise factor that masks the acquisition of genetic data. In this case, there will also be a phenotypic variation for the character (much greater than in the previous case) but this time it will be due almost exclusively to genetic causes ($V_P \approx V_G$), since the design tries to make the environmental variation null ($V_E \approx 0$). Therefore, for this population $H^2 \approx 1$, even if it is the same trait and population under study as in the previous case. In fact, in this scenario H^2 can never reach the limit of 1 although it may get close, given that a total elimination of sources of environmental variation is not possible, as we will see in a later section. This shows that the definition of a quantitative trait includes both its description and the specification of the environment in which it is evaluated.

The previous reasoning reveals one of the possible methods of estimating the degree of genetic determination or broad-sense heritability. Using genetically identical individuals (such as the mentioned clones), an estimate of the environmental variance can be obtained which, subtracted from the phenotypic variance of a genetically variable population, allows for obtaining an estimate of the genotypic variance of this. This can also be done with highly inbred lines (which we will study in Chapter 4) as is the case with species whose natural form of reproduction is autogamy. The phenotypic variance of pure lines (with individuals homozygous for the whole genome and identical to each other), or the crossing between these lines, provides an estimate of the environmental variance. The method, however, has its complications. First, it must be assumed that the sources of environmental variation that affect the genetically variable population are the same as those that affect populations composed of genetically identical, homozygous or heterozygous individuals. Second, highly inbred lines are not a random sample of all possible ones, but of those that are viable. Finally, it has been repeatedly observed that the environmental variance is generally greater for homozygous populations (pure lines) than for heterozygous ones (crossings between lines). This probably occurs because the phenotypic expression of many traits in the face of environmental changes must be different when individuals have only one allelic variant per locus in their genome, instead of having two variants. In the latter case, it is likely that individuals can respond more easily to a wider spectrum of environmental variation without the need to modulate their level of expression, with important phenotypic consequences. In fact, studies of transcriptomic analysis in *Drosophila* (Kristensen et al., 2005; García et al., 2012) show that the variation in expression between inbred lines is greater than that between non-inbred lines of the same population. In addition to the method explained, there are other methods for estimating the broad-sense heritability, which will be studied in Chapter 6, although in general it is a parameter difficult to be obtained with guarantees.

3.4.2 Heritability

The concept of broad-sense heritability, discussed in the previous section, is primarily of interest in providing a measure of the degree to which genetic or environmental determination affects the phenotypic variation of a trait. However, it is generally impossible to estimate and, in addition, it is the additive values of the individuals, and not the genotypic ones, that allow predictions to be made about the expected genotypic value of the offspring, since this is the average of the additive values of their parents (equation (3.7)). The additive values are, therefore, the main factors responsible for the resemblance between parents and progeny, and between relatives in general, and, as we will see later, the response to natural or artificial selection. Therefore, the proportion of the phenotypic variance (V_P) attributable to the additive values (V_A) constitutes a parameter of essential practical interest, called narrow-sense heritability or, simply, heritability:

$$h^2 = \frac{V_A}{V_P}.$$
(3.16)

The role of narrow-sense heritability in determining the resemblance between relatives is made clear with the following argument. The orthogonal decomposition of the genotypic value that we carried out in the previous sections implied that the additive values (A) were independent of the other genetic components. If, in addition, the genotypic values are independent of the environmental deviations, that is, $\text{cov}(G, E) = 0$, then the additive values are independent of the rest (R) of components of the phenotypic value (P). Therefore, $\text{cov}(A, P) = \text{cov}(A, A + R) = V_A$, and the regression of the additive value on the phenotypic value is $b_{AP} = \text{cov}(A, P)/V_P = V_A/V_P = h^2$. In this way, the best linear prediction of the additive value is simply the phenotypic value multiplied by h^2:

$$E[A] = h^2 P.$$
(3.17)

The accuracy of this prediction, in turn, is given by the correlation between A and P, which would be $r_{AP} = \text{cov}(A, P)/\sqrt{V_A V_P} = V_A/\sqrt{V_A V_P} = \sqrt{V_A/V_P} = h$, that is, the square root of the heritability. Given that, as said before, the expected value of the offspring is the additive value (A) of their parents, the regression of the phenotypic value of the offspring on that of their parents is a simple way of estimating the heritability, which we will discuss in detail in Chapter 6. Let us take a simple example to understand the consequences of previous predictions. Suppose that the mean of a population for human height is 160 cm and that the heritability of the trait in the population is $h^2 = 0.6$. Suppose that the height of a man and a woman are 170 and 158 cm, respectively, that is, 10 cm above and 2 cm below the population mean, respectively. The expected additive value of the man (deviated from the population mean) is $10 \times 0.6 = 6$ cm and that of the woman is $-2 \times 0.6 = -1.2$ cm. If the couple has children, their expected height would be $(6 - 1.2)/2 = 2.4$ cm above the population mean, that is, 162.4 cm, although this prediction will have a considerable error.

As already indicated, both H^2 and h^2 depend on the trait analysed, the population under study and the environmental conditions. As a general rule, traits related to reproduction and fitness tend to have lower heritability values than those of morphological traits. Figure 3.4 shows the means of 1120 estimates of heritability, compiled by Mousseau and Roff (1987), obtained in a total of 76 species and classified according to the type of quantitative character. Although there is a lot of variation between estimates (the standard deviations of the estimates are of the order of the value of the means in most cases), the average estimates for

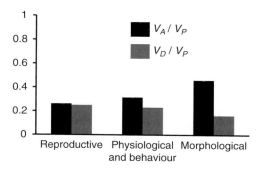

Figure 3.4 Average heritability estimates ($h^2 = V_A/V_P$) of reproductive (341 estimates), physiological and behavioural (209) and morphological (570) traits compiled in 76 invertebrate and vertebrate species (Mousseau and Roff, 1987). Average V_D/V_P estimates collected by Crnokrak and Roff (1995) in 38 wild or domestic species for reproductive (26 averaged estimates), physiological and behavioural (28) and morphological (52) traits.

morphological traits are greater than those of reproductive characteristics, while those of physiological and behavioural traits are intermediate between both. The figure also shows estimates of the V_D/V_P ratio collected by Crnokrak and Roff (1995) in 38 species for different types of characters. Note that the average proportion of dominance variance for reproductive traits is of the same order as the proportion of additive variance, and that, however, V_A is considerably greater than V_D for morphological traits.

Comparisons between h^2 for characters of different types can be misleading, however. Although h^2 is generally lower for fitness traits than for morphological traits, when the additive variance is measured scaled by the mean of the trait, that is, the coefficient of additive genetic variation ($CV_A = \sqrt{V_A}/M$), this is generally greater for reproductive traits than for morphological traits (Houle, 1992). In general, therefore, reproductive characters can also possess a large amount of additive variance, being controlled by many loci, but also possess a substantial amount of dominance, and even epistatic, genetic variance (Figure 3.4), for reasons we will study later, and, above all, they are much more affected by environmental variation than morphological characters. Note, however, that the use of the coefficient of additive genetic variation has a diffuse meaning when the scale of measurement of the trait does not have obvious biological characteristics as is the case, for example, of characters measured as proportions or artificial indexes, such as the coefficient of intelligence.

3.4.3 Genetic Correlation between Traits

So far we have only considered a single quantitative character. When attending to two characters (say X and Y) we can find that some or all genes that contribute to the genetic variation of one of them also contribute to the other, which can be quantified by the additive genetic correlation (r_A) between both. This correlation can appear for two reasons. One is pleiotropy, that is, the existence of genes with effect on the two traits. The second is that, although the genes that control each trait are different, they are physically close in the genome, so that the variation of one of them goes hand in hand with the other due to linkage

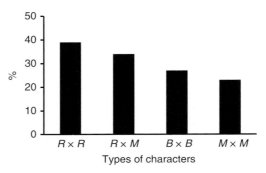

Figure 3.5 Percentage of negative correlations between two reproductive traits ($R \times R$), one reproductive and one morphological trait ($R \times M$), two behavioural traits ($B \times B$) or two morphological traits ($M \times M$). Data collected by Roff (1996) in studies conducted in a total of 51 animal and plant species. The number of estimates is 152 ($R \times R$), 175 ($R \times M$), 166 ($B \times B$) and 1210 ($M \times M$).

disequilibrium. Recombination will break this second type of genetic correlation. A trivial example of genetic correlation, in line with those previously indicated, would be that of height and weight. It seems obvious that many genes will increase height and weight or will decrease both, so a positive genetic correlation between these traits is expected. In other cases, however, there may be a negative genetic correlation, when most of the variation is due to genes that increase the value of one trait and reduce that of the other. These negative genetic correlations can exist or be created by the effect of genetic manipulation, such as artificial selection that we will study later, or natural selection. As it occurs for the components of variance, the phenotypic correlation (r_P) between two traits can also be of environmental origin (r_E), that is, the environmental factors can produce changes in the phenotype of both traits in the same direction (positive environmental correlation) or in opposite directions (negative correlation). For reasons of estimation, which we will see in Chapter 6, the environmental correlation estimate (r_E) also includes the non-additive genetic components and, if the heritability of the two traits is the same (h^2), the contribution of this correlation and the genetic one to the phenotypic correlation (r_P) is given by

$$r_P = r_A h^2 + r_E(1 - h^2). \tag{3.18}$$

Given that natural selection tends to fix those alleles that improve the fitness of individuals and eliminate those that deteriorate it, and that this will occur for the different components of fitness, it is expected that in the case of these types of traits there will be an excess of segregating alleles that increase one and reduce another in relation to those that increase or reduce both. This will mean that the genetic correlations between reproductive traits will often be negative. This prediction will not be fulfilled for traits poorly related to fitness, such as morphological traits. The compilation of correlations obtained by Roff (1996) that includes almost 1800 estimates of genetic correlations in 51 species of animals and plants confirms this prediction. Figure 3.5 shows the percentage of negative genetic correlation estimates considering reproductive (R), behavioural (B) and morphological (M) traits, and shows that negative correlations between reproductive traits are more frequent than those for morphological traits.

3.5 The Environmental Deviation and Its Contribution to Phenotypic Variance

The environmental variation can be due to a multitude of factors. The most important are usually the climatic, nutritional and interaction of individuals with their environment. Errors in the measurement of the characters are also a possible source of environmental variation, although generally small. The maternal effects, which occur particularly in mammals, are also important, and often come from the previous generation (sometimes of two generations) as a consequence of the phenotypic value of the mothers. For example, in many species large mothers have larger litters at the expense of the average body size of the offspring, so it may occur that the genetic factors transmitted by the mother promote a greater size of the offspring, while the maternal environmental factor impacts negatively on the size. Maternal effects are, in fact, a particular type of indirect genetic effects, that is, inheritable effects of an individual that influence the phenotypic value of others. Social interactions among individuals, which may include competition for resources, and social behaviours, such as aggression, social dominance or altruism, have great evolutionary importance (Wolf et al., 1998; Gardner et al., 2007) and can affect the phenotypic expression of individuals. There may also be maternal or paternal effects related to the epigenetic changes that take place in the offspring's genome (Lawson et al., 2013). Environmental variation can have many and very varied consequences on the phenotype and phenotypic variation, as well as on the partition of these into their components. In the context of animal and plant breeding, sometimes macro-environmental or classifiable effects can be considered, such as the year, the season or even the herd, which are usually included as fixed effects in the analyses, while other less specific environmental conditions are included as random effects. In most cases, environmental effects are responsible for random changes in the phenotype, but in some cases they produce directional changes with an adaptive basis, as will be explained below.

3.5.1 Phenotypic Plasticity

The ability of a given genotype to produce different phenotypes in response to environmental variation is called phenotypic plasticity, and it is commonly accepted that this property of responding to the environment has a genetic basis (for a broad review on this phenomenon see Pigliucci, 2005; Nussey et al., 2007). Phenotypic plasticity often occurs as an adaptive response of individuals. There are, for example, many documented cases of phenotypic changes of a defensive type induced in response to interaction with predators. We will use an example to illustrate this phenomenon and its experimental analysis. It has been described that the presence of crabs, common predators of marine snails, induces a thickening of the shell of the latter with a clearly defensive purpose (Trussell, 1996). Figure 3.6 illustrates a schematic representation of the experiment carried out by Brookes and Rochette (2007) in which this plastic response is demonstrated in the marine gastropod *Littorina obtusata*. In this work two contiguous aquariums were used, allowing the flow of water to take place from the lower to the upper. In the upper one, juvenile individuals were introduced and grown for a period of 96 days, and at the end of this period various characters of the shell and soft body, as well as physiological traits, were evaluated. In the lower aquarium a common predator for this species (a green crab) was introduced (treatment with predator) or not (control treatment), along with some congeners of

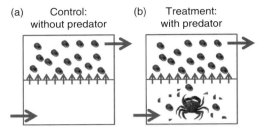

Figure 3.6 Experimental design carried out by Brookes and Rochette (2007) to analyse the phenotypic plasticity in the snail *Littorina obtusata* in response to the presence of chemical substances produced by a predator (crab) and its predatory activity on snails of the same species. In the experiment, aquariums were used that allowed the flow of water in the direction of the arrows. In the upper aquarium, juvenile individuals were grown for a period of 96 days, after which various characters were evaluated. In the lower aquarium, (b) some snails of the same species were introduced in the presence of a predator, whereas in the control treatment, (a) no predator was introduced. The snails evaluated in the upper aquarium never came into direct contact with the predators or snails of the lower aquarium, but the water passed freely from the lower aquarium to the upper one, inducing environmental changes in the development of the snails. Adapted from Brookes and Rochette (2007).

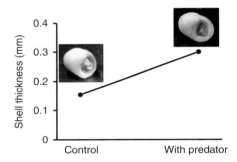

Figure 3.7 Reaction norm for shell thickness in the experiment carried out by Brookes and Rochette (2007) to analyse phenotypic plasticity in *Littorina obtusata*. (The snail photographs do not correspond with the original experiment but are given for illustration.)

the experimental snails. Note that the evaluated snails resident in the upper aquarium never came into direct physical contact with the predators, snails or their remains of the lower aquarium.

The thickness of the shell measured at the end of the growth period in the upper aquariums was approximately double in the snails developed in the presence of predator exudate (0.300 mm) than in the snails of the control treatment (0.154) as illustrated in Figure 3.7. This shows the environmental sensitivity of genotypes to different environments, which is called the reaction norm. Therefore, with this experiment it was clearly shown that the exudate produced by the crabs or their predatory activity stimulated a greater production of calcareous deposits in the shells of the snails.

The analysis of various quantitative traits in the snails provided more information about the changes experienced by those that grew in the presence of the exudate of the predator compared to those of the control treatment (Figure 3.8). The snails exposed to the fluids of the

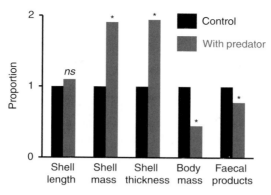

Figure 3.8 Difference (in proportion relative to the control) for different quantitative characters of the snails studied in the experiment carried out by Brookes and Rochette (2007) to analyse phenotypic plasticity in *Littorina obtusata*. The difference found for shell length was not significant (*ns*), while significant changes (*) were found for the other characters.

aquarium with the crab showed a linear growth of the shell similar to that of the control, but the total mass of its shell and its thickness was 91% and 94.5% higher than those of the first, respectively. Phenotypic plasticity, however, has a cost, and these phenotypic changes of a defensive nature were accompanied by parallel changes in other characters. In particular, a significant reduction in body mass (by 55%) and food activity (by 22%, measured as the number of faecal products) in snails exposed to crab exudate with respect to controls. Therefore, the plastic changes propitiated by the environment implied a greater mass and thickness of the shell at the expense of a lower food activity and of gain of corporal mass. This example shows that environmental changes can not only produce arbitrary deviations of the phenotypic value, but sometimes they involve directional phenotypic changes of adaptive nature.

3.5.2 Genotype–Environment Interaction

One of the assumptions assumed in the partition of the phenotypic value (deviated from the population mean) in genotypic and environmental components, $P = G + E$, is that the environmental effects contribute equally or proportionally to the different genotypes. The way to verify this premise consists in the analysis of the reaction norms of different genotypes, as illustrated in Figure 3.9, which presents the reaction norm of two genotypes in two different environments. Suppose, for example, that G_R and G_L represent the genotypes of two plants of the same species, one (G_R) taken from a population that inhabits the bank of a river (E_R environment) and another (G_L) from a population that lives on the side of a mountain (E_L environment). Suppose we now plant cuttings of each plant in the two environments (the bank and the hillside). We phenotypically evaluate a given character (P), such as the number of leaves per plant, and find that the value of this character for the G_R plant is greater than for the G_L plant in the E_R environment (riverbank). That is, the G_R plant has more leaves than the G_L on the riverbank. When we observe the phenotype in the E_L environment (hillside), the number of leaves changes in both plants (the phenotypic plasticity discussed in the previous section). In the case of graph 3.9a, the effect of the environment is the same for the two genotypes, a reduction in the average number of leaves. Therefore, the premise alluded to above for the partition of the phenotypic value in components is fulfilled. However, the situation presented in Figure 3.9b implies that the effect of the environment is different for the two genotypes, in one case increasing and in the other decreasing the phenotypic value. The plant that normally inhabits the riverbank has a greater number of leaves than the one that inhabits the hillside, when it grows on the riverbank, but the opposite occurs when it grows on the hillside and vice versa. It is said then that there is genotype–environment (or genotype by environment, $I_{G \times E}$) interaction. For there to be $I_{G \times E}$ it is not necessary that the reaction norms of the genotypes cross to each other (as in the case of Figure 3.9b), but it is sufficient that their slopes are not parallel. When $I_{G \times E}$ exists, the partition of the phenotypic value is

$$P = G + E + I_{G \times E}, \tag{3.19}$$

and analogously, there is a component of variance attributable to this interaction ($V_{G \times E}$) so that the partition of the phenotypic variance is

$$V_P = V_G + V_E + V_{G \times E}. \tag{3.20}$$

$G \times E$ interaction is very common (Sgrò and Hoffmann, 2004) and responds to the adaptation of the genotypes to the habitats where they are found. In spite of being individuals of the same species, the plant population of the riverbank is better adapted (a concept that will be understood in later chapters) to the environmental conditions (humidity, temperature, etc.) of the riverbank, and the same occurs with the individuals of the mountain population in their environment. To evaluate the reaction norm, it is necessary to test the same genotype in different environments, which is only possible in some situations such as the one described, although it is usually done using different individuals from the same population, as in the experiment presented in Figures 3.6–3.8. The importance of $G \times E$ interaction is fundamental in various fields, such as animal and plant breeding, the conservation of biological resources, or the study of evolution. For example, in conservation genetics it is sometimes necessary to carry out restocking programmes in areas where a certain species of interest has disappeared. Restocking with individuals of populations that live in environments different from those that

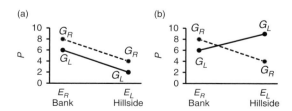

Figure 3.9 Illustration of genotype–environment interaction. The phenotypic value (P) of two genotypes (G_R and G_L) is represented in two different environments (E_R, the bank of a river, and E_L, the hillside of a valley). (a) There is no genotype–environment interaction. (b) There is genotype–environment interaction.

are intended for recovery may not be successful, since the expressions of many quantitative traits, including those of reproductive importance, may have unexpected values in the new environment. An example is the restocking of Atlantic salmon in the rivers of northern Spain carried out in the 1970s–1980s with individuals from northern Europe (Scotland and Norway), which was a fiasco mainly due to the unsuitability of imported salmon to the climatic conditions of the Iberian Peninsula and, in short, to the $G \times E$ interaction for traits related to survival and reproduction (García de Leániz et al., 1989). For this reason, the current restorations are always carried out with specimens from nearby rivers, which share similar environmental conditions. More details on the analysis of $G \times E$ interaction can be found in the book of Mather and Jinks (1983, chapter 6).

3.5.3 Genotype–Environment Correlation

Another assumption we have made in the previous sections about the partition $P = G + E$ is that the two components, G and E, are independent of each other. It is, however, possible that these components are correlated, in which case we speak of genotype–environment correlation, and the phenotypic variance would be decomposed algebraically in

$$V_P = V_G + V_E + 2\text{cov}_{GE}, \tag{3.21}$$

where cov_{GE} is the covariance between genotypic values and environmental deviations. This occurs, for example, in animal breeding when individuals with better genotypic values for a trait of interest are offered more and better resources, which is called differential treatment. Another example is that of some human attributes, particularly those related to intelligence, where there may be a correlation between the inherited genetic constitution and the family and social environment. The covariance component cannot be estimated, being incorporated into one or both types of variation, genetic and environmental. It is often not possible, therefore, to separate components due exclusively to the genotype or to the environment, that is, part of V_G is environmental and part of V_E is due to hereditary causes. Only when the cause that induces cov_{GE} is known can this covariance be included in V_G or V_E. In experimental procedures it is essential to eliminate this source of covariation, randomizing the evaluated individuals in the different available environments, although this is not always possible.

3.5.4 Decomposition of the Environmental Variance: Repeatability

In laboratory experimentation, most of the environmental factors that affect the phenotype can be controlled to a certain extent, and the maximal standardization of the environment is usually pursued to reduce this source of variation. There are, however, uncontrolled environmental factors, such as those that result from errors in the development or environmental internal circumstances of the individual. It is therefore impossible to completely eliminate environmental variation, although it can be greatly reduced.

When a trait can be measured more than once in space or time, it is possible to distinguish between sources of local or temporal variation, and also between causes of non-localized or permanent variation. For example, all morphological structures that are repeated by bilateral symmetry can be measured on each side of the body, and all the characters that are expressed more than once temporarily can be measured at each moment. Examples could be the length of the two wings of an insect, in the first case, or the litter size of a mammal in different births, in the second. If these multiple measurements can be taken, the environmental variation can be decomposed into a component due to non-localized or permanent environmental causes, the general environmental variance, V_{EG}, and another due to internal or temporary causes, the special environmental variance, V_{ES}:

$$V_E = V_{ES} + V_{EG}. \tag{3.22}$$

The variation between measurements taken in the same individual would correspond to V_{ES}, while that observed between the average values of individuals would include the genotypic and the general environmental variation, $V_G + V_{EG}$. It is important to emphasize that, for this partition to be possible, the different measures must have the same genetic basis, that is, the genetic correlation between them must be $r_A = 1$. Only in this case will the observed differences between measures be attributable to environmental factors, such as developmental errors or temporary changes in the expression of the same set of genes. In the first two columns of Table 3.7 we present examples for two traits in *Drosophila melanogaster*, the number of ovarioles measured in the two ovaries (Robertson, 1957), and the number of bristles (sensitive hairs) measured in two abdominal segments (Reeve and Robertson, 1957). The estimated value of the general environmental variance (V_{EG}) is quite small (only 3–4% of the total phenotypic variation), indicating that the experimental study was carried out under very uniform laboratory environmental conditions. The special component, V_{ES}, uncontrollable by the experimenter is, however, very high, of the same order or even higher than the genetic variance.

With this partition of the environmental variance a statistic can be defined, the repeatability (r), which quantifies the degree of similarity between measures within individuals or the degree of differences between individuals:

$$r = \frac{V_G + V_{EG}}{V_P}. \tag{3.23}$$

This statistic allows for estimating the contribution of the special environmental component to the phenotypic variance, $V_{ES}/V_P = 1 - r$. Note that the r statistic is a higher limit of the broad-sense heritability, $H^2 \leq r$, although for this to be true, it is not only necessary that the two measurements refer to traits with the same genetic basis, but also that other circumstances can be excluded, such as genotype–environment interaction between different measures and maternal effects, among others (see Dohm, 2002).

When a trait can be measured repeatedly in each individual and the values of each measurement are averaged, the special environmental variance is reduced proportionally to the number of measures (n) per individual, V_{ES}/n, which implies a reduction of the environmental component of variation, which obviously depends on the magnitude of V_{ES} in relation to the total phenotypic variance, that is, $V_{ES}/V_P = 1 - r$. For example, if the number of abdominal bristles is averaged for two abdominal segments (third column of Table 3.7), the V_{ES} component is reduced by a half (58%/2 = 29%) without changing the other components of variation. By scaling the

Table 3.7 *Example of partition of the environmental variance with repeated measures in each individual*

	Number of ovarioles in the ovary	Number of abdominal bristles in one segment	Average of the number of abdominal bristles in two segments
V_A	23%	33%	46.4%
V_{NA}	27%	6%	8.2%
V_{EG}	4%	3%	4.1%
V_{ES}	46%	58%	41.3%
V_P	100%	100%	100%
r	0.54	0.42	0.59
H^2	0.50	0.39	0.55
h^2	0.23	0.33	0.46

Source: Data from Robertson (1957) and Reeve and Robertson (1954); taken from Falconer and Mackay (1996, table 8.4).

percentages of variation to 100%, the percentages of the third column are obtained, which indicate that when the environmental variance is reduced, the values of heritability and repeatability increase. If more than two measures of the character can be averaged, this will imply greater reductions in the environmental variance and, therefore, the phenotypic variance. However, this reduction of the phenotypic variance will be proportionally lower as the number of measures of the character increases and, in general, considering the workload implied by a greater number of measurements, it will only be worth carrying out two or three measurements per individual (see Falconer and Mackay, 1996, p. 142).

Problems

3.1 The analysis of a *Drosophila melanogaster* population has made it possible to detect an allele with a substantial effect on the number of sternopleural bristles segregating in the population at a frequency of 20%. The effect produced by this allele in the phenotype of the homozygous individual is an average increase of three bristles in relation to non-carriers. In heterozygotes, however, the effect is reduced to an increase of one sterno-pleural bristle. Calculate: (a) The contribution of this gene to the population mean and the average effects of the alleles. (b) The contribution of the gene to the genotypic, additive and dominance variance of the trait.

3.2 Suppose that the average height of a human population is 170 cm and that H^2 and h^2 estimates for this trait and population are, respectively, 0.8 and 0.6. If a man of this population is 185 cm tall, (a) what height would he be if instead of growing in the environment in which he did, he grew in the average environment of the population? (b) If that man has a child with a woman whose height is 165 cm, and the child grows up in the same environment as his father, what will be the expected height of the child when he reaches adulthood?

3.3 In the analysis of the sternopleural bristles in the *Drosophila melanogaster* population mentioned in Problem 3.1, in addition to the locus with alleles A and a, another one was found (with alleles B and b) whose allele b segregated with frequency 10% in the population and produced an increase of five bristles in homozygous individuals and two bristles in heterozygous individuals, in relation to individuals which did not carry that allele (*BB*). When the genotypic value of the different genotypes was estimated for the two loci as a whole, it was found that all genotypic values were the sum of the genotypic values of the loci separately, except for the double homozygote *aabb*, whose genotypic value was 32 bristles instead of the expected value of 8 bristles. It was, therefore, a case of synergistic epistasis, as described in Table 3.6. Calculate: (a) The dominance effects and the average epistatic effects. (b) The effects of epistatic interaction. (c) The contribution of the two loci to the additive, dominance and epistatic variances of the trait.

3.4 Individuals of two varieties of pea, which is a selfing species, were crossed obtaining 50 descendants (F_1 generation), in which the length of the pod was evaluated. The average length of the pod was 9.5 cm and the phenotypic variance 0.12 cm². Then F_1 individuals were selfed to obtain 50 offspring (F_2 generation), whose mean and variance for the length of the pod were 9.3 cm and 0.25 cm², respectively. Is it possible, with these data, to obtain an estimation of the heritability? What are the assumptions that must be made to obtain such an estimate?

3.5 Fifty individuals of a herbaceous plant have been evaluated for the number of seeds per fruit by randomly choosing two fruits per individual. The average number of seeds per fruit is 14.5. The analysis of variance of the data allows for obtaining an estimate of the variance between the average measures of the individuals, $\sigma_B^2 = 145$, and the average variance of the two measures of each individual, $\sigma_W^2 = 64$. (a) Calculate the repeatability of the trait. (b) If four fruits were evaluated per plant instead of two, how would it affect the estimation of environmental variance and repeatability?

Self-Assessment Questions

1 The additive value of an individual is defined as the sum of the average effects of the alleles carried by the individual.

2 The expected additive value of an individual is the average of the additive values of its parents.

3 The dominance deviations are the minimal quadratic differences between the additive values and the genotypic values.

4 Even with full dominance, the additive variance is greater than the dominance variance if the allele frequencies are intermediate.

5 When there is gametic disequilibrium between two loci, the global genotypic variance is always greater than the sum of the variances of each locus separately.

6 Synergistic and antagonistic epistatic interactions always increase the genetic variance.

7 The heritability varies between the different traits and species, but for a given trait and species it has a fixed value, although the estimates may differ due to errors in the estimation.

8 The population mean for a quantitative trait with heritability $h^2 = 0.4$ is 190. The phenotypic value of an individual of that population is 195, so the expected additive value of the individual is 2.

9 The additive genetic variance for a quantitative trait is 2.5, the dominance variance is 0.5 and the environmental variance is 3. Therefore, the heritability (h^2) of the trait is 0.50.

10 Phenotypic plasticity is due to phenotypic changes induced by the environment and, therefore, it has no genetic basis.

4   Inbreeding and Coancestry

Concepts to Study

- Coefficients of inbreeding and coancestry
- Molecular estimates of inbreeding
- Mutation–drift equilibrium
- Purged inbreeding coefficient
- Wright's statistics
- Migration–drift equilibrium
- Estimators of genetic differentiation between sub-populations

Objectives for Learning

- To understand the concepts of inbreeding coefficient and coancestry and to learn how to calculate them from genealogical and molecular markers data
- To learn how to predict inbreeding in the ideal population of Wright–Fisher and to understand its relationship with the process of genetic drift
- To learn how to calculate the coefficient of inbreeding in populations subject to regular inbreeding
- To understand how inbreeding is modulated by the effect of mutation and selection
- To understand the way in which inbreeding is ascribed to different causes in structured populations
- To learn how to predict the magnitude of genetic differentiation among sub-populations
- To identify the problems inherent to the estimation of genetic differentiation among sub-populations using genetic markers

4.1 The Coefficients of Inbreeding and Coancestry

Inbreeding is a consequence of the mating between relatives. This is an inevitable phenomenon in populations of small census size even when crossing between their individuals is 'random'. But inbreeding can also exist in large populations, when relatives mate with each other naturally or forcedly. In this chapter we will analyse the conceptual and mathematical treatment of inbreeding, whose bases are largely due to Sewall Wright. We will address the concepts of coefficient of inbreeding and coancestry, the ways in which these are calculated from genealogical information and genetic markers data, as well as their modulation by the different population forces of change in the allele frequencies that act in the populations.

The inbreeding is quantified by the inbreeding coefficient (F) of Wright (1922, 1969), which is defined as the probability that the two united gametes to form an individual carry alleles at a given locus that are identical by descent. This means that both are copies of

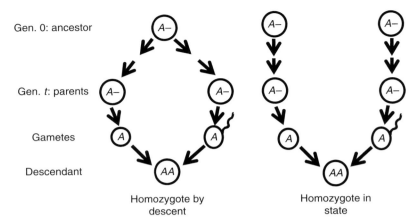

Gen. 0: ancestor

Gen. *t*: parents

Gametes

Descendant

Homozygote by descent

Homozygote in state

Figure 4.1 Difference between homozygotes identical by descent, in which the two copies of the allele come from an ancestor common to the parents of the individual, and homozygotes identical in state, in which the two alleles come from different ancestors.

another allele carried by a common ancestor of the parents of the individual in question or, in other words, the alleles trace their ancestry back (or coalesce) to the same ancestral allele, as illustrated in Figure 4.1. Therefore, the inbreeding coefficient is the probability that the individual is homozygous by descent for that locus. If the individual, even being homozygous, had not received copies of the allele from a common ancestor (Figure 4.1), it is called homozygous in state, and does not compute in the calculation of inbreeding. Wright's original definition (1922), however, was not formulated in terms of probability but of the correlation between the alleles carried by the individual at a given locus. Defined in this way, F could take negative values as will be seen later.

A fundamental aspect of the coefficient of inbreeding is that its value is always relative to that of a starting base population. The distinction between identity by descent and identity in state may disappear since, in principle, all current alleles of the same type are expected to be copies of a single ancestral allele in a more or less remote past. In practice, however, this distinction refers to the aforementioned base population, usually to the most remote generation for which information is available. If, in this initial base population, there were several copies of a given allele, it must be assumed that these are not identical by descent, because we are not aware of it, even if in fact they are. Therefore, by definition, the inbreeding coefficient of individuals of the base population or of individuals for whom information on their ancestry is not available is assumed to be zero, which implies that the value of the coefficient is a downward estimate of its true value, which could only be known if the whole tree of life were available. The inbreeding coefficient, defined as the probability of identity by descent for a given locus, ranges, as such, between 0 and 1, and can be generalized to all the genes of the individual if defined as the proportion of loci of the genome of the individual that are expected to be homozygous by descent. If it refers to a population, it would be obtained as the average inbreeding coefficient of all the individuals of the population.

A coefficient intimately related to that of inbreeding is the kinship coefficient, coancestry coefficient, or simply coancestry (f) defined by Malécot (1948) as the probability that two alleles at the same locus taken at random from two individuals are identical by descent. Note that although the coancestry of two crossing individuals is precisely the coefficient of inbreeding expected in their offspring, the coancestry can be calculated between any two individuals, including an individual with itself, as we shall see later. It is also interesting to note that twice the coancestry between two individuals (if they are not inbred) is the expected proportion of alleles they share. For example, the expected coancestry between full sibs is 1/4, and the percentage of alleles that are expected to share is 1/2. Also, the coancestry between a father or mother and their offspring is also 1/4, although in this case the percentage of alleles they share is exactly 1/2.

4.1.1 Calculation of F and f from Genealogies

The inbreeding and coancestry coefficients can be calculated from genealogical data by two procedures. Wright's paths method (1922) is a direct computation of the probability of identity by descent throughout the pedigree generations. Consider the example in Figure 4.2. To obtain the coefficient of inbreeding of individual X (F_X), we calculate the probability that two copies of an allele from any ancestor of the parents of X will reach that individual. The only known common ancestor of the parents of X is A in this case, and since the others are unknown, as indicated by the question marks, they are all assumed to be different. The probability that A will transmit a copy of the same allele (of the two that it carries in a given locus) to its offspring B and C will depend on the genetic constitution of individual A. If it is homozygous by descent, with probability equal to F_A by definition, the probability that it would transmit the same allele to B and C would be 1. If A were not homozygous by descent (with probability $1 - F_A$), the probability that A would transmit copies of the same allele to B and C would be 1/2. Therefore, the joint probability is $(F_A \times 1) + [(1 - F_A) \times 1/2] = 1/2 \times (1 + F_A)$. Assuming that B and C have received a copy of the same allele from A, the probabilities that these alleles are transmitted from B to F and from F to X, on the one hand, and from C to G and from G to X, on the other, is simply $(1/2)^4$, implying four transmissions with probability 1/2. Multiplying the probabilities, the coefficient of inbreeding of individual X is $F_X = 1/2 \times (1 + F_A)(1/2)^4 = (1 + F_A)(1/2)^5$. If we assume that A is not homozygote identical by descent (when there is no additional information this is the default assumption), $F_A = 0$, and $F_X = (1/2)^5 = 1/32$. A simple way to apply this procedure in practice is to write the path from the common ancestor, located in the centre of the series, to the parents (i.e. FBACG in the example) and compute $(1 + F_{anc})(1/2)^n$, where F_{anc} is the coefficient of inbreeding of the ancestor in question and n the number of letters of the path (5 in the example). This procedure must be repeated, if necessary, for all possible ancestors common to the parents of the individual whose coefficient is to be calculated, as well as for all possible paths, and the probabilities must be added. A practical rule that must be fulfilled is that a path must never have any repeated letter and, of course, it must be accomplished that genes in the path are always transmitted from ancestors to descendants.

The second procedure for calculating the inbreeding coefficient of an individual is based on the coancestry (f). While the method of the path performs the calculation of probabilities from ancestors to descendants, the coancestry method runs through the genealogy in the opposite direction and is based on the application of the following four rules that we will explain using the genealogy of Figure 4.2:

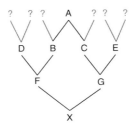

Figure 4.2 Genealogy showing the known ancestors of individual X.

1 The inbreeding coefficient of an individual is, by definition, the coancestry of its parents. For example,

$$F_X = f_{FG}. \qquad (4.1)$$

2 The coancestry between two individuals is the average of the coancestries between one of them and the parents of the other. For example, for individuals F and C or F and E is

$$f_{FC} = \tfrac{1}{2}(f_{DC} + f_{BC})$$
$$f_{FE} = \tfrac{1}{2}(f_{DE} + f_{BE}). \qquad (4.2)$$

3 Applying the same rule to the coancestry between two individuals of the same generation, for example, $f_{FG} = \tfrac{1}{2}(f_{FC} + f_{FE})$, and replacing the equations (4.2), we get

$$f_{FG} = \tfrac{1}{4}(f_{DC} + f_{DE} + f_{BC} + f_{BE}). \qquad (4.3)$$

4 Finally, the coancestry of an individual with itself, or self-coancestry, is, for example for individual F,

$$f_{FF} = \tfrac{1}{2}(1 + F_F). \qquad (4.4)$$

Note that the probability of self-coancestry (equation (4.4)) is that two alleles taken from the same individual in a given locus (with replacement) are identical by descent and, therefore, the result is identical to that obtained by calculating the probability that an individual transmits a copy of the same allele to two of its offspring in the path method. Note also that if individual F is not inbred ($F_F = 0$) the self-coancestry of F is $f_{FF} = 0.5$.

Let us apply the previous rules to obtain F_X in the genealogy of Figure 4.2. With the rule (4.3), $F_X = f_{FG} = \tfrac{1}{4}(f_{DC} + f_{DE} + f_{BC} + f_{BE}) = \tfrac{1}{4}(f_{BC})$ since the coancestry between unknown individuals, or between these and other individuals in the genealogy, is assumed to be zero, so $f_{DC} = f_{DE} = f_{BE} = 0$. Analogously, $f_{BC} = \tfrac{1}{4}(f_{?A} + f_{??} + f_{AA} + f_{A?}) = \tfrac{1}{4}(f_{AA})$. Therefore, substituting and using equation (4.4), $F_X = f_{FG} = \tfrac{1}{4}[\tfrac{1}{4}(f_{AA})] = (1/16)[\tfrac{1}{2}(1 + F_A)]$, if we assume $F_A = 0$, it results in $F_X = 1/32$, as was obtained with the path method.

Note that, from equations (4.1), (4.2) and (4.4) applied, for example to individual F, it can be reached algebraically to

$$f_{FF} = \tfrac{1}{4}(f_{DD} + f_{DB} + f_{BD} + f_{BB}) + \tfrac{1}{4}[1-(F_D + F_B)/2]. \qquad (4.5)$$

Figure 4.3 Genealogy used for the computation of inbreeding in Table 4.1.

The first term of equation (4.5) is equivalent to equation (4.3), relative to the flow of genes from parents to offspring. The second term refers to the Mendelian gene segregation characteristic of heterozygotes (1/4), with a correction $[1 - (F_D + F_B)/2]$ that expresses the reduction of that term due to the inbreeding of the parents.

The coancestry method, although apparently more complex than the paths method, is much simpler to apply in complex genealogies using the so-called tabular method. In this method, the coancestry coefficients of each individual are calculated with all the others in an $N \times N$ matrix, where N is the total number of individuals in the genealogy and, simply, each element of the matrix is calculated from the average of the elements with the coancestry between the parents of the individuals in question. The application of the paths method and the coancestry method, as well as the tabular version of the latter, is illustrated with the genealogy of Figure 4.3 and Table 4.1. For the analysis of large pedigrees in animal breeding and conservation genetics fast and efficient algorithms for the computations of inbreeding are used (Sargolzaei et al., 2005).

4.1.2 Molecular Coancestry and Inbreeding Coefficients

The use of molecular genetic markers allows for the calculation of the coefficients of molecular inbreeding and coancestry, as an alternative when there is no genealogical information or as a complement to it. The calculation is immediate and is illustrated in Table 4.2. The coefficient of molecular inbreeding of an individual for a given locus is 0 if it carries different alleles for the marker and 1 if these are identical, and the molecular coancestry coefficient between two individuals is calculated by comparing two by two alleles of each individual. Note that using markers cannot make a distinction between identity by descent and identity in state, that is, the molecular coefficients of inbreeding (F_M) and coancestry (f_M) are generally higher than those obtained from genealogies because they include both types of identity. Thus, they are not proper inbreeding and coancestry coefficients and very often they are called molecular homozygosity and similarity coefficient (Malécot, 1948; Walsh and Lynch, 2018, p. 693), respectively. However, we will keep the denominations of molecular inbreeding and molecular coancestry in analogy to the genealogical ones. When more than one locus is used in the calculation, molecular inbreeding and coancestry values are obtained by averaging them over loci.

Table 4.1 *Example of estimation of the inbreeding coefficient of individual X in Figure 4.3*

Paths method

Ancestor	Path	Probability
A	ECBAD	$(1/2)^5 (1 + F_A) = (1/2)^5$
A	ECAD	$(1/2)^4 (1 + F_A) = (1/2)^4$
D	DE	$(1/2)^2 (1 + F_D) = (1/2)^2$
		$F_X = 11/32 = 0.34375$

Coancestry method

$F_X = f_{ED} = \frac{1}{2}(f_{CD} + f_{DD})$ (average of the coancestry between D and the parents of E)
We calculate each of the elements in brackets: $f_{CD} = \frac{1}{4}(f_{BA} + f_{B?} + f_{AA} + f_{A?})$ (average of the coancestry
     between the parents of C and D)
The terms, in turn, are
$f_{B?} = f_{A?} = 0$ (unrelated individuals)
$f_{AA} = \frac{1}{2}(1 + F_A) = 1/2$ (self-coancestry of A)
$f_{BA} = \frac{1}{2}(f_{A?} + f_{AA}) = \frac{1}{2}(f_{AA}) = \frac{1}{2}(1/2) = 1/4$ (coancestry between A and the parents of B)
So that $f_{CD} = \frac{1}{4}(1/2 + 1/4) = 3/16$
Finally, $f_{DD} = \frac{1}{2}(1 + F_D) = 1/2$ (self-coancestry of D)
Therefore
$F_X = f_{ED} = \frac{1}{2}(3/16 + 1/2) = 11/32 = 0.34375$

Tabular method

Parents	? ?	A ?	A B	A ?	C D	E D
Individuals	A	B	C	D	E	X
A	0.5	0.25	0.375	0.25	0.3125	0.28125
B	0.25	0.5	0.375	0.125	0.25	0.1875
C	0.375	0.375	0.625	0.1875	0.40625	0.296875
D	0.25	0.125	0.1875	0.5	0.34375	0.421875
E	0.3125	0.25	0.40625	0.34375	0.59375	0.46875
X	0.28125	0.1875	0.296875	0.421875	0.46875	0.671875

Note: For the tabular method, each cell is built with the previous ones starting by the oldest individual
(A). The diagonal (self-coancestries) is obtained as $f_{ii} = \frac{1}{2}(1 + F_i)$, where F_i is the coefficient of
inbreeding of individual i, which is the coancestry between its parents. For example, $f_{CC} = \frac{1}{2}(1 + F_C) =$
$\frac{1}{2}(1 + f_{AB}) = \frac{1}{2}(1 + 0.25) = 0.625$. Values outside the diagonal are obtained by averaging the coancestry
values of each individual with the parents of the other. For example, $f_{CE} = \frac{1}{2}(f_{CC} + f_{CD}) = \frac{1}{2}(0.625 +$
$0.1875) = 0.40625$. Note that the inbreeding coefficient of X is obtained from $f_{XX} = \frac{1}{2}(1 + F_X)$, that is,
$F_X = 2f_{XX} - 1 = 2 \times 0.671875 - 1 = 0.34375$, as in the other methods. The resulting matrix is called the
coancestry matrix (matrix **A**).

The average molecular coancestry of all individuals of a population is the complement to the unit of the expected frequency of heterozygotes or expected heterozygosity for the
marker (H_M), that is, $\bar{f}_M = 1 - H_M$. To illustrate this let us consider a simple example. Let us
assume a population formed by $N = 4$ individuals whose genotypes for a locus with three
alleles are A_1A_1, A_1A_3, A_2A_3 and A_2A_2. The frequencies of the three alleles A_1, A_2 and A_3 are p_1
$= 3/8, p_2 = 3/8$ and $p_3 = 2/8$, respectively. Using equation (2.2) of Chapter 2, heterozygosity is
$H_M = 1 - p_1^2 - p_2^2 - p_3^2 = 21/32$. On the other hand, the molecular coancestry matrix of the
individuals can be obtained (Table 4.3), and if the 16 values of the table are averaged

Table 4.2 *Coefficients of molecular inbreeding (homozygosity) of an individual and molecular coancestry (similarity) between a pair of individuals for a multiallelic genetic marker*

Genotype of individual	Molecular inbreeding (or molecular homozygosity) (F_M)
A_1A_1	1
A_1A_2	0

Genotype of a pair of individuals	Molecular coancestry (or molecular similarity) (f_M)
A_1A_1, A_1A_1	1
A_1A_1, A_1A_2	0.5
A_1A_2, A_1A_2	0.5
A_1A_2, A_1A_3	0.25
A_1A_2, A_3A_4	0

Table 4.3 *Coefficient of molecular coancestry between four individuals with the genotypes shown for a multiallelic genetic marker*

Genotype	A_1A_1	A_1A_3	A_2A_3	A_2A_2
A_1A_1	1	0.5	0	0
A_1A_3	0.5	0.5	0.25	0
A_2A_3	0	0.25	0.5	0.5
A_2A_2	0	0	0.5	1

$(\bar{f}_M = \sum_{i=1}^{N}\sum_{j=1}^{N} f_{M,i,j}/N^2 = (11/2)/16 = 11/32)$, where i, j denote the different individuals and $N = 4$, the result is $H_M = 1 - \bar{f}_M = 21/32$.

If the pedigrees of the individuals were available to calculate the average genealogical coancestry of the population ($\bar{f}$), this is equal to the expected frequency of homozygotes by descent. The relation between molecular and genealogical coancestry in a given population comes from the relationship

$$1 - \bar{f}_M = \left(1 - \sum_{k=1}^{a} p_k^2\right)(1 - \bar{f}),$$ (4.6)

where p_k is the frequency of allele k of the marker and the sum is for all its alleles (a). Solving, we obtain an estimate of the genealogical coancestry from the molecular coancestry:

$$\bar{f} = \frac{\bar{f}_M - \sum_{k=1}^{a} p_k^2}{1 - \sum_{k=1}^{a} p_k^2}.$$ (4.7)

However, in order to correctly infer $\bar{f}$ from $\bar{f}_M$ it is necessary that the values of the allele frequencies (p_k) in equation (4.7) are those of the base population from which the current one originates. Unfortunately, in most cases these original frequencies are not available and only the current ones can be used, which may differ from the initial ones due to genetic drift and other factors, so the inference of $\bar{f}$ from $\bar{f}_M$ will be biased. Although there are several methods to correct these estimates, none is exempt of problems (Toro et al., 2002). The bias is obviously due to the difference between identity by descent associated with $\bar{f}$ and identity in general associated with $\bar{f}_M$. Note that, rearranging equation (4.7) we obtain $\bar{f}_M = \bar{f} + \left(\sum_{k=1}^{a} p_k^2\right)\left(1 - \bar{f}\right)$, where the first term at the right of the expression is the probability that two alleles taken from the population are identical by descent, and the second is the probability that these alleles are alike in state. If in the initial population all the individuals carried different alleles, that is, there were $2N$ different alleles in N individuals, the two identities would coincide and $\bar{f} = \bar{f}_M$. This assumption that is accepted as a procedure in many computer simulations is, however, very unlikely in real situations and, consequently, the two parameters will generally differ. Fortunately, molecular coancestry and inbreeding tend to be highly correlated with their corresponding genealogical values (Toro et al., 2002), which allows the first to be used instead of the second when the latter are not available.

Equation (4.7) can also be used to estimate the genealogical coancestry coefficient of a particular pair of individuals, or the coefficient of inbreeding of an individual. For example, if data of L biallelic markers are available, the estimate of the coefficient of inbreeding of individual i, substituting the terms of coancestry by those of inbreeding and the frequencies of the markers in the current generation, is

$$\hat{F}_i = \frac{\sum_{s=1}^{L} F_{M,i,s} - \sum_{s=1}^{L}\sum_{k=1}^{a} p_{s,k}^2}{L - \sum_{s=1}^{L}\sum_{k=1}^{a} p_{s,k}^2},$$

(4.8)

where $F_{M,i,s}$ is the coefficient of molecular inbreeding of individual i for marker s (1 if the individual is homozygous for the marker and 0 otherwise), and $p_{s,k}$ is the frequency of allele k of marker s in the population. Taking into account that the allele frequencies used will generally be those of the current population (instead of those of the original base population), equation (4.8) provides a measure of the excess in homozygosis with respect to that expected in panmixia. Therefore, its value can be positive or negative and will be close to zero when the population is panmictic.

Analogously, it is also possible to estimate the coancestry between a particular pair of individuals from equation (4.7). However, an alternative procedure frequently used with data from high-density molecular markers, such as SNPs, is made in terms of molecular covariance. Cockerham (1969) showed that the covariance between the gene frequencies of the alleles of two individuals for a biallelic locus is $\text{cov}_{M,i,j} = f_{M,i,j} q(1-q)$, where q is the frequency of one of the alleles, resulting that $f_{M,i,j} = \text{cov}_{M,i,j}/[q(1-q)]$. If we consider L biallelic loci ($s = 1 \ldots L$) and N sampled individuals, the estimate of the molecular coancestry between individuals i and j is then

$$\hat{f}_{M,i,j} = \frac{\sum_{s=1}^{L}(q_{i,s} - \bar{q}_s)(q_{j,s} - \bar{q}_s)}{\sum_{s=1}^{L}\bar{q}_s(1 - \bar{q}_s)}$$

(4.9)

(Van Raden, 2008), where $q_{i,s}$ is the frequency of marker s in individual i (equal to 0, 1/2 or 1 if the individual carries none, one or two copies of the allele), and $\bar{q}_s = \frac{1}{N}\sum_{i=1}^{N} q_{i,s}$ is the average frequency of the marker s for all individuals (N). Equation (4.9) can also be obtained from the average frequency of markers for each individual and then averaging over markers, that is, substituting in equation (4.9) $\bar{q}_s$ for $\bar{q}_i = \frac{1}{L}\sum_{s=1}^{L} q_{i,s}$, with slightly different results (Toro et al., 2011).

The genealogical inbreeding indicates the expected homozygosity, but there is a variation of the actual homozygosity around that expected. For example, the proportion of the genome that is expected to be homozygous in the progeny of full sibs is a variable whose mean value is 0.25 but the standard deviation of that proportion is 0.18 (Hill and Weir, 2011). In some cases, inbreeding estimates based on genomic data from genetic markers may be more useful than genealogical inbreeding measures, particularly when the markers are very numerous and cover a large part of the genome, as is the case of SNPs, which can be analysed by tens of thousands (Kardos et al., 2015).

The possibility of determining the genotype of individuals for high-density genetic markers has also allowed us to discern between recent and ancestral inbreeding by means of the so-called regions or runs of homozygosity (ROH) of the genome, that is, portions of the genome in which all or the vast majority of the bases are homozygous. It is expected that inbreeding will generate large homozygous genomic regions by descent but also that these will be broken over time by recombination producing homozygous fragments of smaller size. Therefore, the largest ones will be considered due to recent inbreeding while the smallest ones are expected to come from more distant inbreeding. Following theory developed by Fisher (1949), the length of the ROH segments of identity by descent in Morgans (1 centimorgan, cM, implies 1% of recombination) follows an exponential distribution with mean $1/2g$, where g is the number of generations passed from the closest common ancestor. For example, it is expected that ROH with a common ancestor 10 generations ago have average lengths of the order of $1/20 = 5$ cM or approximately 5 Mb, assuming the common rule in mammals that 1 cM equals approximately 1 Mb of the genome (Dumond and Payseur, 2008). Likewise, ROH with a common ancestor 50 generations ago would have an expected length of 1 cM or 1 Mb. With these arguments, the inbreeding coefficient of the ROH of individual i ($F_{ROH,i}$) is then defined as the proportion (0 − 1) of the genome found in these segments:

$$F_{ROH,i} = \frac{\sum_{k} \text{length}(ROH_k)}{L} \tag{4.10}$$

(McQuillan et al., 2008). That is, $F_{ROH,i}$ is the sum of the lengths (in base pairs) of all the ROH of the individual's genome, divided by the total length of the genome (L, in base pairs). Equation (4.10) can be applied arbitrarily to different intervals of fragment lengths. For example, if only ROH greater than, say, 5 Mb (long ROH) are considered, the inbreeding produced recently (in the last 10 generations) will be more probably investigated. If, on the other hand, sections larger than, say, 0.5 Mb (short ROH) are considered, there will be an excess of small fragments due to ancient inbreeding, generated over 100 generations ago. In practice, the definition of the ROH when data of SNPs are available (instead of a complete sequence) is done by accepting a series of more or less arbitrary rules referred to the minimum number of SNPs that should constitute a ROH, the maximum number of heterozygous SNPs that are allowed in the region to take into account the possibility of mutation, the minimum density of markers in the region to be considered as such, and the maximum distance between consecutive ROH. Analogous to obtaining ROH inbreeding it is also possible to obtain the corresponding coancestry coefficients between pairs of individuals.

4.1.3 The Expected Inbreeding in the Ideal Population

So far we have studied how inbreeding and coancestry coefficients are calculated in real populations, either through a genealogy or with molecular marker data. Suppose now that we want to make predictions about the expected average values of these coefficients in a given population. The simplest way to do this is to assume the simple assumptions of the Wright–Fisher ideal population, described in Chapter 1 in the context of genetic drift. Suppose an ideal population of census size N in which initially all the alleles present ($2N$ alleles) are different and, therefore, the average inbreeding coefficient of the population in generation 0 is $F_0 = 0$. Since in this population the individuals produce a large number of gametes with equal probability, and these are randomly united to form the N individuals of the next generation, the expected coefficient of inbreeding in the first generation will be $F_1 = 1/2N$, that is, the probability of choosing twice a given allele between the $2N$ possible. When this happens it will necessarily be by self-fertilization of a given individual. This inbreeding, which is due to the union of two copies of the same allele of generation 0, may also occur in generation 2 due to self-fertilization of an individual of generation 1, that is, again $1/2N$. However, in generation 2 there will also be the possibility that copies of an allele of generation 0 will be united by cross-fertilization between different individuals of generation 1. This probability will be $(1 - 1/2N)F_1$, so the expected global inbreeding in generation 2 will be $F_2 = 1/2N + (1 - 1/2N)F_1$. In successive generations the same thing will occur, and the coefficient of inbreeding in any generation t will be determined by

$$F_t = \frac{1}{2N} + \left(1 - \frac{1}{2N}\right)F_{t-1}. \tag{4.11}$$

The first term ($1/2N$) is what is called new inbreeding and it can be generated in all generations by self-fertilization of the individuals of the previous generation ($t - 1$). The second is the ancient inbreeding, accumulated throughout generations, that occurs by crossbreeding between different individuals and is due to the union of alleles whose common ancestor belongs to generations prior to $t - 1$. This separation in new and old cumulative inbreeding illustrates a very important practical aspect. In the absence of other factors, the inbreeding accumulated in a population cannot be removed even if the generation of new inbreeding is avoided. Suppose that the population size has been N up to generation t, accumulating an inbreeding whose value is given by equation (4.11) and that from generation $t + 1$ the population size increases enormously to a new size N', so large that the new inbreeding ($1/2N'$) is negligible. Under these conditions the inbreeding will no longer increase but the value reached in generation t (F_t) will not be eliminated. Only in the case in which other forces, such as mutation or migration, act will it be possible for the accumulated inbreeding to be reduced or eliminated.

If the $1/2N$ term is solved from equation (4.11) it is found to be equal to

$$\Delta F = \frac{1}{2N} = \frac{F_t - F_{t-1}}{1 - F_{t-1}}, \tag{4.12}$$

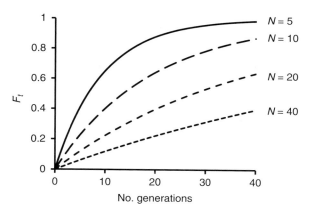

Figure 4.4 Expected inbreeding coefficient in each generation (t) in populations of size N.

that is, $1/2N$, the rate of increase in inbreeding (ΔF), is the increase in inbreeding from one generation to the next, relative to the precise change so that inbreeding takes the maximal value of one.

Equation (4.11) enables a prediction of the average inbreeding in the ideal population to be obtained in any generation based on the previous one, but it can be rearranged so that it is not necessary to make a recurrent calculation. Calling $P = 1 - F$ and substituting in equation (4.12), we have $1 - \Delta F = P_t/P_{t-1}$, so $P_t = (1 - \Delta F) P_{t-1}$ and, by recursion, $P_t = (1 - \Delta F)^t P_0 = (1 - \Delta F)^t$. Therefore, $F_t = 1 - (1 - \Delta F)^t$ and

$$F_t = 1 - \left(1 - \frac{1}{2N}\right)^t. \tag{4.13}$$

Note that equation (4.13) implies that the coefficient of inbreeding increases progressively depending on the population size until reaching the maximum value of one (Figure 4.4).

In Chapter 1 the process of genetic drift, or random change in allele frequencies due to the random sampling of gametes in a finite population, was explained and the corresponding predictive equations were presented. Genetic drift and inbreeding are two intimately related processes, since in populations of reduced size, where genetic drift has a greater impact, mating between relatives and, therefore, inbreeding, are inevitable processes with the course of generations. Thus, genetic drift involves the loss or fixation of alleles, which implies a reduction in the expected frequency of heterozygotes. Substituting equation (4.13) in equation (2.9), we obtain the relation between the variance of allele frequencies and the inbreeding coefficient, as a function of the initial allele frequencies in the population in the case of a biallelic locus:

$$\sigma_{q_t}^2 = p_0 q_0 F_t. \tag{4.14}$$

Substituting expressions (2.10) to (2.12), we obtain the expected frequencies of the three genotypes of the locus as a function of inbreeding:

$$AA : p_0^2 + p_0 q_0 F_t = p_0 F_t + p_0^2 (1 - F_t), \tag{4.15}$$

$$Aa : 2p_0 q_0 - 2p_0 q_0 F_t = 2p_0 q_0 (1 - F_t), \tag{4.16}$$

$$aa : q_0^2 + p_0 q_0 F_t = q_0 F_t + q_0^2 (1 - F_t). \tag{4.17}$$

The homozygotes frequencies expressed by equations (4.15) and (4.17) include both homozygotes identical by descent and homozygotes identical in state and the expressions to the right of the equality illustrate this difference. For example, the first term $p_0 F_t$ is the probability of being homozygote AA by descent (the probability of obtaining a homozygote by descent AA would be equal to the probability of choosing an allele A, whose probability is p_0, and finding an identical copy by descent of that same allele in the population, whose probability is F_t, by definition). And the second term, $p_0^2 (1 - F_t)$, is the probability of being homozygote in state. This partition is equivalent to the one performed previously to distinguish between genealogical and molecular coancestry from equation (4.6).

Equations (4.15) to (4.17) show that inbreeding implies the increase in the frequency of homozygotes at the expense of heterozygotes and, from equation (4.16), the expected heterozygosity in generation t (H_t) in relation to that of the initial generation (H_0) turns out to be

$$H_t = H_0(1 - F_t) = H_0 \left(1 - \frac{1}{2N} \right)^t. \tag{4.18}$$

That is, the heterozygosity is reduced in each generation by a rate of $1 - (1/2N)$:

$$H_t = H_{t-1} \left(1 - \frac{1}{2N} \right). \tag{4.19}$$

Note that, after a number of generations, we will arrive at $F_{t=\infty} = 1$, which implies that the expected frequency of heterozygotes will be $H_{t=\infty} = 0$, and therefore it is expected that finally there will only be homozygotes, either of genotype AA with probability p_0 or of genotype aa with probability q_0.

4.2 Populations with Regular Inbreeding

4.2.1 Highly Inbred Lines

In populations with random mating, inbreeding is produced by random mating between relatives, which will be more probable the lower the population size. In the most extreme case of populations with only one or two individuals, all matings are obviously consanguineous. The creation of populations with these very small sizes, called highly inbred lines, is a usual procedure to obtain pure lines, that is to say, homozygous for the whole genome, with diverse practical and experimental interests. Figure 4.5 shows the expected genealogy in both cases, considering two or three consecutive generations. For $N = 1$, which implies necessary self-fertilization, applying the expressions (4.1), (4.2) and (4.4) we have $F_X = \frac{1}{2}(1 + F_A)$ and, since A and X represent generic individuals of generation $t - 1$ and t, respectively:

Table 4.4 *Inbreeding coefficient (F_t), rate of increase in inbreeding (ΔF) and expected heterozygosity (H_t) in each generation (t) in populations of size N = 1 (selfing line) and N = 2 (brother–sister mating)*

N = 1				N = 2			
t	F_t	ΔF	H_t	t	F_t	ΔF	H_t
0	0.000		1.000	0	0.000		1.000
1	0.500	0.5	0.500	1	0.000	0.000	1.000
2	0.750	0.5	0.250	2	0.250	0.250	0.750
3	0.875	0.5	0.125	3	0.375	0.167	0.625
4	0.938	0.5	0.063	4	0.500	0.200	0.500
5	0.969	0.5	0.031	5	0.594	0.188	0.406
6	0.984	0.5	0.016	6	0.672	0.192	0.328
7	0.992	0.5	0.008	7	0.734	0.190	0.266
8	0.996	0.5	0.004	8	0.785	0.191	0.215
9	0.998	0.5	0.002	9	0.826	0.191	0.174
10	0.999	0.5	0.001	10	0.859	0.191	0.141

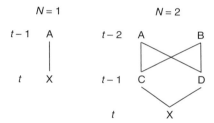

Figure 4.5 Genealogies with $N=1$ (self-fertilization line) and $N=2$ (line of mating between full sibs) in consecutive generations (*t*).

$$F_t = \frac{1}{2}(1 + F_{t-1}) = 1 - (\tfrac{1}{2})^t. \tag{4.20}$$

The expected inbreeding values, the rate of increase in inbreeding and the expected heterozygosity corresponding to the case of $N=1$ are presented in Table 4.4. The rate of increase in inbreeding, which can be obtained by substituting (4.20) in (4.12), is simply $\Delta F = 1/2N = 1/2$, which takes a constant value from the first generation. Being an ideal population, the inbreeding expected in this case can also be obtained by means of equation (4.13). Note that inbreeding grows very rapidly, as expected, and will be close to 100% after less than 10 generations.

In the case of an ideal population with $N=2$ where individuals can self-fertilize at random, inbreeding and ΔF are obtained again through expressions (4.11) and (4.13). However, in the case of species that cannot self-fertilize, as in most animals, these expressions do not give accurate predictions. The genealogy of Figure 4.5 refers to three consecutive generations of brother–sister mating lines without self-fertilization. Using the expressions (4.1), (4.2) and (4.4) we have that $F_X = f_{CD} = \frac{1}{4}(f_{AA} + f_{BB} + 2f_{AB}) = \frac{1}{4}[1 + 1/2$

$(F_A + F_B) + 2f_{AB}]$. Since A and B are individuals of generation $t-2$, their average inbreeding is equal to F_{t-2} and as the coancestry between A and B is equal to the inbreeding of their offspring in generation $t-1$, $f_{AB} = F_{t-1}$, so that we can get the recursion expression:

$$F_t = \frac{1}{4}(1 + F_{t-2} + 2F_{t-1}). \tag{4.21}$$

The corresponding inbreeding values and the rate of increase in inbreeding are shown in Table 4.4. Note that, initially, $\Delta F = 1/4$ (the expected value in an ideal population of size $N = 2$) but gradually decreases to a value of 0.191 due to the absence of self-fertilization.

A more general expression excluding self-fertilization is $F_t = F_{t-1} + (1 - 2F_{t-1} + F_{t-2})/2N$ (Crow and Kimura, 1970, p. 102), which holds for any value of N. If $N = 2$, equation (4.21) is recovered, as expected.

4.2.2 Inbreeding in Large Populations

Consider now the case of a very large population but in which a given proportion of the matings are inbred and the rest are produced by random breeding. For a self-fertilizing species where the selfing rate is β and the crossbreeding rate is $1 - \beta$, assuming that the population is infinitely large ($N = \infty$) and that, therefore, there is no inbreeding due to the reduced census size, the proportion β of matings produces an amount of inbreeding that can be predicted using equation (4.20), while the $1 - \beta$ proportion of matings does not produce inbreeding, that is, $F_t = \beta[\frac{1}{2}(1 + F_{t-1})] + 0(1-\beta)$. As the population is infinite, the inbreeding produced by self-fertilization is diluted by cross-fertilization and an equilibrium inbreeding is reached in which $F_{t-1} = F_t = \hat{F}$. Substituting, we have $\hat{F} = \beta \frac{1}{2}(1 + \hat{F})$, from where $\hat{F}$ can be obtained:

$$\hat{F} = \alpha = \frac{\beta}{2 - \beta}. \tag{4.22}$$

This theoretical coefficient of inbreeding is usually referred to as α instead of F, and it represents the disequilibrium factor with respect to the Hardy–Weinberg equilibrium due to mating between relatives, so that the observed frequency of heterozygotes in a population (H_o) is a fraction $(1 - \alpha)$ of the expected one, that is, $H_o = 2pq(1 - \alpha)$. Thus, in a population with a very large size where, for example, 50% ($\beta = 0.5$) of matings were by self-fertilization, we would expect an equilibrium inbreeding of $\hat{F} = \alpha = 1/3$, and a defect of observed heterozygotes of 2/3 of those expected if mating were random. Obviously, if the population is not infinite, there would be a progressive increase in inbreeding above the theoretical value indicated by equation (4.22).

The same argument can be applied to an infinite population with separate sexes and with a certain percentage (β) of mating between full sibs. Multiplying the right part of equation (4.21) by β and replacing $F_{t-2} = F_{t-1} = F_t = \hat{F} = \alpha$, we obtain

$$\hat{F} = \alpha = \frac{\beta}{4 - 3\beta}. \tag{4.23}$$

In an infinite population with 50% of matings between full sibs, $\hat{F} = \alpha = 0.2$, there would be a defect of observed heterozygotes equal to 80% of those expected if mating were random. A generalization of equations (4.22) and (4.23) for any type of mating between relatives is given by

$$\alpha = \frac{\beta}{2^j - (2^j - 1)\beta}, \tag{4.24}$$

where j determines the degree of coancestry ($j = 1$ for self-fertilization, 2 for mating between full sibs, 3 for mating between half sibs, etc.), and it is also possible to obtain expressions for combinations of different types of matings between relatives (Hedrick and Cockerham, 1986). If the percentage of mating between relatives is 100%, an equilibrium inbreeding coefficient of one will be reached, except when the coancestry is equal to or less than that of second cousins. In an infinite population in which all matings are between second cousins, the equilibrium inbreeding coefficient will be $\hat{F} \approx 0.016$.

We saw in the previous section that the inbreeding accumulated in a population due to its finite size cannot be removed even if the generation of new inbreeding is avoided by greatly increasing the population size. This is not the case, however, when the population has a very large size and inbreeding is produced by a percentage of inbred matings, such as implicit in equations (4.22) to (4.24). If this population were to be maintained by panmictic mating, the previous inbreeding produced would disappear completely.

We have also seen in previous sections that the coefficient of coancestry of the parents is equal to the coefficient of inbreeding of their offspring. Therefore, in a panmictic population the average inbreeding in a given generation $(\overline{F})$ will be equal to the average coancestry $(\overline{f})$ in the previous generation, that is, there is a delay of one generation between both averages, and their difference is generally small in each generation. When there are more matings between relatives than expected by pure chance, the gap is extended for more generations and the difference between averages increases, so that $\overline{F}$ may be much higher than $\overline{f}$ in a certain generation. In contrast, if matings between relatives are deliberately avoided, we will have that $\overline{F} < \overline{f}$, although this difference will always be very small. Nevertheless, despite the lag, except in particular situations (such as the permanent subdivision of the population), the rate of increase in inbreeding will be equal to the rate of increase in coancestry, that is, $\Delta F = \Delta f$.

4.3 Modulation of Inbreeding by the Effect of Mutation and Selection

The expressions obtained previously for the coefficient of inbreeding imply the absence of forces of change of the allele frequencies, such as mutation, selection and migration between sub-populations. In the presence of these factors, the inbreeding coefficient is expected to reach certain equilibrium values, as we shall see below (the effect of migration will be seen later in the context of subdivided populations).

4.3.1 Mutation–Drift Equilibrium

In the absence of mutation, inbreeding leads to allelic fixation and complete loss of heterozygosis. If the possibility that allelic variants mutate is considered, the eventual absolute homozygosis is not guaranteed. Consider equation (4.11) that gives the probability of identity by descent of two alleles taken at random from the population. If one of those alleles mutated (with probability u) there would no longer be identity by descent. Therefore, the probability of identity by descent must be multiplied by the probability that neither of the two alleles mutates, that is, $(1 - u)^2$. Thus, the inbreeding coefficient would be $F_t = [(1/2N) + (1 - 1/2N)F_{t-1}](1 - u)^2$. Given that the finite population size increases inbreeding and that mutation acts against this increase, a balance called mutation–drift equilibrium is reached. The coefficient of inbreeding in the equilibrium between mutation and drift is obtained by replacing $F_{t-1} = F_t = \hat{F}$ in the previous expression and solving for $\hat{F}$ we obtain $\hat{F} = (1 - u)^2/[2N - (2N - 1)(1 - u)^2]$. If we assume that u is small enough, that is, the terms in u^2 are negligible, we arrive at $\hat{F} \approx (1 - 2u)/[4Nu + 1 - 2u]$ or, approximately,

$$\hat{F} \approx \frac{1}{4Nu + 1}. \qquad (4.25)$$

Note that if we assume a reasonable mutation rate per locus and generation of $u = 10^{-6}$, the population size must be very high for the inbreeding coefficient of equilibrium to be much lower than one. For example, if $N = 10^6$, the expected inbreeding in the mutation–drift equilibrium is $\hat{F} \approx 0,2$ but if $N_e = 10^4$, $\hat{F}$ would be very close to unity (0.96). In general, if $Nu \ll 1$, $\hat{F}$ tends to 1.

4.3.2 The Purged Inbreeding Coefficient

As we will see in later chapters, the vast majority of mutations that affect fitness are deleterious. In large populations, natural selection efficiently removes mutations with effect on the heterozygote but acts less efficiently with respect to recessive mutations (Chapter 2), which therefore are accumulated in the populations. Inbreeding, by increasing the frequency of homozygotes, exposes many of these mutations in homozygosis, so that natural selection acts more efficiently, which is called purging or purge selection. The removal by selection of individuals carrying deleterious recessive mutations causes the inbreeding coefficient for these genes and, to a certain extent, for those linked to them, to be lower than that expected with the neutral model (equations (4.11)–(4.13)). Thus, a purged inbreeding coefficient is produced. There are two concepts in this sense. The first is the ancestral inbreeding coefficient (F_a) (Ballou, 1997), which would be the proportion of the genome of an individual that is expected to have been exposed to homozygosity by descent in at least one of its ancestors. The idea is that an inbred individual with consanguineous ancestry is more likely to survive the action of natural selection than an individual with the same inbreeding as the first but without consanguineous ancestry, since in the former recessive deleterious alleles would have been purged in its ancestors.

The second concept is that of the purged inbreeding coefficient (g), which would be approximated by

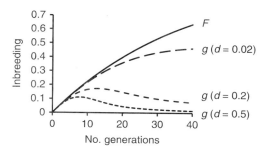

Figure 4.6 Change of the inbreeding coefficient (F) through generations in an ideal population of size $N = 20$ and purged inbreeding coefficient (g) corresponding to a recessive deleterious gene with purging coefficient $d = 0.5$ (lethal), 0.2 and 0.02.

$$g_t = \left[\frac{1}{2N} + \left(1 - \frac{1}{2N}\right)g_{t-1}\right](1 - 2dF_{t-1}) \tag{4.26}$$

(García-Dorado, 2012), where F_{t-1} is the neutral inbreeding coefficient represented by equations (4.11) to (4.13) and d is the dominance effect (equation (2.2)) which is the part of the effect on fitness of a deleterious allele totally or partially recessive that remains hidden in heterozygosis but is expressed in homozygosis. For deleterious mutations this value represents the effect on fitness attributable to a copy of the allele. For example, for a recessive lethal allele ($s = a = 1$ and $h = 0$ in expression (2.2)), $d = 0.5$ (changing the sign for convenience), so that in the homozygote the two copies of the allele produce its effect in homozygosis ($s = 1$). If the gene is partially recessive ($h = 0.2$), the value of d would be lower ($d = 0.3$) and for an additive allele ($h = 0.5$), $d = 0$. In this context, the value of d is called the purging coefficient. The higher this coefficient, the more effective will be the purging of deleterious alleles by natural selection in inbred populations, and to a greater extent the purged inbreeding coefficient will be reduced in relation to the neutral inbreeding coefficient. It has been estimated that the genomic purging coefficient takes values under competitive conditions of the order of 0.3 including all types of mutations, and 0.2 exclusively for non-lethal mutations (López-Cortegano et al., 2016). Figure 4.6 illustrates the purged inbreeding coefficient in comparison to the neutral one for a population with size $N = 20$ and three values of purging coefficients. Note that purging can significantly slow down the increase in inbreeding.

4.4 Inbreeding in Subdivided Populations

4.4.1 Wright's *F* Statistics

The majority of natural and domesticated populations do not form a single reproductive unit, but are subdivided into reproductive groups among which there may be a greater or lesser flow through migration. In this context, three inbreeding coefficients can be defined, called Wright's (1943) *F* statistics. The explanation of these statistics can be done both in terms of genealogical coancestry and inbreeding, and in terms of observed and expected heterozygosity referring to loci with known frequencies. The most correct definition is the former, referring to genealogical data, but perhaps the simplest way to understand the concepts is

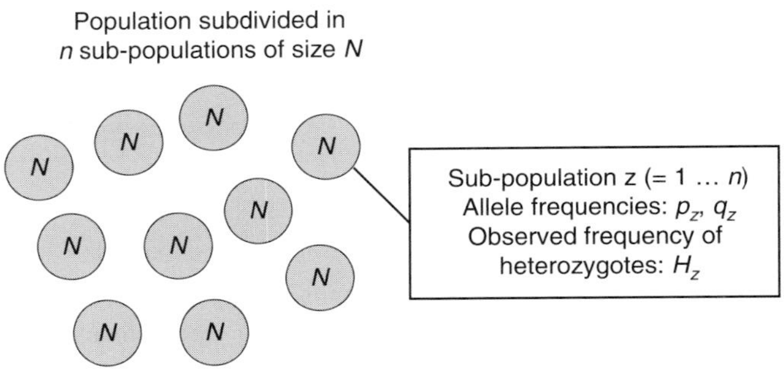

Population subdivided in
n sub-populations of size N

Sub-population z (= 1 ... n)
Allele frequencies: p_z, q_z
Observed frequency of
heterozygotes: H_z

Figure 4.7 Scheme for a population subdivided into n sub-populations of census size N. In sub-population z, the allele frequencies for a biallelic locus are p_z and q_z, and the observed frequency of heterozygotes is H_z.

by the latter, in terms of allele frequencies. This is Nei's (1973) approach, developed for multiallelic loci in terms of expected and observed heterozygosity, and analogous to F statistics if heterozygosities are genealogical ones (in Section 4.4.4 we will discuss the differences between both concepts). Consider a population subdivided into n sub-populations of the same size, as represented in Figure 4.7. Assume a locus, biallelic for simplicity, with allele frequencies p_z and q_z, in sub-population z (where $z = 1 \ldots n$). Let us also assume that the frequency of heterozygotes observed in sub-population z is H_z. The average heterozygosity observed in the whole population (H_I; I from individual) would be, therefore,

$$H_I = \frac{1}{n}\sum_{z=1}^{n}H_z.$$

(4.27)

Now we can obtain the expected frequency of heterozygotes in each sub-population assuming Hardy–Weinberg equilibrium (Chapter 2), that is, panmixia in each sub-population. In sub-population z would be $2p_zq_z$ and, if we average those for all sub-populations, we obtain the expected average heterozygosity in the whole set of sub-populations (H_S; S from sub-population),

$$H_S = \frac{1}{n}\sum_{z=1}^{n}2p_zq_z.$$

(4.28)

If there were more than two alleles, the calculation would be $H_S = \frac{1}{n}\sum_{z=1}^{n}\left(1 - \sum_{k=1}^{a}p_k^2\right)$, where a is the number of alleles per locus. Finally, we can obtain an expected heterozygosity assuming that the entire population is panmictic, that is, if there were no subdivision. In this case, the average allele frequency of the whole set would be $\bar{p} = \frac{1}{n}\sum_{z=1}^{n}p_z$ and $\bar{q} = 1 - \bar{p}$, and the expected total heterozygosity (H_T) would be

$$H_T = 2\bar{p}\bar{q}.$$

(4.29)

Using the above expressions, it is possible to obtain the three statistics. First,

$$F_{IS} = \frac{H_S - H_I}{H_S}$$

(4.30)

indicates the inbreeding due to the possible non-random mating within each sub-population, and is equivalent to the term α defined in expressions (4.22) and (4.24) when referring to populations without subdivision. Note that the numerator is the difference between the expected heterozygosity in the Hardy–Weinberg equilibrium and the observed one, and the denominator aims to standardize this difference.

Therefore, if mating between relatives were greater than that expected by chance $F_{IS} > 0$. On the other hand, if random mating between relatives were avoided, the coefficient could be negative. The range of theoretical variation is $-1 < F_{IS} < 1$, but in general, the negative values obtained by avoiding mating between relatives are small. The F_{IS} statistic can be defined as the probability of identity by descent of the two alleles carried by an individual considering the sub-population to which it belongs as the base population. It can also be defined as a correlation, hence the possibility of taking negative values, instead of a probability (in fact, the original definition of inbreeding given by S. Wright was made in terms of correlation). In this case, F_{IS} would be the correlation between the two alleles of an individual relative to that of two alleles taken at random from the sub-population.

Finally, if there were random mating in the sub-populations, we would expect $F_{IS} = 0$. In reality, it is not exactly zero, but takes a negative value more or less close to zero due to the binomial sampling of gametes. To generate a population of size N, $2N$ gametes should be sampled each generation. If p and q are the frequencies of two alleles in a sample of size $2N$, the proportion of observed heterozygotes obtained by sampling without replacement is

$$\frac{(2Np)(2Nq)}{\binom{2N}{2}} = 2pq\left(\frac{2N}{2N-1}\right), \tag{4.31}$$

which, equalizing to the expected value of heterozygotes $2pq(1 - F_{IS})$, leads to

$$F_{IS} = \frac{-1}{2N-1}, \tag{4.32}$$

that is, a value very close to zero unless N is very small.

The second statistic that can be obtained is

$$F_{ST} = \frac{H_T - H_S}{H_T}, \tag{4.33}$$

which is called the fixation index and determines the inbreeding due to the subdivision of the population, reflecting the nearness to fixation of alleles in each sub-population. Note that the numerator is the difference between the expected heterozygosity without or with subdivision. The range of variation of this coefficient is $0 < F_{ST} < 1$, that is, it only takes positive values, since the heterozygosity of a subdivided population is always less than or equal to that of the same population without subdivision, what is called the Wahlund's (1928) effect. Due to this effect, when Hardy–Weinberg disequilibrium with excess of homozygotes is observed in a sample, this may be indicative that the sample corresponds to a subdivided population instead of a panmictic one. F_{ST} indicates the correlation between alleles taken at random from a sub-population relative to that of two alleles taken at random from any sub-population. It would also indicate the probability of identity by descent of two alleles taken at random from a sub-population considering the set of all sub-populations as the base population.

Another indication given by the F_{ST} statistic, for which it is also often called the genetic differentiation index, is an idea of the diversification in allele frequencies between sub-populations. It is precisely this characteristic that makes the F_{ST} index a parameter of universal application in population and conservation genetics. Note that if all sub-populations had the same allele frequencies, $F_{ST} = 0$. An extreme value of $F_{ST} = 1$ would indicate that the differentiation between sub-populations in allele frequencies is total and would occur when only one allele of the considered locus was present in each sub-population and this was different in each sub-population. In fact, the numerator of equation (4.33) is a measure of the distance in allele frequencies between sub-populations. Nei's (1987) minimum distance between two sub-populations z and w is $D_{z,w} = \sum_k \frac{1}{2}(p_{z,k} - p_{w,k})^2$, where $p_{z,k}$ is the average frequency of allele k in sub-population z, and the average of all distances between sub-populations is

$$D_{ST} = \frac{1}{n^2} \sum_{z,w=1}^{n} D_{z,w} = H_T - H_S. \tag{4.34}$$

Compilations of F_{ST} estimates reveal that 26%, 29% and 55% of the estimates in vertebrates, invertebrates and plants, respectively, show values higher than 0.2 (Frankham et al., 2014), which indicates that genetic differentiation among sub-populations is very common in natural populations.

Finally, the third statistic, which combines the two previous sources of inbreeding, that due to non-random mating within sub-populations (F_{IS}) and that due to population subdivision (F_{ST}) is

$$F_{IT} = \frac{H_T - H_I}{H_T}, \tag{4.35}$$

which takes values in the range $-1 < F_{IT} < 1$ and is defined as the correlation between the two alleles carried by an individual in a locus relative to that of two alleles taken at random from the population as a whole. From expressions (4.30), (4.33) and (4.35) it follows that $1 - F_{IS} = H_I/H_S$, $1 - F_{ST} = H_S/H_T$ and $1 - F_{IT} = H_S/H_T$, so the relationship between the three coefficients is

$$(1 - F_{IT}) = (1 - F_{IS})(1 - F_{ST}). \tag{4.36}$$

An example of a calculation of Wright's statistics from the allele frequencies of a locus is presented in Table 4.5. If the calculations are made for several loci, the procedure consists of calculating the values of H_I, H_S and H_T for each locus and averaging them, using those averages to obtain the corresponding statistics. Alternatively, Wright's statistics can be estimated for each locus and averaging them, weighting by the H_T value of each locus, the same global estimates are obtained (Weir and Cockerham, 1984).

The previous statistics can also be obtained from the coancestry and inbreeding coefficients. Let us consider again the case of Figure 4.7 with n sub-populations of census size N each and, therefore, with total size Nn. Let f_{zz} be defined as the average coancestry between all possible pairs of individuals in sub-population z, that is, $f_{zz} = (1/N^2) \sum_{i,j=1}^{N} f_{ij}$, where f_{ij} is

Table 4.5 *Example of calculation of Wright's statistics with data of allelic frequencies for a locus with three alleles in four sub-populations*

	Sub-pop. I	Sub-pop. II	Sub-pop. III	Sub-pop. IV	Average frequency
Allele A_1	0.3	0	0.6	0	$p_1 = 0.225$
Allele A_2	0.3	0.9	0.1	0	$p_2 = 0.325$
Allele A_3	0.4	0.1	0.3	1	$p_3 = 0.450$
Observed frequency of heterozygotes	0.46	0.20	0.34	0	$H_I = 0.25$
Expected frequency of heterozygotes	0.66	0.18	0.54	0	$H_S = 0.345$
					$H_T = 0.641$
					$F_{IS} = 0.275$
					$F_{ST} = 0.462$
					$F_{IT} = 0.610$

Note: The observed frequency of heterozygotes (H_I) is the average of the frequencies in the four sub-populations (equation (4.27)). The expected frequencies per sub-population (H_S) are calculated with equation (4.28). For sub-population I, for example, $1 - [(0.3)^2 + (0.3)^2 + (0.4)^2] = 0.66$. H_S is the average for the four sub-populations. The expected frequency in the total population (H_T) is calculated from the average frequencies of each allele, that is, $1 - [(0.225)^2 + (0.325)^2 + (0.45)^2] = 0.641$ (equation (4.29)). Wright's statistics are obtained with equations (4.30), (4.33) and (4.35). The results indicate a large inbreeding due to an excess of mating between relatives ($F_{IS} = 0.275$) and a notable genetic differentiation between sub-populations ($F_{ST} = 0.462$).

the coancestry between any two individuals i and j of sub-population z. Notice that the summation also includes the diagonal, the self-coancestries of the individuals, which are $f_{ii} = 1/2(1 + F_i)$, where F_i is the inbreeding coefficient of individual i that can be expressed as $F_i = 2f_{ii} - 1$. If we average the mean coancestries of all sub-populations we obtain $\widetilde{f} = \frac{1}{n}\sum_{z=1}^{n} f_{zz}$. This term is the probability of identity by descent of two alleles taken at random from a given sub-population. Likewise, we can obtain the average of all the inbreeding coefficients of the population as a whole, averaging the values of F_i to obtain $\overline{F}$, which would indicate the probability of identity by descent of the two alleles carried by each individual. Finally, the average coancestry among all individuals in the population can be obtained as $\overline{\overline{f}} = [1/(Nn)^2]\sum_{z,w=1}^{n} \sum_{i,j=1}^{N} f_{iz,jw}$, which is the probability of identity by descent of two alleles taken at random from the total population. With these definitions, we have that Nei's minimum distance between sub-populations z and w is $D_{z,w} = \frac{1}{2}(f_{zz} + f_{ww}) - f_{zw}$. Likewise, $H_I = 1 - \overline{F}$, $H_S = 1 - \widetilde{f}$ and $H_T = 1 - \overline{\overline{f}}$, and

$$F_{IS} = \frac{\overline{F} - \widetilde{f}}{1 - \widetilde{f}}, \quad F_{ST} = \frac{\widetilde{f} - \overline{\overline{f}}}{1 - \overline{\overline{f}}}, \quad F_{IT} = \frac{\overline{F} - \overline{\overline{f}}}{1 - \overline{\overline{f}}}. \tag{4.37}$$

4.4.2 Change of Base Population in the Estimation of Inbreeding

The partition of inbreeding due to mating between relatives in each sub-population (quantified by F_{IS}) and that due to the finite size of sub-populations (quantified by F_{ST}) can also be applied to populations without subdivision, if one wants to distinguish between both types of inbreeding. To

Gen. 0   Gen. t

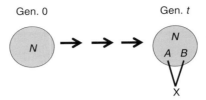

Figure 4.8 Scheme representing a population of census size N maintained under panmixia during t generations. In the last generation, individuals A and B, who are full sibs, are crossed to generate offspring X.

illustrate this let us consider Figure 4.8, which represents the situation in which an ideal population of census size N is maintained for t generations with panmixia. In generation t, two full sibs (A and B) are crossed and we want to know the inbreeding coefficient of their descendant X. Given that X is the offspring of full sibs, we know that its inbreeding, relative to that of its parents (or, in general, to that of the population from which its parents belong, in generation t) is 0.25. This would be the inbreeding coefficient of X if its parents were not inbred, but we know that they are so since they belong to a population of finite size maintained for t generations. The inbreeding due to mating between full sibs (0.25) is equivalent to F_{IS}, while the inbreeding due to the finite population size is equivalent to F_{ST}. The latter can be calculated from equation (4.13). For example, if $N = 20$ and $t = 10$, $F_{t=10} = 0.224$. Therefore, following the analogy and using equation (4.36), $(1 - F_X) = (1 - 0.25)$ $(1 - 0.224)$, from which the absolute inbreeding of individual X (referring to the population at generation 0) is $F_X = 0.418$.

In general, equation (4.36) can be used for any calculation of inbreeding that involves a change in the base population of reference. As we have indicated in previous sections, the inbreeding coefficient measures identity by descent from a starting point, which is the base population. In that population, it is assumed that homozygotes are not so by descent, but only in state. However, if we could go back further in time, we might find that the homozygotes of that base population are also homozygous by descent. In this case, if we consider F_{IS} (F_{new}) as the new inbreeding that has recently occurred with respect to the base population considered, and F_{ST} the inbreeding accumulated by this base population with respect to an ancestral state, the total inbreeding relative to the furthest moment (F_{old}) can be obtained from $(1 - F_{old}) = (1 - F_{new})(1 - F_{ST})$, where $F_{old} = F_{new} + (1 - F_{new})F_{ST}$ (Powell et al., 2010). This partition is equivalent to the one which was carried out between molecular coancestry (identity of alleles) and genealogical coancestry (identity by descent) (equation (4.6)), from which it is obtained that $\bar{f}_M = \bar{f} + (1 - \bar{f})(\sum_{k=1}^{a} p_k^2)$. Thus, F_{new} would indicate the identity by descent created recently in the current base population (genealogical coancestry) and F_{old} would be the molecular identity (whether it is by descent or not) generated from the ancestral state, the ancestral frequency of homozygotes being an estimate of the accumulated inbreeding between both states of the base population.

4.4.3 Migration–Drift Equilibrium

Using an argument analogous to that used in Section 4.3.1 to determine the mutation–drift equilibrium, one can deduce the equilibrium inbreeding value between genetic drift and migration. Let us assume a subdivided population following the model of infinite islands explained in Chapter 2 (Figure 2.6b). Each sub-population is an ideal population of census size N and receives and sends migrants to the whole population with a rate m per generation. Assuming that any allele from immigration is not identical by descent to any of the alleles present in the sub-population, the probability of identity by descent is, using equation (4.11), $F_t = [(1/2N) + (1 - 1/2N)F_{t-1}](1 - m)^2$, and the equilibrium value will be approximately equal to

$$\hat{F} \approx \frac{1}{4Nm + 1}.$$ (4.38)

The coefficient $\hat{F}$ in expression (4.38) is the total inbreeding expressed by Wright's F_{IT}. With panmixia within each sub-population ($F_{IS} \approx 0$), however, it is equated to the coefficient F_{ST}, with the value of Nm being the absolute number of migrants entering the population each generation. In contrast to mutation (equation (4.25)), the force of migration to reduce inbreeding is formidable, even if reasonably low migration values are considered. For example, if we assume a population of arbitrary size N, and a migration rate such that the average number of immigrants per generation is $Nm = 1$, this rate is sufficient to restrict the inbreeding of the population to a value of only $\hat{F} \approx 0.2$ (Figure 4.9). With a larger number of migrants, the expected value of $\hat{F}$ gradually approaches, albeit slowly, to zero. Note that the value that determines the equilibrium inbreeding is the product Nm, that is, for a given value of Nm, populations differing in N and/or m reach the same equilibrium inbreeding, except for second order terms. This occurs because in a larger population there will be less genetic drift than in a smaller one and, therefore, a smaller migration rate will be necessary in the first than in the second to compensate for the increase in inbreeding.

In the case of the stepping-stone model of migration (Chapter 2, Figure 2.6c), where the exchange of individuals occurs only between adjacent sub-populations, the effect of migration in restricting inbreeding is obviously lower, so that in order to achieve a similar effect to that of one migrant per generation in the infinite islands model, about twice as many migrants would be needed in the stepping-stone model.

The equilibrium $\hat{F}$ (or F_{ST} when there is panmixia in each sub-population) due to the joint action of migration and mutation can be obtained simply by combining equations (4.25)

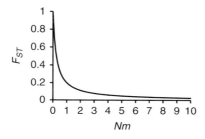

Figure 4.9 Expected value of the F_{ST} statistic based on the number of migrants (Nm) per generation and sub-population in an infinite islands model (equation (4.38)).

and (4.38), $F_{ST} \approx 1/[4N(m + u) + 1]$ that for the model with a finite number of islands (n islands) should be corrected as follows (Takahata, 1983):

$$F_{ST} \approx \frac{1}{4N\left[m\left(\frac{n}{n-1}\right)^2 + u\left(\frac{n}{n-1}\right)\right] + 1}. \tag{4.39}$$

4.4.4 Estimation of F_{ST} with Genetic Markers

The F_{ST} statistic (as well as other Wright's statistics) provides measures of the probability of identity by descent when obtained with genealogical coancestries and inbreeding. In this sense, the definition of F_{ST} is equivalent to another expressed in terms of coalescence time (the time between the generation in which two alleles of a population are sampled and the closest generation of an ancestral copy of them), so that $F_{ST} = (\bar{t}_T - \bar{t}_S)/\bar{t}_T$ (Whitlock, 2011), where $\bar{t}_T$ is the average time elapsed back to the most recent ancestor of two alleles taken at random from the global population, and $\bar{t}_S$ is the corresponding time when the two alleles are taken from a given sub-population. However, when F_{ST} is calculated from the data obtained with genetic markers, there may be differences with respect to the genealogical or coalescence F_{ST}, depending on the characteristics of these markers, due to the fact that the estimates of heterozygosity are limited between 0 and 1 and it is not usually possible to distinguish between identity by descent and identity in state with markers.

Wright's (1943) original definition of F_{ST} was established in terms of the variance of allele frequencies for a biallelic locus, analogous to equation (4.9), so that $F_{ST} = \sigma_q^2/\bar{p}\bar{q}$, with the numerator being the variance of the allele frequencies of the sub-populations and $\bar{p}$ and $\bar{q}$ the mean frequencies in the global population. In fact, the estimation of F_{ST} is usually obtained in terms of variances of allele frequencies (Weir and Cockerham, 1984). Note that expressed in this way F_{ST} also describes the proportion of the variation in allele frequencies that is attributed to differences between sub-populations. The expression of F_{ST} that we have obtained with the heterozygosity procedure (equation (4.33)) was developed for multiallelic loci and is called G_{ST} (Nei, 1973) instead of F_{ST}. Nevertheless, as mentioned before, Wright's F_{ST} definition and Nei's G_{ST} are analogous when referring to genealogical data and we maintain Wright's F original nomenclature for simplicity. When allelic data rather than genealogical data are considered, however, the genealogical or coalescence F_{ST} can be different from its estimate obtained from allelic data (G_{ST}), as explained below.

In practice, F_{ST} can be estimated from the analysis of variance of the allele frequencies or the heterozygosity values, with the necessary corrections to take into account the finite size of the samples used in their estimation. In any case, the estimates obtained from genetic markers depend on the characteristics of these. In general, if the markers used have a low mutation rate, the molecular F_{ST} estimates do not deviate much from the theoretical ones. However, for highly mutable markers, such as microsatellite loci used in many analyses, the biases may be very high and the F_{ST} statistic may not give a very accurate idea of the differentiation in allele frequencies. This can be easily seen by means of expression (4.33). Microsatellite loci are usually very polymorphic, so sometimes sub-population heterozygosity (H_S) can be close to one, but since total heterozygosity (H_T) also has a maximal limit of one, the numerator is necessarily small and, therefore, the value of F_{ST} will be close to zero even when there is a great differentiation in allele frequencies between sub-populations. In order to be able to compare F_{ST} estimates of different markers, Hedrick (2005) proposed a correction similar to the one proposed for linkage disequilibrium (Chapter 2), consisting of relativizing the estimate of F_{ST} to the highest value that could be obtained assuming the same global allelic frequencies,

$$F'_{ST} = \frac{F_{ST}}{\left(\frac{(n-1)(1-H_S)}{n-1+H_S}\right)}, \tag{4.40}$$

so that F'_{ST} always has a range between 0 and 1 regardless of the H_S value. For the example in Table 4.5, where an estimate of $F_{ST} = 0.46$ was found using equation (4.33), we obtain that $F'_{ST} = 0.79$, that is, 79% of the maximum possible value (0.59), as a result of including the denominator of equation (4.40).

Other alternative statistics of differentiation between sub-populations are based on the allelic differences between them rather than on differences in frequencies. For example, the statistic

$$D = \frac{H_T - H_S}{1 - H_S}\left(\frac{n}{n-1}\right) \tag{4.41}$$

(Jost, 2008) is based on the effective number of alleles (Kimura and Crow, 1964), that is, the number of alleles that there would be if all of them were segregating at the same frequency. For the example in Table 4.5, using equation (4.41), $D = 0.60$, greater than F_{ST} (0.46). In the finite island model (n islands) with migration m, D can be predicted as

$$D \approx \frac{1}{1 + \frac{m}{u(n-1)}} \tag{4.42}$$

(Jost, 2008). Note that D does not depend on the population size (N) and can be approximated simply by $(n-1)(u/m)$ if $u \ll m$ (Whitlock, 2011). That is, D provides a measure of the differentiation between sub-populations that will increase the greater the mutation rate of the loci considered and will decrease if migration between sub-populations increases.

Figure 4.10 shows the expected values of F_{ST} in a model of n sub-populations obtained by means of equation (4.39), considering migration and mutation (the two terms of the equation, or only migration). In the latter case, F_{ST} (m) indicates the theoretical value of F_{ST} that provides the level of differentiation between sub-populations that would be obtained with genealogies or coalescence and, therefore, is independent of the type of genetic marker used. The value of (F_{ST} ($m + u$)) is dependent on the mutation rate (u) of the genetic marker (this is more precisely the concept of Nei's G_{ST}). Note that if u is small (panel a) there is not much difference between the two values of F_{ST}. However, when the marker has a high mutation rate (panel b), the estimate that would be obtained with that marker (F_{ST} ($m + u$)) would be considerably lower than the theoretical one (F_{ST} (m)) when the migration rate m is low. The D statistic (equation (4.42)) also depends on the marker used and may be lower or higher than the theoretical F_{ST} as a function of the relationship between m and u. Depending on the type of marker and demographic scenario, D and F_{ST} can be very different, the former giving a more appropriate measure of allelic differentiation between sub-populations than F_{ST} when referring to a particular set of markers (Jost et al., 2017).

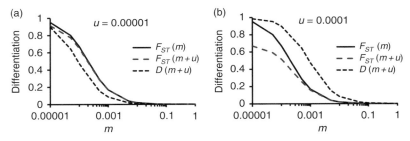

Figure 4.10 Evolution of the statistics of differentiation between sub-populations, assuming a finite island model ($n = 10$ islands) with sub-population size $N = 100$, as a function of the migration rate (m) between islands. The solid line indicates the prediction of F_{ST} from equation (4.39), ignoring mutation ($u = 0$), and therefore indicates the genealogical or coalescence value of F_{ST}. The broken lines indicate the prediction of F_{ST} from a genetic marker with mutation rate u, obtained from equation (4.39), including m and u. Finally, the dashed line indicates the prediction of D (equation (4.42)) also assuming the particular values of m and u.

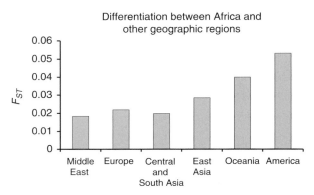

Figure 4.11 Differentiation in allele frequencies (F_{ST}) between Africa and other geographical regions obtained from 377 microsatellite loci in 1052 humans from different geographical regions. (Data from Rosenberg et al. (2002) analysed with the METAPOP programme (Pérez-Figueroa et al., 2009a).)

Finally, Figure 4.11 presents an example of F_{ST} values obtained from genetic markers. The data correspond to the analysis of 377 microsatellite loci in 1056 individuals from diverse human populations grouped in seven large regions of the world (Rosenberg et al., 2002). The figure represents the value of F_{ST} between the region of Africa and the other geographical areas. The average estimate of F_{ST} considering the seven regions is 0.047, indicating that of all variation in allele frequencies, only 4.7% is ascribed to differences between regions and, therefore, most of these differences occur within them. The results are compatible with the hypothesis of a recent African origin, according to which the human populations originating in Africa colonized progressively the Middle East and Europe, to later colonize Asia, Oceania and America, passing through successive population bottlenecks (see Reich, 2018 for more information on human origins). This hypothesis would imply a greater differentiation between the region of origin (Africa) and regions geographically further away from it in the various colonizations, following approximately the abscissa axis of the figure

(Handley et al., 2007). The hypothesis is also compatible with the fact that the greatest genetic diversity is assigned to Africa (mean heterozygosity $H = 0.79$ and average number of alleles per locus 9.13) and the lowest to America ($H = 0.66$ and mean number of alleles per locus 5.79).

Problems

4.1 In a set of populations with 20 reproductive individuals each, the initial allele frequencies of a biallelic locus are $p_0 = 0.3$ and $q_0 = 0.7$. If the populations are kept under ideal conditions, what will be the expected values of the heterozygosity, the variance of allele frequencies and the inbreeding at generations $t = 5$, 20 and 100?

4.2 A plant species reproduces naturally with autogamy and allogamy in equal proportions. From a population of large census size of this species a line with size $N = 20$ is founded and maintained for 10 generations in the greenhouse by cross-fertilizing reproduction. What will be the expected value of the inbreeding coefficient of the population in generation 10?

4.3 In Problem 4.1 it was assumed that the locus considered is neutral. Suppose that one of the alleles of the locus is deleterious with a purging coefficient of $d = 0.2$. What would then be the expected values of the heterozygosity, the variance of allele frequencies and the inbreeding for this locus in generation $t = 20$?

4.4 In a population subdivided into $n = 10$ sub-populations of constant census size $N = 100$, among which migration takes place with rate $m = 0.01$ per generation following an island model, a molecular marker whose mutation rate is $u = 10^{-6}$ is analysed. (a) What is the value of the differentiation indices F_{ST} and D, considering or not mutation for the marker? (b) What would these values be if the marker mutation rate were $u = 10^{-3}$?

4.5 In an experimental fish population maintained in 10 tanks, diversity for a group of molecular markers (SNPs) is analysed obtaining an average expected heterozygosity in each tank of $H_S = 0.21$ and a total heterozygosity for the whole set of tanks of $H_T = 0.32$. (a) Calculate the differentiation index F_{ST}, the corrected index relative to its maximum value and the D index. (b) The analysis is repeated with microsatellite markers obtaining values of $H_S = 0.83$ and $H_T = 0.89$. Are the results of both types of markers consistent with each other?

Self-Assessment Questions

1 The expected coefficient of inbreeding of the progeny of first cousins is 1/16.

2 The coancestry coefficient between two half sibs is 1/8, so they are expected to share 1/8 of their alleles.

3 The coefficient of molecular inbreeding is generally an underestimate of the coefficient of genealogical inbreeding.

4 It is expected that homozygous fragments (ROH) of 4 Mb had a common ancestor about 25 generations ago.

5 In an infinite population where 30% of autogamy is carried out, a heterozygosity of 82% is expected.

6 A coefficient $F_{IS} < 0$ indicates a defect of heterozygotes in the sub-populations of a structured population.

7 An F_{ST} coefficient close to zero indicates that the variance of allele frequencies between sub-populations is small.

8 A defect of heterozygotes with respect to the equilibrium value of Hardy–Weinberg may be indicative of the structuring of the population.

9 The number of migrants per generation that produces a certain equilibrium value of inbreeding is lower for an island model than for a stepping-stone model.

10 The values of F_{ST} obtained by highly polymorphic molecular markers, such as micro-satellite loci, are underestimates of the genealogical or the coalescence values.

5 Effective Population Size

Concepts to Study

- Effective population size
- Population bottleneck
- Variance of contributions from ancestors to descendants
- Discrete generation equivalent
- Cumulative effect of selection on effective size
- Selective sweep
- Allelism of lethals

Objectives for Learning

- To understand the general concept of effective population size
- To distinguish the different types of effective population size (variance, inbreeding, mutation, eigenvalue and coalescence)
- To understand the impact of a population bottleneck on the effective size
- To know the concept of discrete generation equivalent for the estimation of effective size in populations with overlapping generations
- To understand the cumulative effect of selection on effective population size and the concept of selective sweep
- To know the applications that derive from the theory of effective population size for the conservation of genetic resources
- To learn how to estimate the effective size of a population using demographic data and by the method of allelism of lethals
- To understand the basis of the procedures used to estimate the effective population size using molecular markers

5.1 Definition and Methods for the Prediction of the Effective Population Size

In Chapter 2 we analysed the process of genetic drift, or random change of allele frequencies in populations of small size due to the sampling of gametes, and in Chapter 4 we considered the inbreeding generated in these populations by the inevitable mating between relatives. We studied these phenomena under the simplified conditions of the ideal population of Wright–Fisher, which are described in Section 2.5. Under this simple model, we derived the expressions of the expected variance of allele frequencies by genetic drift (equations (2.8) and (2.9)) and the expected inbreeding coefficient and its rate of increase per generation (equations (4.12) and (4.13)), all of them being a function of the population census size, N. However, real populations may fail to meet one or more of the ideal conditions, so that the mentioned expressions would no longer hold. For example, separate sexes may exist rather than hermaphroditic individuals, the census size may vary with generations rather than being constant, mating between individuals may not be random, or individuals may possess different reproductive capacities and not the same. The expressions obtained in Chapters 2 and 4, however, can still be used as predictive tools in real situations using the concept of effective population size, N_e, introduced by Wright (1931, 1933). This is defined as the size that would have an ideal population that would give rise to the drift or inbreeding observed in the real population. That is, if we could obtain a measure of the variance of allele frequencies observed from genetic markers, or the rate of increase in inbreeding per generation from the pedigrees, the effective population size would be the size of an ideal population in which the variance of allele frequencies or the increase in inbreeding were those observed. Thus, if $\sigma^2_{q_1 OBS}$ is the variance of the allele frequencies observed in the first generation for a set of loci with initial allele frequency q_0, we infer from equation (2.8) that $N_e = p_0 q_0 / (2\sigma^2_{q_1 OBS})$. Likewise, if ΔF_{OBS} is the inbreeding rate observed per generation, from equation (4.12) we obtain $N_e = 1/(2\Delta F_{OBS})$. The effective sizes to which these expressions refer to are called, respectively, variance effective size and inbreeding effective size.

In practice, the problem is often the opposite. There are no empirical estimates of drift and inbreeding in the population and it is intended to deduce what their magnitude would be by predicting the corresponding effective size, which is often possible. Considering the possible deviations of the ideal population, one by one or in combinations, it is possible to predict the effective size of a population based on criteria of drift or inbreeding, as we will see in the next sections. But there are other procedures to calculate the effective size. In Chapters 2 and 4 we also saw that drift and inbreeding produce a reduction in the frequency of heterozygotes, and the rate at which this occurs can be obtained from the largest non-unit eigenvalue of the transition matrix (**P**) of a Markov chain, whose elements determine the frequencies of the $2N + 1$ possible allelic states in a population of size N,

$$p_{ij} = \binom{2N}{j} \left(\frac{i}{2N}\right)^j \left(1 - \frac{i}{2N}\right)^{2N-j}, \tag{5.1}$$

where p_{ij} is the probability that there are j copies of an allele in a population at some generation given that there were i copies of the allele in the previous generation. It can be shown that the main eigenvalue of this matrix is $\lambda = 1 - [1/2N] = 1 - \Delta F$. If the matrix is constructed with models other than those corresponding to the ideal population, the effective size can be predicted as $N_e = 1/[2(1 - \lambda)]$, what is called the eigenvalue effective size.

Another less used method is that of mutation effective size, which is defined from the probability of identity in state of a gene, instead of the probability of identity by descent, under a mutation model with a defined rate.

Finally, a procedure increasingly used to predict the effective size, particularly in the evolutionary context, is the coalescence effective size. The coalescence time of two given alleles is the number of generations that must be traced back in the genealogy to the ancestor from which they come. The probability of coalescence of two alleles t generations ago in an ideal population is $[1 - (1/2N)]^{t-1}(1/2N)$, therefore, the mean time of coalescence is $T = (1/2N)\sum_{t=1}^{\infty} t[1 - (1/2N)]^{t-1} = 2N$ generations, and if we can estimate the mean time of coalescence (T) in a real population we can predict the effective size as $N_e = T/2$.

Except in very specific situations, the predictions of the effective population size obtained with the above criteria coincide, or only differ in second order terms. When subdivided populations are considered, the variance and inbreeding effective sizes differ substantially during the first generations, but if the population structure remains constant, both sizes tend towards the same asymptotic effective value. In the next sections we will make reference only to these asymptotic values, obtained by means of one or more of the previous procedures.

The effective size is a fundamental parameter in population and quantitative genetics, and has implications in evolutionary biology, conservation and animal and plant breeding, since the rates of increase in genetic drift and inbreeding depend on its magnitude, as well as the efficiency of processes such as natural and artificial selection and migration. In the next sections we will study the most important predictive formulas and the methods for their estimation from empirical data. The theoretical developments initiated by Sewall Wright were continued by James Crow and Motoo Kimura (Kimura and Crow, 1963a; Crow and Kimura, 1970) and later by other authors (see reviews by Crow and Denniston, 1988; Caballero, 1994; Wang and Caballero, 1999; Charlesworth, 2009; Wang et al., 2016). Most estimation methods based on empirical data are more recent and can be consulted in the reviews by Luikart et al. (2010) and Wang et al. (2016).

5.2 Prediction of N_e in Unselected Populations

In this section we will explain how to predict N_e for different cases in which one or more of the characteristics of the ideal population are not met. In all of them we will assume that the population in question is not subject to selection, since the predictions are more complex in that case and will be explained in a later section.

5.2.1 Absence of Self-Fertilization

Suppose, to begin with, that in the population under study all the characteristics of the ideal population are met except that individuals are unable to self-fertilize. In this case, two alleles taken at random from generation t that are identical by descent cannot be a copy of the same allele of generation $t - 1$ but, at least, of one allele of generation $t - 2$, that is, the absence of self-fertilization delays the appearance of inbreeding in a generation and reduces its rate of increase. It can be shown that the latter is approximately $\Delta F \approx 1/(2N + 1)$ (Wright, 1969) and the effective size

$$N_e \approx N + \frac{1}{2}.$$

(5.2)

Unless N is very small (e.g. for full-sib lines), the impact of random self-fertilization is modest. Therefore, we will ignore this effect in the following sections and, in general, the terms of order $1/N^2$, which complicate the expressions and are generally negligible if $N \geq 10$.

5.2.2 Different Number of Males and Females

Let us now consider the situation in which, once again, all the conditions of the ideal population are fulfilled but, in addition to no self-fertilization, there are parents of separate sexes that can differ in number, being N_m and N_f, the number of male and female reproducers, respectively, and $N = N_m + N_f$, the total. Since half of the genes at generation $t - 1$ come from males and half from females from generation $t - 2$, the probability that two alleles carried by an individual of generation t are identical by descent and come from males of generation $t - 2$ is 1/4, and the probability that they come from the same male is $1/4N_m$. The same can be argued with females. Therefore, the probability that the two alleles carried by an individual in generation t come from the same individual, male or female, of generation $t - 2$ is $1/4N_m + 1/4N_f$. In the ideal population this probability is simply $1/N$. If both are equated and N is replaced by N_e, we obtain a prediction of the effective size

$$\frac{1}{N_e} = \frac{1}{4N_m} + \frac{1}{4N_f}; \quad N_e = \frac{4N_mN_f}{N_m + N_f} \tag{5.3}$$

(Wright, 1931). If the number of males and females is identical ($N_m = N_f = N/2$), the effective size is equal to the number of reproducers, $N_e = N$, as it occurs in the ideal population, although remember that this is only approximate since we ignore the absence of self-fertilization expressed in equation (5.2).

Note that N_m and N_f are not the total number of males and females present in the population, but the number of individuals of each sex that can intervene in reproduction. An unbalanced sex ratio is very common in nature where only a few dominant males mate with females in many species and, particularly, in domestic animal populations, where the number of sires is generally much smaller than the number of dams. For example, an analysis of pedigrees in the Holstein-Friesian breed of cattle showed that 5 bulls were the grandparents of more than 60,000 registered cows, and the first in the ranking had 17,000 granddaughters. Suppose, as an illustration, that the reproduction is produced by crosses between $N_m = 5$ males and $N_f = 45$ females. Applying equation (5.3) we obtain $N_e = 18$. That is, it is expected that, if the population is maintained in successive generations with these numbers of breeding animals, the rate of inbreeding and genetic drift are the same as would occur in an ideal population (that is, without separate sexes) of size $N = 18$, instead of $N = 50$, which is the total number of reproducers. Note that equation (5.3) provides twice the harmonic mean of the sizes of each sex. Given that the impact of the population size on drift and inbreeding occurs in the denominator of the equations (see, for example, (2.8) and (4.12)), the harmonic mean is the one that should be used whenever we want to average population sizes or effective population sizes for the purpose of predicting their impact on drift or inbreeding. In a harmonic mean the small values are those that have the greatest relevance and, therefore, whenever the sex ratio is not 50%, it will occur that $N_e < N$, and if one of the sexes has disproportionately low values, the reduction of N_e will be considerable.

5.2.3 Variable Population Size across Generations

Imagine now that another of the conditions of the ideal population is not met, the constancy of the population size throughout generations, keeping the rest. We saw in Chapter 4 (equation (4.19)) that, in a population of size N, heterozygosity is reduced each generation at a rate $H_t/H_{t-1} = (1 - 1/2N)$. Therefore, in a population of variable size, with $N_0, N_1, N_2, \ldots$ N_{t-1} reproductive individuals in consecutive generations, the cumulative reduction will be $H_t/H_0 = (1 - 1/2N_0)(1 - 1/2N_1)(1 - 1/2N_2) \ldots (1 - 1/2N_{t-1})$. In the ideal population (constant size equal to N), the corresponding drop is $H_t/H_0 = (1 - 1/2N)^t$, which can be equated to the previous one, and substitute N for N_e to obtain the prediction of the effective size, that is, $\Pi(1 - 1/2N_i) = (1 - 1/2N_e)^t$. If the sizes are not too small, both expressions can be approximated by $1 - \Sigma(1/2N_i) \approx 1 - (t/2N_e)$, from where we get

$$\frac{1}{N_e} \approx \frac{1}{t}\sum_{i=0}^{t-1}\frac{1}{2N_i} \tag{5.4}$$

(Wright, 1938), that is to say, the effective size during the period considered is the harmonic mean of the population sizes in each generation. Again, low sizes in some generations determine reduced global values of N_e. For example, if the population sizes in five consecutive generations are 2500, 2000, 50, 1500 and 3000, applying equation (5.4) we obtain $N_e \approx 228$. Note that, again, the smallest size is the one that has the greatest impact in N_e. The illustrated situation is what is called a population bottleneck, a frequent event in natural settings. Let us notice that although the size in the last generation is very high (3000 individuals), the drift or inbreeding generated by the bottleneck are appreciable, since they would be, in the same period of five generations, those corresponding to an ideal population of size $N = 228$. This explains why on some occasions the genetic diversity observed in a population is lower than what is expected with its current size.

We explained at the beginning of the chapter that the variance and inbreeding effective sizes coincide in most cases, but they differ when the population size changes with generations. The drift of allele frequencies depends on the number of offspring, while the inbreeding in the offspring depends on the number of parents. To illustrate this let us consider an extreme case. If in generation $t - 1$ there is a single heterozygous individual that reproduces by self-fertilization and generates infinite progeny in generation t, the allele frequency does not change when passing from one generation to the next (0.5), that is to say, there is no genetic drift and, therefore, the variance effective size is infinite. However, half of the progeny will be homozygous by descent, that is, an inbreeding of 0.5 will have been generated, and the inbreeding effective size will be equal to one.

In situations where genealogical or molecular marker information is available, the inbreeding effective size should be estimated from the rate of increase in inbreeding ($1/2\Delta F$; see Chapter 4) while the variance effective size should be estimated from the rate of increase in coancestry ($1/2\Delta f$), an indicator of genetic drift (Caballero and Toro, 2000), although both are expected to be generally the same.

5.2.4 Non-Random Contribution of Parents to Offspring

One of the key assumptions of the ideal population is that all their individuals have the same probability of contributing with descendants to the next generation. This is a condition that, for various reasons, is frequently violated. Suppose that in a population of constant size N, each individual i contributes to the next generation with k_i gametes, the mean contribution being $\bar{k} = \Sigma k_i/N = 2$, where the summation refers to the N individuals in the population, and the total number of gametes contributed is $2N$. The probability that two gametes taken at random from the population come from the same individual is

$$\frac{\Sigma \binom{k_i}{2}}{\binom{2N}{2}} = \frac{\Sigma \frac{k_i(k_i-1)}{2}}{\frac{2N(2N-1)}{2}} = \frac{\Sigma k_i^2 - \Sigma k_i}{2N(2N-1)}. \tag{5.5}$$

Since the variance of the k_i values is $S_k^2 = (\Sigma k_i^2/N) - \bar{k}^2$, we have that $\Sigma k_i^2 = N(\bar{k}^2 + S_k^2) = N(4 + S_k^2)$ and, since $\Sigma k_i = N\bar{k} = 2N$, substituting in (5.5), the probability is reduced approximately to $(S_k^2 + 2)/4N$. Equating this to the corresponding probability in the ideal population $(1/N)$ and substituting N for N_e in the last one we obtain

$$N_e \approx \frac{4N}{2 + S_k^2} \tag{5.6}$$

(Wright, 1938). The two terms of the denominator in equation (5.6) represent the two sources of variation of the allele frequencies in the population. The term S_k^2 is the variation due to the different contributions from individuals, and the term '2' indicates the source of variation due to the Mendelian segregation of the heterozygotes, that is, the variation in allele frequencies that arises as a consequence of the fact that the progeny of a heterozygote may receive one or the other allele. In the ideal population, in which all individuals are equally likely to contribute offspring to the next generation, the distribution of the number of gametes (that is, of progeny) contributed by the parents is Poisson, with which $S_k^2 = 2$ and $N_e = N$. If the individuals have, as usually occurs, different reproductive capacities, so that $S_k^2 > 2$, the effective size will be lower than the number of parents.

Equation (5.6) can be extended to the case where the numbers of males and females differ,

$$N_e = \frac{(16N_mN_f)/(N_m + N_f)}{2 + S_k^2}, \tag{5.7}$$

being

$$S_k^2 = \frac{N_f}{N_m + N_f} \left[S_{mm}^2 + 2\left(\frac{N_m}{N_f}\right)S_{mm,mf} + \left(\frac{N_m}{N_f}\right)^2 S_{mf}^2 \right]$$

$$+ \frac{N_m}{N_m + N_f} \left[S_{ff}^2 + 2\left(\frac{N_f}{N_m}\right)S_{fm,ff} + \left(\frac{N_f}{N_m}\right)^2 S_{fm}^2 \right], \tag{5.8}$$

where S_{xy}^2 is the variance of the number of offspring of sex y contributed by parents of sex x (x and y being equal to m or f), and $S_{xm,xf}$ is the covariance of the number of sons and daughters of parents of sex x. In the case that the contribution of offspring has Poisson distribution ($S_{xy}^2 = N_y/N_x$ and $S_{xm,xf} = 0$), equations (5.7) and (5.8) are reduced to (5.3), and if the number of individuals of each sex is the same, they are reduced to (5.6), as expected. Equations (5.7) and (5.8) can also be extended to the case in which the size varies with generations, but the equations are more complex and are not presented here.

5.2.5 Partial Mating among Relatives

In the ideal population, mating between individuals is random. The above equations can easily be extended to the case where a proportion of the matings takes place between relatives (see Section 4.2.2 of Chapter 4). For example, many plant species self-fertilize, either 100% or less. We saw in Chapter 4 that the effect of mating between relatives is the generation of Hardy–Weinberg disequilibrium measured by the parameter α, which is the correlation between the alleles carried by an individual. If there is a proportion β of selfing in the population and the rest of the pairings are panmictic, $\alpha = \beta/(2 - \beta)$ (equation (4.22)). Hardy–Weinberg disequilibrium implies a decrease in the frequency of heterozygotes and a corresponding increase of homozygotes. The decrease of heterozygotes reduces the importance of the source of drift due to their Mendelian segregation, that is, the '2' of equation (5.6) is reduced by a fraction α. In contrast, the increase of homozygotes increases in the same fraction as the source of drift due to the variation between parental contributions, since homozygotes only contribute a single allele to their offspring, generating a greater variation of allele frequencies. Therefore, the general prediction of the effective size is

$$N_e \approx \frac{4N}{2(1 - \alpha) + S_k^2(1 + \alpha)} \tag{5.9}$$

(Crow and Morton, 1955). When the number of descendants by self-fertilization and cross-fertilization are distributed according to independent Poisson laws, $S_k^2 \approx 2 + 2\beta$, and equation (5.9) is reduced to $N_e = N/(1 + \alpha)$. For a population of a 100% autogamous species, $\beta = \alpha = 1$ and $N_e = N/2$, that is, the effective size is equal to half the number of breeders.

In the case of mating between full sibs, instead of self-fertilization, α is given by equation (4.23) and the variance of contributions not only increases by α, the correlation between alleles of the same individual, but also by the correlation between alleles of the individuals that make up the couple. The two correlations, intra- and inter-individuals, are very similar, so that the expression of the effective size in the case of partial full-sib mating is equal to (5.9) except that the second term of the denominator is $S_k^2(1 + 3\alpha)$ (Caballero and Hill, 1992a).

As we saw in Chapter 4 (Section 4.4.1), with random mating the value of α, which is equivalent to F_{IS} in a subdivided population, is not exactly equal to zero but rather $\alpha \approx -1/(2N - 1)$ (equation (4.32)). However, as we have already said, in this chapter we ignore the second order terms. Therefore, if mating between individuals is random, $\alpha \approx 0$ and equation (5.9) is reduced to equation (5.6).

Equation (5.9), which is expressed as a function of the variance of the number of descendants contributed in one generation, can also be expressed in terms of the variance of contributions of genes from ancestors to long-term descendants (V_∞),

$$N_e \approx \frac{2N}{(1 + V_\infty)(1 - \alpha)} \tag{5.10}$$

(Woolliams and Thompson, 1994; Caballero and Toro, 2000). With Poisson distribution of the number of offspring per parent and random mating, $V_\infty = 1$, $\alpha = 0$ and $N_e = N$.

5.2.6 Overlapping Generations

Another characteristic of the ideal population is that generations do not overlap, that is, the individuals of a given generation do not mate with those of other generations. In practice, the generations can be imbricated and there exists a complex structure of ages with differential survivals and fecundities in each age class. This generational overlap is common in nature and very frequent in animal breeding programmes. Fortunately, if the age structure can be assumed constant over time, N_e can be predicted with a small modification of equation (5.6),

$$N_e \approx \frac{4N_a I_g}{2 + S_k^2} \tag{5.11}$$

(Hill, 1979), where N_a is the number of reproductive individuals that enter the population in each cohort, suppose that in each year, I_g the generational interval in years, that is, the average age of parents when their offspring are born, and S_k^2 the variance of the contributions of individuals to progeny throughout their lives. This is obtained as the sum of the variance of the contributions in each age class ($V_{k(a)}$) and the variance of the contributions between different classes ($V_{k,a}$), that is, $S_k^2 = V_{k(a)} + V_{k,a}$. An example of the required calculation is explained in Table 5.1. A constant structure is assumed where there are three age classes, with individuals of one, two or three years (column a). We assume that each year $N_a = 100$ individuals, of which all survive at the age of one year, half at two years and 1/4 at three years (column b). The total number of gametes contributed each year is 200. The one-year class is supposed to contribute 1/4 of the gametes, the two-year class with 1/2 and the three-year class with 1/4 (column c), so the expected number of gametes (offspring) from each class is 50, 100 and 50, respectively (column d). Therefore, the generational interval is $I_g = 2$ years and $N_a I_g = 200$. The average number of offspring per parent in each class is 50/100 = 1/2 in the one-year class, 100/50 = 2 in the two-year class and 50/25 = 2 in the three-year class (column e). To obtain the contribution of descendants of each age class we note the following. Of the 100 individuals that are born each year, 50 go on to be two years, then 50 (1/2) will only reach one year (column f) and their vital proportional contribution to the offspring will be 1/2 (column g). Likewise, of the 50 that reach the age of two years, 25 reach three, then only 25 (1/4) will have been two years old and their vital contribution will be 1/2 + 2 = 5/2. Finally, the 25 who have reached the age of three years (1/4) will have had a vital contribution of 1/2 + 2 + 2 = 9/2.

Table 5.1 *Example of the calculation of the effective size of a population with overlapping generations*

(a) Age classes	(b) Proportion of surviving parents	(c) Contribution of each class	(d) Expected number of offspring	(e) Mean number of offspring per parent	(f) Proportion of individuals	(g) Total contribution to each class
1 year	1 × 100 = 100	1/4	50	1/2	50/100 = 1/2	1/2
2 years	1/2 × 100 = 50	1/2	100	2	25/100 = 1/4	1/2 + 2 = 5/2
3 years	1/4 × 100 = 25	1/4	50	2	25/100 = 1/4	1/2 + 2 + 2 = 9/2

Note: Each year, 100 individuals are born, and the generation interval is $I_g = 2$.

Therefore, the average contribution of the three age classes is equal to the variance of the contributions given a certain age ($V_{k(a)}$), because it is supposed to be distributed as a Poisson, and it will be (columns $f \times g$): $V_{k(a)} = (1/2) \times (1/2) + (1/4) \times (5/2) + (1/4) \times (9/2) = 2$. Likewise, the variance of contributions between the different age groups ($V_{k,a}$) will be (column $f \times$ quadratic deviations of column g to the mean): $V_{k,a} = (1/2)[(1/2) - 2]^2 + (1/4)[(5/2) - 2]^2 + (1/4)[(9/2) - 2]^2 = 11/4$, so $S_k^2 = V_{k(a)} + V_{k,a} = 2 + 11/4 = 19/4$. Substituting $N_a = 100$, $I_g = 2$ and $S_k^2 = 19/4$ in equation (5.11) we obtain a prediction of $N_e = 118.52$.

In the analysis of populations with overlapping generations where there is no constant age structure, as in Table 5.1, but the genealogical relationships are available and the inbreeding coefficients of the individuals can be calculated, the effective size can be computed directly from the rate of increase in inbreeding (Gutiérrez et al., 2008). From equation (4.13) of Chapter 4, $F_t = 1 - (1 - \Delta F)^t$, we obtain $\Delta F = 1 - \sqrt[t]{1 - F_t}$ which can be computed for each individual separately. Since generations are not discrete, t is replaced by the discrete generation equivalent (DGE$_i$),

$$\text{DGE}_i = \sum_{j=1}^{a_i} \frac{1}{2^{g_{ij}}}, \tag{5.12}$$

a_i being the number of ancestors of individual i and g_{ij} the number of generations between individual i and its ancestor j. By averaging the values $\Delta F_i = 1 - {}^{\text{DGE}_i}\sqrt{1 - F_i}$ for all individuals, and obtaining the average $(\overline{\Delta F})$, the effective size can be estimated as $N_e = 1/2\overline{\Delta F}$.

5.2.7 Different Models of Inheritance and Reproduction

Table 5.2 shows some of the predictions of N_e for different models of inheritance and reproduction. For a haploid species, the predictive equation is substantially simplified. If the distribution of the number of descendants per individual is Poisson with mean one, $S_k^2 = 1$ and $N_e = N$, but if we refer to a diploid ideal population, $N_e = N/2$, that is, the effective size of a haploid population is half that of a diploid one. Crow and Morton (1955) proposed a general prediction for different ploidies, where j is the number of haploid sets. If $j = 1$ ($\bar{k} = 1$), the equation for haploids is obtained and, if $j = 2$ ($\bar{k} = 2$), that of diploids (equation (5.9)) is recovered.

The expression of genes linked to the X chromosome (Wright, 1933; Caballero, 1995), applicable to the whole genome in haplo-diploid species, gives greater weight to the number of males (if this is the heterogametic sex). With an equal number of males and females, the effective size is 3/4 of that corresponding to autosomal genes in diploid species.

When reproduction occurs through harems where the most successful males mate with most of the females and these only mate generally once, the number of males is also given more weight (Nomura, 2002). If females only mate once, the effective size is 2/3 of that corresponding to panmictic mating.

Finally, Yonezawa (1997) provides predictions for monoecious species that carry out asexual propagation with proportion δ and variance of asexual contributions S_c^2. If $\delta = 0$, the equation is reduced to (5.9). If all reproduction were asexual ($\delta = 1$) and all individuals were

Table 5.2 *Predictions of effective population size for different models of inheritance and reproduction*

Model	Prediction	Simplification
Haploid species	$N_e = N/S_k^2$	$N_e = N/2$ [a]
Different ploidies	$N_e = \dfrac{4N}{4\left(\frac{j-1}{j}\right)(1-a)+4\left(\frac{S_k^2}{k}\right)\frac{1+(j-1)a}{j}}$ [b]	
Sex-linked genes or haplo-diploid species	$N_e = \dfrac{9N_m N_f}{4N_m+2N_f}$ [c]	$N_e = \dfrac{3N}{4}$ [d]
Reproduction by harems	$N_e = \dfrac{4N_m N_f}{2N_m+N_f}$ [c]	$N_e = \dfrac{2N}{3}$ [d,e]
Sexual reproduction and asexual propagation	$N_e = \dfrac{4N}{[2(1-a)+S_k^2(1+a)](1-\delta)+4\delta S_c^2(1+a)}$ [f]	$N_e = \dfrac{N}{S_c^2(1+a)}$ [g]

[a] Poisson distribution of the number of descendants and referring to a diploid ideal population. [b] Variable j is the number of haploid sets. [c] A Poisson distribution of the number of descendants is assumed. [d] It is assumed that the number of males is equal to the number of females. [e] It is assumed that females mate only once. [f] Variable S_c^2 is the variance of asexual contributions, with mean 1, and δ is the proportion of reproduction by asexual propagation. [g] It is assumed that all reproduction is asexual ($\delta = 1$).

homozygotes ($\alpha = 1$), the effective size is $N_e = N/2$, equal to that of a population with autogamous sexual reproduction. There are also predictions for other models of inheritance and reproduction, as well as extensions of the previous ones to other situations (see reviews by Caballero, 1994; Wang et al., 2016).

5.3 Prediction of N_e in Selected Populations

5.3.1 Cumulative Effect of Selection

The equations of the previous section assume that the variation in parental contributions to offspring (S_k^2) is due to non-heritable causes and, therefore, changes in allele frequencies are not correlated in consecutive generations. When the variation in contributions is due to heritable causes, such correlations exist because a fraction of the individual's selective advantage is transmitted to its progeny and that of the progeny to its corresponding one, and so on. Therefore, the frequency of a neutral allele carried by a successful breeder will tend to increase in successive generations and those changes in frequency will be greater and more directional than those due exclusively to drift. This is what has been called genetic draft (Gillespie, 2000), which makes a play on words with genetic drift. The effective size can be obtained from the variance of the change in allele frequency of a neutral gene that may or may not be linked to the selected genes. In this section we will consider the case in which there is no linkage, and in the following we will extend the predictions to the linkage situation.

Robertson (1961) introduced the concept of the cumulative effect of selection, whereby the average selective value of an individual is reduced by half each generation, since it is assumed that this and its descendants are paired with individuals taken at random from the population, whose average selective value is zero. The cumulative series is then $Q = 1 + 1/2 + 1/4 + 1/8 + \ldots = 2$. If the variance of the selective advantages is C^2 (the average selective advantage being equal to unity), for the purpose of drift and inbreeding, the variance of parental contributions due to non-heritable causes (S_k^2) is increased by a factor $4Q^2C^2$ (the cumulative effect is $2Q$, as the average contribution of each parent is of two individuals and must be squared since the variance is a mean quadratic value). Equation (5.9) then becomes

$$N_e \approx \frac{4N}{2(1-\alpha) + (S_k^2 + 4Q^2C^2)(1+\alpha)} \tag{5.13}$$

or its equivalent with $(1 + 3\alpha)$ in the case of mating between full sibs or other types of biparental inbreeding. Therefore, selection reduces the effective size with the consequent increase in genetic drift and inbreeding. This decrease occurs progressively when a selection process begins, following the cumulative terms of Q, so that equation (5.13) determines the asymptotic effective size.

The variance of the selective advantages, C^2, is the genetic variance of the selected trait (standardized to an average of one), which can be any quantitative trait, or fitness itself, in the case of natural selection. Since the genetic variance is reduced by the effect of drift and selection (the latter will be discussed in Chapter 9), the cumulative factor Q is also reduced. On the other hand, the genetic variance can also be increased if there is a correlation between the selective advantages of the crossing parents. If we call G the proportion of the remaining variance after its decay by drift and selection, and r the aforementioned correlation, the cumulative factor is

$$Q = 1 + \frac{G(1+r)}{2} + \left(\frac{G(1+r)}{2}\right)^2 + \left(\frac{G(1+r)}{2}\right)^3 + \cdots = \frac{2}{2 - G(1+r)} \tag{5.14}$$

(Santiago and Caballero, 1995). When mating between individuals is random $r \approx 0$, but if mating occurs between relatives (for example self-fertilization or mating between full sibs) with probability β, the correlation will tend to be $r \approx 1$ for the fraction of matings that occur between relatives, and approximately 0 in the remainder, so $r \approx \beta$. With these disquisitions, equation (5.13) can be applied to a scenario of partial selfing, and its analogue with $(1 + 3\alpha)$ to one of breeding between full sibs or other types of biparental inbreeding.

There are extensions of equation (5.13) to various scenarios, such as different numbers of males and females, overlapping generations, artificial selection by selection indexes or BLUP (which we will study in Chapter 9), and selection assisted by genetic markers, among others (see Wang et al., 2016).

Equation (5.13) applies to any trait subject to selection. If the trait is fertility, a main component of fitness, an interesting way to present the equation is in terms of its heritability (Nei and Murata, 1966). Let us call V_k the observed variance of the contributions of individuals to progeny, or variance of family sizes. If the decay in variance with selection and drift is ignored, and random mating of individuals is assumed, that is, $G = 1$, $r = 0$ and $Q = 2$, we have $V_k = S_k^2 + 4C^2$. The first term $S_k^2 = V_k(1 - h^2)$ is the non-heritable component of the variance, where h^2 is the heritability of fertility. The second, $4C^2 = V_k h^2$, is the heritable component. Substituting in (5.13) with $\alpha = 0$, we obtain

$$N_e = \frac{4N}{2 + (1 + 3h^2)V_k}. \tag{5.15}$$

5.3.2 The Impact of Linkage

The previous predictions assume that neutral loci, whose drift or inbreeding is intended to be quantified by N_e, are not linked to the genes that control the selected trait. If there is linkage between neutral and selected genes, the association that produces the cumulative effect of selection is extended over time, which leads to an even greater reduction in the effective population size. Although the argument can be developed for any character subject to selection, the application of greatest interest refers to the case of natural selection against deleterious mutations, which is called background selection (Charlesworth and Charlesworth, 2010, Chapter 8). If we consider fitness and a random mating population, equation (5.13) is best expressed exponentially, since the allelic effects on fitness are represented by a multiplicative model. That is, from equation (5.13) with $\alpha = 0$, $N_e = N/(1 + Q^2 C^2) = N\exp(-Q^2 C^2)$.

As we will see in Chapter 7, most of the allelic variants that are generated by mutation are deleterious (reduce fitness) and are continuously eliminated by natural selection. If the rate at which these mutations appear per gamete and generation is U for the whole genome and its deleterious effect on homozygosis is s (see Chapter 2), the average fitness of a population of large size is reduced each generation by a Us factor. In an equilibrium population, natural selection compensates for this decline and, as will be seen in Chapter 10, the response of natural selection is equal to the genetic variance for fitness and, therefore, $C^2 = Us$.

On the other hand, the cumulative effect of selection produced by that mutation on a neutral gene depends on the frequency of recombination (c) between them, that is, their genetic distance. For this pair of genes, the cumulative effect is expressed by

$$Q_c = \sum_{i=0}^{\infty} [(1-s)(1-c)]^i \approx \frac{1}{s+c} \qquad (5.16)$$

(Santiago and Caballero, 1998). Note that if the two loci are not linked ($c = 0.5$) and the effect of the selective gene is very small ($s \rightarrow 0$), $Q_c \approx 2$, as was deduced in the previous section. The cumulative effect of selection can be seen in Figure 5.1, which illustrates $N_e = N\exp(-Q_c^2 C^2)$ for a neutral allele where Q_c is obtained with equation (5.16) and $C^2 = Us$. In the presented cases, deleterious mutations with effect $s = 0.1$ that appear with rate $U = 0.1$ per generation are considered, linked with different magnitudes to the neutral allele. When there is no linkage ($c = 0.5$) the reduction of N_e is small and occurs in very few generations. As linkage between the neutral and selective alleles is greater (lower value of c), the reduction of the effective size is greater and the asymptotic effective size is reached in a longer period of time.

Assuming that the neutral locus is located in a central position of a chromosome or chromosomal segment of length L (in Morgans) and that the selected genes are evenly distributed in the segment, an average of Q_c can be obtained for this segment by integrating equation (5.16), giving $Q^2 \approx 1/[s(s + L/2)]$. Substituting this value and $C^2 = Us$ in the expression of the effective size, we obtain

$$N_e = N\exp(-Q^2 C^2) \approx N\exp\left[\frac{-U}{s+(L/2)}\right]. \qquad (5.17)$$

Note that the higher the linkage in the chromosomal segment in question (lower L), the lower the asymptotic effective size indicated by equation (5.17). The time necessary to

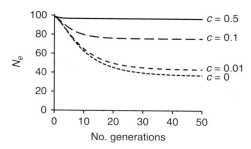

Figure 5.1 Prediction of the effective population size (N_e) of a neutral allele as a consequence of the action of selection on deleterious genes with effect $s = 0.1$ that appear by mutation with rate $U = 0.1$ per haploid genome and generation in a population of size $N = 100$. The recombination rate between the neutral allele and the selective gene is c. Results obtained by equation $N_e = N\exp(-Q_c^2 C^2)$, with $C^2 = Us$ and Q_c from equation (5.16).

reach this maximum reduction of N_e will be greater the tighter the linkage (Figure 5.1). Equation (5.17) requires selective effects to be large in relation to the effective size, in particular, $s > 1/N_e$.

5.4 Prediction of N_e in Subdivided Populations

5.4.1 Prediction with Different Models of Population Structure

Most populations, both wild and domestic, are spatially structured. Structure is particularly important in endangered species, where habitat fragmentation accentuates the division of the population into small and often isolated reproductive groups. In addition, the populations of species preserved in captivity, such as those kept in zoos, natural parks, botanical gardens or germplasm banks, are also subdivided, and the estimation of their effective size is essential for their management.

Table 5.3 presents some of the predictive equations of N_e for the most common models of population structure, the island and stepping-stone models represented in Figure 2.6 of Chapter 2, consisting of n sub-populations of census size N with migration rate m. For both models, if the migration rate is high, $N_e \approx Nn$, the total number of individuals of the population as a whole. The lower the migration rate, the greater will be N_e in relation to Nn.

5.4.2 General Model

In Chapter 4 we studied the partition of inbreeding in a structured population in terms of Wright's statistics, where F_{IS} indicates the inbreeding due to non-random mating within each sub-population, which is equivalent to the term α in expression (5.13), F_{ST} indicates the inbreeding due to the subdivision of the population, and F_{IT} is the total inbreeding. The effective size of a structured population can be obtained in a general way in terms of these statistics. Let us assume a population formed by n sub-populations of size N as illustrated in Figure 4.7 of Chapter 4. The effective size can be approximated by

Table 5.3 *Predictions of effective population size for two models of population structure*

Model	Prediction
Island model (Figure 2.6b; Chapter 2)	$N_e = Nn\left(1 + \frac{(n-1)^2}{4Nmn^2}\right)$
Circular stepping-stone model (Figure 2.6c with circular structure; Chapter 2)	$N_e = Nn + \frac{n^2}{2m\pi^2}$

Note: It is assumed that each sub-population is ideal except by receiving migrants from other sub-populations. Variable n is the number of sub-populations; N is the number of reproductive individuals in each sub-population, constant for all of them; m is the rate of migration per generation and sub-population.

$$N_e = \frac{Nn}{\frac{1}{4}(1 - F_{ST})[2(1 - F_{IS}) + S_k^2(1 + F_{IS}) + 4V] + 2NVF_{ST}\left(\frac{n}{n-1}\right)} \tag{5.18}$$

(Wang and Caballero, 1999), where the first two terms within brackets correspond to the denominator of equation (5.13) substituting α for F_{IS}, so that S_k^2 is the variance of the contributions of the parents to their progeny in the sub-populations, and the new term, V, is the variance of the contributions of each sub-population to the next generation.

If we assume, to simplify, that the parental contributions are distributed according to a Poisson ($S_k^2 = 2$) and mating within sub-populations is random ($F_{IS} \approx 0$), and if the number of sub-populations is not very small, equation (5.18) is reduced to

$$N_e \approx \frac{Nn}{(1 - F_{ST})[1 + V] + 2NVF_{ST}} \tag{5.19}$$

(Whitlock and Barton, 1997). Suppose now that all sub-populations contribute equally to the offspring in each generation, in which case $V = 0$. Equation (5.19) is then simplified to

$$N_e = \frac{Nn}{1 - F_{ST}}, \tag{5.20}$$

which was deduced by Wright (1943). Note, that the effective size increases with the differentiation between sub-populations. With total differentiation ($F_{ST} = 1$) N_e is, in theory, infinite. As mentioned at the beginning of this chapter, this is one of the few situations in which variance and inbreeding effective sizes differ. The idea is that, if the sub-populations remain isolated indefinitely, the different allelic variants present in the population could become fixed in the different sub-populations without the possibility of being lost. The approach to this limiting situation would imply little or no genetic drift, producing a very high variance effective size. On the contrary, the situation would generate a rapid increase in the frequency of homozygotes and, therefore, in the inbreeding of the sub-populations, reflected in a very low inbreeding effective size. If there is genetic exchange between the sub-populations by migration, the difference between both effective sizes is only temporary, converging towards the same asymptotic N_e.

Equation (5.20) implies, in any case, an increase of N_e with an increase in differentiation. However, the scenario to which this expression refers assumes that the sub-populations contribute identically to the offspring in each generation ($V = 0$). If this is not the case, the result can change drastically. In fact, if there is a variation of the contributions between sub-populations equal to $V = 1/N$, and assuming that $1/N \ll 1$, substituting in (5.19), we obtain

$$N_e \approx \frac{Nn}{1 + F_{ST}},\tag{5.21}$$

which is just the opposite result to that of equation (5.20), implying that N_e is reduced with an increase in differentiation, which may be important in practice.

A drastic case of disproportionate contribution of the sub-populations occurs in species composed of metapopulations, that is, groups of populations governed by their continuous extinction and recolonization. This is frequent in numerous species of insects in which populations are an ephemeral and variable concept in time and space. Suppose that, maintaining the characteristics already described of a structured population, we add that in each generation the sub-populations can be extinguished with probability e, and that each extinct sub-population is recolonized immediately by individuals from any other sub-population. In that case, the contribution of descendants from each sub-population to the next generation is 0 with probability e and $N/(1 - e)$ with probability $(1 - e)$, so that the average contribution is $[N/(1 - e)] \times (1 - e) = N$, and the variance of the contributions of the sub-populations is $[N/(1 - e)]^2 \times (1 - e) - N^2 = N^2 e/(1 - e)$ that, scaled to the unit, is $V = e/(1 - e)$. Substituting in equation (5.19), we obtain

$$N_e \approx \frac{Nn(1 - e)}{1 - F_{ST} + 2eNF_{ST}}\tag{5.22}$$

(Whitlock and Barton, 1997). If there were no extinction ($e = 0$), equation (5.20) would be recovered. In short, a frequent extinction entails a drastic reduction of the effective population size.

5.5 Applications of the Theory of Effective Population Size to Conservation

5.5.1 Contributions with Minimal Variance

The concept of effective size is fundamental in the preservation of genetic diversity in conservation programmes, since it provides a summary of the past history of the population regarding its inbreeding and drift and provides the basis for achieving its future sustainability. The previous sections allow us to formulate general recommendations for the preservation of diversity. The general idea is that the greater the effective population size, the lower the loss of diversity due to genetic drift and the lower the increase in inbreeding, whose negative consequences will be explained in Chapter 8. The predictions of N_e presented in Section 5.2 for unstructured populations indicate that this parameter is reduced with a decrease in the census size of breeders (all expressions), with a disequilibrium between the frequency of males and females (equation (5.3)), with temporal changes in the census size (equation (5.4)), with a variation in parental contributions greater than that expected by chance, particularly with consanguineous matings (equation (5.9)), and with short generational intervals (equation (5.11)). Therefore, conservation practices will require us to maintain populations with the maximum possible number of reproducers, with equal number of parents of each sex, constant throughout generations, to reduce the variance of parental contributions, to avoid consanguineous matings and to lengthen the generational interval.

One of the previous recommendations with more obvious consequences is to try to equalize the parental contributions, that is, that each pair of individuals contributes two descendants to the next generation. In that case, $S_k^2 = 0$ and from equation (5.6), $N_e \approx 2N$ is obtained, which implies that drift and inbreeding rates are half those which would occur in an ideal population of size N. This procedure is the basic method traditionally recommended for the conservation of genetic resources and is the one that is usually applied, for example, for the collection of plants in germplasm banks and their subsequent regeneration (Vencovsky et al., 2012). The extension of this procedure to a different number of males and females was designed by Gowe et al. (1959). With this design each father has a son and N_f/N_m daughters, and each mother has a daughter and a probability of N_m/N_f of having a son. Therefore, in this case, $S_{fm}^2 = (N_m/N_f)[1 - (N_m/N_f)]$ and all the other variances and covariances in equation (5.8) are zero, so that equation (5.7) is reduced to

$$N_e = \frac{16N_m N_f}{3N_f + N_m}. \tag{5.23}$$

If $N_m = N_f$, the former expression is reduced to $N_e = 2N$, as expected. A better alternative than the previous one was proposed by Wang (1997) and is that, among the N_f/N_m females mated with each male, one is selected at random to contribute one son, another to contribute two daughters and the rest contribute one daughter each. In this scenario, $S_{ff}^2 = 2N_m/N_f$, $S_{fm}^2 = (N_m/N_f)[1 - (N_m/N_f)]$, and a negative covariance is generated between the number of sons and daughters of mothers, $S_{fm,ff} = -N_m/N_f$. Substituting in (5.7) and (5.8),

$$N_e = \frac{16N_m N_f^2}{3N_f^2 + 2N_m^2 - N_m N_f}, \tag{5.24}$$

which can produce effective sizes up to 17% greater than with the design corresponding to equation (5.23). The extension of the method to consecutive generations further improves the expectations of increasing N_e (Sánchez et al., 2003). The previous methods do not require genealogical information but, if it is available or estimated through genetic markers, a more general and efficient method of deducing the contributions that maximize N_e is to find, through iterative heuristic methods, those that minimize the coancestry in the offspring, which is called minimum coancestry contributions (Ballou and Lacy, 1995; Fernández et al., 2003). It has been possible to demonstrate that this method maximizes the variances of the contributions of ancestors to descendants and, therefore, maximizes the effective population size (Caballero and Toro, 2000, 2002).

The above procedures are aimed at reducing or eliminating the variance of parental contributions. If this is achieved, the only source of drift is the Mendelian segregation of heterozygotes, represented by the term '2' in equation (5.6). But it is also possible to reduce this source of variation. One possibility is to use marker-assisted selection (Wang and Hill, 2000). This can be achieved by reducing the probability of 1/2 that two alleles from the father (or mother) are identical by descent. Success will depend on the amount of information from the markers, the size of the genome and the number of offspring that are genotyped for the markers in each family. Another possibility is to use reproductive technologies for the manipulation of meiosis, such as the *in vitro* culture of premeiotic germ cells and the microinjection of primary spermatocytes into oocytes (Santiago and Caballero, 2001). Using more than one gamete from a single meiosis, the drift generated by Mendelian segregation can be partially or completely removed. Thus, for example, if the parental contributions are equalized and the gametes produced by parents are manipulated so that they come from the same meiosis in each case, the resulting effective size is $3N$, instead of the typical $2N$.

5.5.2 Mating Systems

The control of inbreeding in conservation programmes can also be carried out, to a lesser extent, with the design of the mating system. Let us start from the basis that parental contributions are equalized ($S_k^2 = 0$), equation (5.9) is then reduced to $N_e = 2N/(1 - \alpha)$. This indicates that, if mating between relatives is forced ($\alpha > 0$) in a scenario of equal parental contributions, the effective size will increase. The situation is similar to that discussed above for a subdivided population with equal contributions from parents and sub-populations to the next generation (equation (5.20)). In fact, exactly the same situation occurs if we assume that sub-populations are in this case families and migration is the crossing between individuals of different families. If mating between relatives is complete ($\alpha = 1$), which would correspond to the scenario in which isolated lines are highly inbred (for example by self-fertilization or brother–sister mating), the effective size would, in theory, be infinite, as explained in the case of structured populations with no migration between sub-populations. As also mentioned, in this situation the variance and inbreeding effective sizes differ. In the case of independent lines of self-fertilization or mating between full sibs, the variance effective size is infinite while the inbreeding effective size is 1 or 2, respectively.

From the point of view of the maintenance of genetic diversity, in a structured population with no migration between sub-populations (or families in the previous case) the differentiation that is reached between sub-populations implies the maintenance of a maximum allelic diversity in the population since, in theory, the different alleles of a locus could be conserved in the different sub-populations. However, heterozygosity in each sub-population would be minimal, since isolation would lead to total homozygosis. From a practical point of view, it could then be considered that sub-population isolation is useful in maintaining allelic diversity, expressed by the variance effective size. However, the inbreeding generated in each sub-population is a negative factor because, as we will see in Chapter 8, homozygosity brings undesirable consequences with respect to the reproductive capacity and viability of individuals. Therefore, a maximum inbreeding forced by the total isolation of sub-populations is discouraged in the field of conservation.

A less extreme scenario, which has often been considered, is to force inbreeding only partially, in which case the variance and inbreeding effective sizes coincide even if the drift takes place one or more generations later than inbreeding. Circular mating is one of these cases. We can divide the population into numbered pairs that have a son and a daughter each. In each generation the son of one family is crossed with the daughter of the next to give birth to a son and a daughter, and this is repeated in a successive and sequential manner until the son of the last family is crossed with the daughter of the first. With this, the parental contributions are equalized generating a certain inbreeding, but continuous gene flow is maintained between families. The effective size for circular pair mating can be approximated by

$$N_e \approx \frac{(N + 12)^2}{2\pi^2} \tag{5.25}$$

(Kimura and Crow, 1963b). So, for example, if $N = 20$, $N_e \approx 52$, while it would be $N_e = 2N = 40$ with random pairing.

The proposal to force consanguineous matings of some kind in conservation programmes is based on two premises. First, the increase in the effective size mentioned. Second, the possible purging effect of inbreeding on deleterious mutations, which we will study in Chapter 8. However, the effective size to which the predictions refer is always the asymptotic one, which is reached after several or perhaps many generations depending on the cases. The rate of decay in the frequency of heterozygotes with forced inbreeding is only less than that corresponding to random mating after that initial period, during which the increase in inbreeding will be higher. As Robertson (1964) demonstrated, circular mating and other designs of inbred matings result in lower long-term inbreeding rates at the expense of producing higher rates in the short term. For this reason, consanguineous matings are not usually recommended in conservation programmes and, in fact, are generally discouraged.

Avoiding consanguineous matings that occur by chance, however, usually has a tiny effect on the effective size, unless the population size is very small. If, in a given population, mating between relatives is avoided to the maximum, as proposed by Wright (1921), it is possible to delay inbreeding substantially. For example, if the population size is $N = 2^n$, matings can be designed in a way that avoids crosses between individuals that have a common ancestor in the last n generations. As mentioned above, inbreeding depends on the parental generation, while genetic drift depends on the progeny generation, so that with random mating, inbreeding has a generation of delay with respect to drift. With the maximum avoidance of inbreeding system of mating the delay will increase in n generations and the effective size is

$$N_e \approx 2N - \frac{n + 1}{2}. \tag{5.26}$$

For example, if $N = 16 = 2^4$, N_e would be approximately 29.5 instead of 32, which illustrates the low impact of avoiding mating between relatives. The most important effect of the delay of inbreeding is to avoid the negative impact of this on fitness.

The availability of genetic markers allows mating between individuals to be carried out with minimal coancestry even when genealogical information is not available. Decisions on the design of matings can be made after those corresponding to the parental contributions of minimum coancestry have been taken, or both simultaneously, which is called mate selection (Toro and Pérez-Enciso, 1990).

Finally, there is a method of mating applicable to artificial selection programmes, compensatory mating (Santiago and Caballero, 1995), whose objective is to try to partially eliminate the cumulative effect of selection on the effective size expressed in equation (5.14). The method consists of crossing individuals from large families with others from small families, with the idea of producing negative correlations between the drift caused by selection and that caused by sampling. This system can also be combined with minimum coancestry matings with good results (Caballero et al., 1996).

5.6 Estimation of Effective Population Size Using Demographic Methods

So far we have studied how to predict the effective size of a population. In this section we will explain the different procedures for estimating N_e from empirical data. As we discussed earlier, if the genealogical relationships between the individuals of a population are known, the effective size can be estimated directly from the rate of increase in inbreeding. However, genealogies are usually only available in domestic animals, and captive and experimental populations. When genealogies are not available, N_e can be estimated using the previous predictive formulas if it is possible to obtain empirical estimates of the demographic data required by the equations, such as the number of breeding individuals of each sex, the type of mating, the variance of the parental contributions to the offspring, the generation interval, and so on. This is what is called the demographic estimate of N_e.

Consider an illustrative example of this form of estimation with Atlantic salmon data from River Lérez (Pontevedra, Spain) (Saura et al., 2008). During the period from March to December of the years 2004 and 2005 all individual adults returning to the river were collected for mating after their stay at sea (14 males and 20 females in 2004 and 14 males and 34 females in 2005). With such a large collection period, it was guaranteed to capture practically all population breeders. After the breeding season (between July and September 2005 and 2006, respectively), 91 and 89 juveniles were sampled, respectively. Both parents and juveniles were genotyped for six microsatellite loci that were used to make paternity tests, which allowed the identity of the parents of practically all juveniles to be determined. Molecular data were also used to estimate the effective population size by genetic methods that we will study later. With the data of the mean $(\bar{k})$ and the variance of progeny number (S_k^2), the effective size can be estimated for each sex and year separately, using equations (5.3) and (5.6), as indicated in Table 5.4. The estimated

Table 5.4 *Example of estimation of the effective size by the demographic method*

	Year 2004		Year 2005	
	Males	Females	Males	Females
N_s	14	20	14	34
$\bar{k}$	2.43	1.7	2.67	1.07
V_k	10.11	8.12	27.33	2.75
S_k^2	7.20	10.89	15.84	7.87
$N_{e,s}$	6.1	6.2	3.1	13.8
N_e	12.3		10.2	

Note: N_s is the number of parents of each sex ($s = m$ males or f females); $\bar{k}$ is the mean number of offspring per parent. The variance of family sizes (S_k^2) is obtained from the observed one (V_k) adjusted by assuming a mean of two progeny per parent and random survival of juveniles, by means of the relation $[(S_k^2/2) - 1] \approx (2/\bar{k})[(V_k/\bar{k}) - 1]$ (Crow and Morton, 1955). The effective population size of each sex, $N_{e,s}$, is obtained by equation (5.6), and the global effective size, N_e, is obtained by equation (5.3) using $N_{e,s}$ instead of N_s.
Source: Data from Saura et al. (2008) corresponding to the Atlantic salmon population or River Lérez (Pontevedra, Spain).

effective size was similar in the two years, $N_e \approx 10\text{--}12$ individuals. The demographic estimates have the advantage of allowing for an understanding of the factors involved as a source of genetic drift and inbreeding, in this case a smaller number of breeding males than females and a variance of contributions much higher than that expected in an ideal population, since $(S_k^2/\bar{k}) > 1$ in all cases.

5.7 Estimation of Effective Population Size Using the Allelism of Lethals Method

In the case of some species in which chromosome manipulation can be carried out by crosses, as for example in *Drosophila*, it is possible to estimate the effective population size from the frequency of lethal alleles of the same locus. The idea is the following. As we saw in Chapter 2, lethal alleles are removed by natural selection, so their population frequency will generally be very low. Therefore, if a sample of individuals is taken from a large population, it will be very unlikely to find two or more carriers of the same lethal allele or different lethal alleles of the same locus. This probability, however, will not be negligible if the effective population size is small. This allows us to design a procedure to calculate N_e from an estimate of the allelism rate, that is, the frequency of lethal genes found in a sample that are alleles of the same gene (Nei, 1968). The procedure to detect the presence of lethal alleles on chromosome II of *Drosophila melanogaster* is described in Figure 5.2. Males of a strain carrying multiple inversions that prevent recombination are used. These chromosomes also carry visible marker genes for easy identification, such as *Cy* ('*Curly*'), which produces curved wings and *If* ('*Irregular facets*') that produce irregular eye facets, both lethal in homozygosis. By crossing individuals of the population (+/+) to individuals *Cy*/*If*, it is possible to isolate a single chromosome II from the population and detect the presence of a lethal gene on that chromosome (Figure 5.2a). Once the lethal alleles present in the sample are detected, it can be verified if some of them are alleles of the same locus (Figure 5.2b).

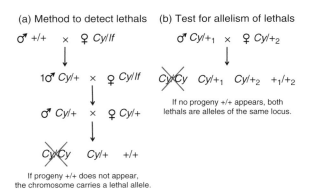

Figure 5.2 (a) Crosses in *Drosophila melanogaster* to detect lethals in a chromosome II sampled from the population. Strains are used with balanced chromosomes carrying multiple inversions that prevent recombination and carry visible dominant markers (*Cy*, *If*), lethal in homozygotes (+ indicates absence of visible marker). (b) Test of allelism of lethals to find out if two chromosomes of the population ($+_1$ and $+_2$) with a lethal allele carry that lethal in the same locus.

Table 5.5 *Example of estimation of the effective population size by the method of allelism of lethals*

	Number	Calculation
Chromosomes evaluated	549	
Chromosomes carrying a lethal	96	$q_L = 96/549 = 0.1749$
Crosses for test of allelism of lethals	275	
Allelic lethals found	10	$I_{a,obs} = 10/275 = 0.0364$
Estimations	Equation	Result
Previous estimate of U_L [a]		$U_L = 0.0032$
Previous estimate of u_L [a]		$u_L = 2.3 \times 10^{-6}$
Allelism rate	(5.29)	$\hat{I}_a = 0.0301$
N_e assuming recessive lethals	(5.27)	$N_e = 1603$
N_e assuming partially recessive lethals	(5.28)	$N_e = 2579$

Note: Data correspond to a population of *Drosophila melanogaster* maintained with $N \approx 3000$ individuals.
[a] Previous estimates from García-Dorado et al. (2007); 142 lethals were found in 4431 chromosomes in a period of 10 generations, so that $U_L = 142/(4431 \times 10) = 0.0032$. Assuming a total of 1400 vital loci in chromosome II of *Drosophila*, the mutation rate per locus and generation is $u_L = U_L/1400 = 2.3 \times 10^{-6}$.
Source: Data from López-Cortegano et al. (2016).

The estimate of N_e from these data is

$$N_e = \frac{1 - I_a}{2\pi U_L I_a} \tag{5.27}$$

if the lethal alleles are assumed to be completely recessive or

$$N_e = \frac{1 - I_a}{4(U_L I_a - u_L)} \tag{5.28}$$

if they are assumed to be partially recessive, where I_a is the lethal allelism rate, U_L is the lethal mutation rate per chromosome and generation and u_L the corresponding rate per locus. For the estimations, it is therefore necessary to have previous estimates of U_L and u_L.

To estimate I_a it is necessary to know the observed frequency of lethal chromosomes in the analysed sample (q_L) and the observed frequency of lethals found that are allelic with one another ($I_{a,obs}$), such that

$$\hat{I}_a = \frac{-\ln\left(1 - I_{a,obs} q_L^2\right)}{[\ln(1 - q_L)]^2}. \tag{5.29}$$

Consider the example illustrated in Table 5.5, which presents the estimates obtained by López-Cortegano et al. (2016) for a laboratory population of *Drosophila melanogaster* maintained with census size $N \approx 3000$ individuals. A total of 549 chromosomes were sampled, of which 96 were found to carry a lethal allele. Then 275 crosses of allelism test were carried out with the lethals found, obtaining that in 10 of the crosses the lethals were alleles of the same locus. The estimated effective size would oscillate between 53% and 86% of the census size of reproducers depending on the recessivity of the lethal alleles.

5.8 Estimation of Effective Population Size with Molecular Markers

The incessant development of methods to obtain and analyse markers in almost every species with a lower cost and time means that genetic methods are applied more and more frequently in conservation and animal and plant breeding studies. There are several methods for estimating the effective population size using molecular genetic markers. The procedures described below generally refer to the current effective size or that referring to a few past generations, which has the greatest interest in the fields of study mentioned, although some methods can be used to obtain estimates of N_e in different periods of time. In all cases they apply to situations in which the markers are neutral, they are not linked to selected genes and there are no sources of change in allele frequencies other than genetic drift. The violation of these premises will lead to biased estimates of N_e. All the equations shown below must be corrected for sample size and other second order terms, which will be ignored for simplicity.

5.8.1 Heterozygosity Excess Method

In an infinite population, with random mating and in the absence of forces of change in the allele frequencies, it can be expected that alleles segregate independently both within the locus (Hardy–Weinberg equilibrium) and between loci (gametic equilibrium). In a finite population, however, genetic drift entails associations between intralocus alleles, which results in an excess of heterozygotes, and interlocus alleles, resulting in linkage disequilibrium. In this section we will study how N_e can be estimated in the first case, and in the next we will focus on the second.

In a population with a reduced number of males (N_m), females (N_f), or both, random differences between the allele frequencies of both sexes will be expected, and this will generate an excess of observed heterozygotes (H_o) with respect to those expected with Hardy–Weinberg equilibrium (H_e) in such a way that $H_o = H_e(1 - \alpha)$, where α can be approximated by

$$\alpha = -\frac{1}{8N_m} - \frac{1}{8N_f} = -\frac{1}{2N_e} \tag{5.30}$$

(Robertson, 1965), and N_e is given by equation (5.3). Expression (5.30) is equivalent to the generic (4.32) of Chapter 4. Therefore, if α can be estimated as $\hat{\alpha} = (H_e - H_o)/H_e$, the effective size can be approximated by $\hat{N}_e \approx 1/(2\hat{\alpha})$ (Pudovkin et al., 1996). This method has low precision and a lot of error, often giving infinite estimates of N_e. However, it is useful for very small size populations or when biallelic markers (such as SNPs) are used in very high numbers.

5.8.2 Linkage Disequilibrium Method

Genetic drift generates linkage disequilibrium between loci. For neutral genes unlinked to selective genes in an isolated population of effective size N_e, the expected value of the squared correlation between the allele frequencies of a pair of loci (r^2; equation (2.7), Chapter 2), which is also the variance of r since $E[r] = 0$, can be approximated by

$$E[r^2] = V(r) \approx \frac{1}{4N_e c + 1} \tag{5.31}$$

(Sved, 1971; Hill, 1981), where c is the recombination frequency between the two loci. Therefore, having markers whose genetic distances can be estimated, the effective size can be calculated by $N_e \approx 1/[4V(r)c]$. This is one of the most popular methods to estimate N_e, both for its simplicity and robustness against complications such as population structure and non-random mating, and for providing estimates with great precision. The possibility of discerning between markers with different levels of linkage may also allow the inference of N_e at different times in the past to be obtained (Hayes et al., 2003). The idea is that, if the recombination frequency between two markers is c, the expected time to their nearest common ancestor is $1/2c$ generations (see Section 4.1.2 of Chapter 4). Therefore, markers separated by this distance will give an idea of the effective size $1/2c$ generations ago. If a large number of markers (for example SNPs) are available at different genetic distances, they can be grouped according to these and obtain estimates of N_e for different time periods (Tenesa et al., 2007). However, this approach can only be applied to scenarios of constant effective size or linear changes, such as steady growth or decline (Walsh and Lynch, 2018, p. 97) and is not appropriate to detect sudden changes such as population bottlenecks. A development of the method accounting for these drastic changes has been given by Santiago et al. (2019), who also showed that temporal changes in N_e will have an impact on pairs of SNPs at all genetic distances, so that the estimates of N_e at any given time period ($1/2c$ generations) should consider changes in linkage disequilibrium from all pairs of SNPs and not only from those at a distance c.

5.8.3 Temporal Method

As we already know, the allele frequencies in a population of finite size change by genetic drift. Therefore, if we can estimate the allele frequencies in two or more moments in time separated by at least one generation, we can estimate N_e (Krimbas and Tsakas, 1971; Waples, 2005). The idea can be understood from equation (4.13) of Chapter 4, which gives the expected value of inbreeding after t generations, $F_t = 1 - (1 - 1/2N_e)^t$. If $t \gg N_e$, it is approximated by $F_t \approx t/2N_e$. Recalling now the relation between inbreeding and genetic drift, $\sigma_{q_t}^2 = p_0 q_0 F_t$ (equation (4.14)), we have that the effective size can be calculated as $N_e \approx t/2F_t$, where $F_t = \sigma_{q_t}^2/p_0 q_0$ is evaluated with the so-called moment estimator $\hat{F}$, which calculates the standardized variance of the changes in allele frequencies. In essence, this estimator is analogous to F_{ST} (Chapter 4) except that the former measures temporal, rather than spatial, differentiation. There are several types of estimators $\hat{F}$ depending on whether the sampling is done with replacement or not of individuals. For example, one of them is

$$\hat{F} = \frac{\sum\limits_{i=1}^{a} (x_i - y_i)^2}{\sum\limits_{i=1}^{a} z_i(1 - z_i)}, \tag{5.32}$$

where x_i and y_i are the observed frequencies of the i allele at the two sampling moments (a is the total number of alleles in the locus in question) and z_i is the average of both frequencies. The estimate obtained by (5.32) must be corrected for sample size.

The temporal method has the disadvantage, with respect to other methods, that it requires at least two samplings in time separated by at least one generation. It is, therefore, less appropriate in species with long generational intervals or when sampling is expensive. However, it makes fewer assumptions than other methods and is quite robust to the complications of real populations, such as population structure and overlapping generations. In the example of Table 5.3 with data on Atlantic salmon, in addition to the demographic estimates, others were also carried out using the temporal method based on the changes in allele frequency of the six microsatellites analysed in the two successive generations (parental and juveniles). The estimates were $N_e = 12.0$ for year 2004 and 9.2 for year 2005, values practically identical to those obtained with the demographic method.

The moment estimator method $\hat{F}$ (equation (5.32)) is very simple and straightforward, but N_e can also be estimated with the temporal method by means of maximum likelihood or Bayesian procedures. These have in general greater precision and accuracy, especially with markers with rare alleles, but require a much higher computing time.

5.8.4 Coancestry and Sib Frequency Methods

Given that the lower the effective size, the greater the coancestry between individuals, Nomura (2008) proposed a method for estimating N_e that uses the increase in molecular coancestry between two successive generations. It has been shown, however, that this procedure usually provides very skewed estimates and with a great error. The main problem is that, to be able to apply it, it is necessary to have some pairs of unrelated individuals that serve as a reference, which sometimes is not possible, in addition that a total lack of relatedness between individuals is somewhat arbitrary. A much more robust alternative method is to estimate the frequency of full-sib (Q_{FS}) and half-sib (Q_{HS}) pairs and estimate N_e with the expression

$$\frac{1}{N_e} = \frac{1+3\alpha}{4}(Q_{HS} + 2Q_{FS}) - \frac{\alpha}{2}\left(\frac{1}{N_m} + \frac{1}{N_f}\right) \tag{5.33}$$

(Wang, 2009), which also has the advantage of allowing N_e to be estimated with non-random mating ($\alpha \neq 0$). The difficulty lies in estimating the census sizes of reproducers N_m and N_f but, if mating is random ($\alpha \approx 0$), the problem disappears. The idea of the method is that in populations of reduced effective size it will be more likely to find pairs of full sibs or half sibs, which must be assigned probabilistically according to the information provided by the markers.

5.8.5 Methods with Multiple Information Sources

It is also possible to estimate N_e with approximate Bayesian computation (ABC) methods that use a statistic based on a variety of parameters, such as the number of alleles per locus, the expected heterozygosity, the linkage disequilibrium, the estimated value of α, and so on. By means of simulations, the value of N_e can be found, which gives a simulated statistic as close as possible to that observed with the data. However, it is not clear how the different parameters of the statistic should be weighted, which are also correlated and reflect different time scales that correspond to the processes of drift and inbreeding.

Of all the methods mentioned above, the most robust, accurate and with less error are those of linkage disequilibrium, the temporal method and the frequency of sibs method. It is common, however, to estimate N_e by several procedures, since they may reflect different time scales and different sources of drift. For example, if there are changes in the population size over generations, the actual variance and inbreeding effective sizes may differ. The methods of excess of heterozygotes and coancestry estimate the inbreeding effective size, while the temporal method estimates the variance effective size. The linkage disequilibrium method and the ABC methods probably estimate a combination of both effective sizes.

5.9 Estimates of N_e/N in Natural Populations

As we have seen in previous sections, most of the sources of drift that occur in natural populations produce a reduction of N_e in relation to the number of breeding individuals. Although there are thousands of estimates of N_e published, there are many less that provide the ratio N_e/N, basically because of the difficulty of obtaining estimates of N that, in many cases, especially in very numerous species such as insects or fish, are more difficult to obtain than those of N_e.

In general, demographic estimates apply to species of low reproductive capacity, such as mammals, birds or reptiles, on which it is often possible to obtain demographic data. Demographic estimates tend to be overestimates of N_e since they may not consider all sources of drift that act on the population. However, they have the advantage of allowing a correct understanding of the sources involved. The values of N_e/N for demographic estimates of the mentioned species are around 0.4, but with a great variation (Frankham, 1995). The genetic estimates with markers have the advantage of taking into account all sources of drift involved but are subject to more possible biases, such as those produced by migration and population structure and, in addition, do not provide indications of the sources of drift. The most typical values of N_e/N are of the order of 0.1–0.2 (Frankham, 1995; Palstra and Ruzzante, 2008). Figure 5.3 shows the distribution of N_e/N ratios obtained with the temporal method and compiled by Palstra and Ruzzante (2008), distinguishing natural populations of stable size, with an average of 0.18, conserved populations, with a mean of 0.27, and commercially exploited populations, with an average of 10^{-4}, this last result due to the enormous selective pressure exerted on them.

Problems

5.1 In Problem 4.1, we considered a set of ideal populations of census size $N = 20$ individuals each. Suppose now that the populations are maintained with 4 males and 16 females each generation. What would be the expected values of the heterozygosity, the variance in the allele frequencies and the inbreeding at generations $t = 5$, 20 and 100?

5.2 In Problem 4.2 of Chapter 4, we considered a plant species whose natural reproduction is a combination of autogamy and allogamy in equal proportions. From a large population of this species a line with census size $N = 20$ was founded, which was maintained for 10

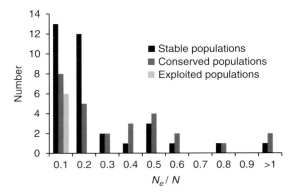

Figure 5.3 Distribution of the estimates of the ratio between N_e and N obtained with the temporal method and compiled by Palstra and Ruzzante (2008), distinguishing stable natural populations (33 estimates with mean $N_e/N = 0.18$), conserved populations (26 estimates with mean $N_e/N = 0.27$) and commercially exploited populations (6 estimates with mean $N_e/N = 10^{-4}$).

generations in the greenhouse by allogamous reproduction. Suppose now that the line is maintained with the natural system of reproduction of the species instead of forcing allogamy. What would then be the expected value of the coefficient of inbreeding of the population in generation 10?

5.3   In a population that is maintained with 16 males and 32 females in panmixia, what is the effective size corresponding to autosomal, X-linked and Y-linked loci?

5.4   Consider the lines of size $N = 20$ of Problem 5.2, which are reproduced with 50% autogamy, and calculate their effective size taking into account the effect of natural selection acting on deleterious mutations. Assume that the haploid genomic rate of deleterious mutation is $U = 0.2$, that the effect of mutations is $s = 0.1$, constant for all of them, and that there is no linkage.

5.5   Suppose that a population is maintained with $N = 16$ individuals that mate in pairs and the following numbers of descendants per pair are obtained: 1, 3, 4, 0, 2, 0, 5 and 1. (a) What would be the effective size of the population? (b) If the contributions of the pairs were equalized so that each couple contributed a male and a female to the offspring, what would then be the effective size? What would it be if the contribution of four of the pairs were of two males and that of the other four pairs of two females?

Self-Assessment Questions

1   The different types of effective size (variance, inbreeding, eigenvalue, coalescence, etc.) usually coincide exactly or approximately in their asymptotic value.

2   To average different population sizes, the harmonic mean is usually used, because the effective size usually affects the denominator of the expressions in which it is found.

3   In populations with a certain percentage of matings between relatives, the variance of the contributions of parents to progeny decreases with respect to that corresponding to a panmictic population.

4 The effective size decreases as the generation interval increases in populations with overlapping generations.

5 The magnitude of the genetic drift that affects the genes of the X chromosome in XX-XY species or the Z chromosome in ZZ-ZW species is 25% less than that of autosomal genes.

6 The effective size referring to neutral genes is drastically reduced when there is linkage between these and other loci subjected to selection.

7 The effective size of a subdivided population always increases with the differentiation in allele frequencies between sub-populations.

8 With equal contributions from parents to offspring, if mating between relatives is forced, the long-term effective size increases in comparison with the panmictic scenario.

9 The demographic methods of estimating the effective size tend to produce underestimates, by not taking into account all possible sources of genetic drift in the population.

10 The larger the effective population size, the larger is the expected linkage disequilibrium between two closely linked loci.

6 Estimation of Genetic Values, Variances and Covariances

Concepts to Study

- Covariance between relatives
- Coefficients of additive and dominance relationships
- The animal model
- REML estimation
- BLUP prediction and BLUE estimation

Objectives for Learning

- To understand the relationship between the phenotypic covariance among relatives and the components of genetic variance and covariance
- To learn how to estimate the heritability of a character and the genetic correlation between two characters from data of parents and offspring or groups of sibs
- To understand the sources of bias in the estimates of heritability
- To know the animal model and its applications in the estimation of genetic variance components and the prediction of additive values
- To learn how to use genetic markers for the estimation and prediction of genetic values and variances

6.1 Estimation of Heritability with Simple Experimental Designs

In Chapter 3 we studied the partition of the phenotypic value of a quantitative trait, deviated from the population mean, in its genetic and environmental components, and the first in its additive, dominance and epistatic components. The corresponding partition was extended to the components of the phenotypic variance. It was also indicated that the additive values of the individuals are the main values responsible for the resemblance between relatives and that this latter, therefore, can be quantified by the heritability, that is, the ratio of the variance of the additive values and the phenotypic variance, $h^2 = V_A/V_P$. This intimate relationship between additive variance and resemblance between relatives is what allows us to estimate the first one from the phenotypic values of related individuals.

In this section and the next we will deal with the simplest cases of estimation, which allow us to understand the basic concepts and are applicable in simple experimental designs. In the latter, we will consider more complex situations that require more elaborate procedures.

In simple designs, the data may correspond to pairs of parents and offspring or to groups of full sibs or half sibs. In the first case, the estimates are obtained by means of the coefficient of regression of the values of the offspring (O) on that of the parents (P),

$$b_{OP} = \frac{\text{cov}(O, P)}{\sigma_P^2},$$ (6.1)

whose rank with biological interpretation is $0 \le b_{OP} \le 1$, where σ_P^2 is the phenotypic variance observed in the parents. If the similarity between the values of the parents and the offspring is total, $O = P$, $\text{cov}(O, P) = \sigma_P^2$, and $b_{OP} = 1$. If, on the contrary, the similarity between parents and offspring is that between pairs of individuals taken at random from the population, $\text{cov}(O, P) = 0$ and $b_{OP} = 0$.

In the case of groups of full sibs or half sibs, the estimate is obtained from the intraclass correlation coefficient (t) of the groups of relatives considered, which is the proportion of the phenotypic variance between groups with respect to the total, that is,

$$t = \frac{\sigma_B^2}{\sigma_W^2 + \sigma_B^2},$$ (6.2)

where σ_W^2 and σ_B^2 are the intra- and intergroup observed components of variance, respectively. In what follows, the bases of the estimation with these simple designs are explained.

6.1.1 Estimation Based on the Degree of Resemblance between Parents and Offspring

Assuming there is Hardy–Weinberg equilibrium (with random mating with respect to the quantitative trait considered) and in the absence of epistasis, a direct relationship can be established between the regression estimate (6.1) and the causal components of genetic variation. If the character can only be evaluated in one sex, that is, the data pairs include the phenotypic value of a parent (P) and the average of its offspring ($\overline{O}$), equation (6.1) is $b_{\overline{O}P} = \text{cov}(\overline{O}, P)/\sigma_P^2$. Recall that in Chapter 3 we indicated that the phenotypic value P of an individual can be broken down into various genetic and environmental components or, in a simplified way, $P = A + R$, where A is its additive value and R is the rest of its genotypic value plus the environmental deviation, supposedly not correlated with A ($\text{cov}(A, R) = 0$). We also established that the mean of the trait in the offspring of that individual is, by definition, half of its additive value (A). Therefore, $\text{cov}(\overline{O}, P) = \text{cov}(A/2, \ A + R) = \text{cov}(A/2, \ A) + \text{cov}(A/2, \ R) = {}^1\!/_2\text{cov}(A, \ A) = {}^1\!/_2 V_A$ and, since the observed value of the phenotypic variance (σ_P^2) is an estimate of V_P, it follows that

$$b_{\overline{O}P} = \frac{{}^1\!/_2 V_A}{V_P} = {}^1\!/_2 h^2,$$ (6.3)

that is, the regression of the mean of the phenotypic values of the offspring on that of one of their parents is an estimate of half the heritability of the evaluated trait. Note that if $h^2 = 1$, $b_{\overline{O}P} = 1/2$, since only one of the parents is considered in the estimate. An alternative deduction of equation (6.3) can be obtained from the expressions of A and D values and genotype frequencies of Table 3.2 (Chapter 3). The phenotypic value of the parents, deviated from the population mean, is equal to their genotypic value (if the environmental deviations are 0), that is, the sum of $A + D$ values in Table 3.2. On the other hand, the average value of the offspring is $A/2$. Multiplying the values $(A + D) \times (A/2) \times$ genotype frequencies of Table 3.2 gives $\mathrm{cov}(\overline{O}, P) = V_A/2$. Note that equation (6.3) is valid regardless of the number of offspring available per parent, although the accuracy of the estimate will decrease as that number decreases.

If we have the phenotypic values of both sexes and calculate average values of the mother and father, $\overline{P} = (P_m + P_f)/2$, equation (6.1) is $b_{\overline{OP}} = \mathrm{cov}(\overline{O}, \overline{P})/\sigma_{\overline{P}}^2$. Analogously to the previous case, $\mathrm{cov}(\overline{O}, \overline{P}) = \mathrm{cov}(\overline{O}, [P_m + P_f]/2) = \mathrm{cov}(\overline{O}, P_m/2) + \mathrm{cov}(\overline{O}, P_f/2)$ and, assuming that there are no differences in variance between males and females, $V(P_m) = V(P_f) = V_P$, results in $\mathrm{cov}(\overline{O}, \overline{P}) = \frac{1}{2}[2\,\mathrm{cov}(\overline{O}, P)] = \frac{1}{2}V_A$, as was found in the case of a single parent. But now $\sigma_{\overline{P}}^2 = V[(P_m + P_f)/2] = \frac{1}{4}V(P_m) + \frac{1}{4}V(P_f) = \frac{1}{2}V_P$ and, therefore,

$$b_{\overline{OP}} = \frac{\frac{1}{2}V_A}{\frac{1}{2}V_P} = h^2, \tag{6.4}$$

that is, the regression of the mean of the offspring on the parental average provides a measure of heritability. Figure 6.1 presents an example of estimation of the heritability of human height by regression of the value of an individual on the average value of their parents, providing an estimate of $h^2 = 0.71$, in accordance with other estimates obtained for this trait that range between 0.6 and 0.8 (Zaitlen et al., 2013).

It must be taken into account that the previous estimates (equations (6.3) and (6.4)) assume absence of sources of environmental resemblance between parents and offspring. For certain traits this assumption does not hold since parents and offspring share a common environment, and estimates will be overestimations of heritability. In mammals it is common for many traits to have an important cause of resemblance between parents and offspring, which are the maternal effects. At least during the first stages of their lives, individuals are influenced by the maternal environment for characters such as body weight, since larger mothers generally have bigger wombs and produce more milk, contributing to a larger weight of the offspring. In this case, the source of bias can be avoided by using regression data of the offspring on the father, or by randomizing the offspring among the mothers so that feeding is

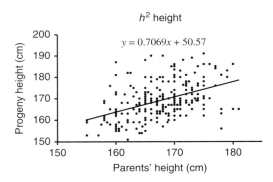

Figure 6.1 Estimation of the heritability for human height from the regression of the value of an individual on the average of their parents. Data of 252 students of biology of the University of Vigo. The heritability estimate is $h^2 = 0.71$.

carried out with nursing mothers, as sometimes done in laboratory experiments. In addition to its interest in the case of traits of economic importance in mammals, the environmental resemblance between parents and offspring acquires a notable influence in humans, especially for behavioural traits.

The error of the estimates of heritability by regression is given by the estimate of the coefficient of regression, whose sample variance can be approximated by

$$\sigma_b^2 \approx \frac{k[1 + (j-1)t]}{nj}, \tag{6.5}$$

where n is the number of data pairs with $k = 1$ or 2 parents and j offspring, and t is the intraclass correlation coefficient for full sibs (equation (6.2)) (Falconer and Mackay, 1996, p. 178). If the total number of offspring evaluated (jn) is fixed, σ_b^2 is minimal if only one offspring per family is evaluated ($j = 1$), in which case $\sigma_b^2 \approx k/n$. Thus, if a single parent were considered ($k = 1$), $\sigma_b^2 \approx 1/n$, and the standard error of the heritability estimate would be $SE[h^2] = 2\sigma_b \approx 2/\sqrt{n}$. If there are two parents ($k = 2$), $SE[h^2] = \sigma_b \approx \sqrt{2/n}$. Therefore, the accuracy of the estimate is greater using two parents than only one. For example, if we want that $SE[h^2] = 0.1$, we would need 400 pairs with one parent, but only 200 with two parents. In the case of Figure 6.1, $k = 2$, $j = 1$, $n = 252$, and the approximate standard error of h^2 would therefore be $SE[h^2] = \sqrt{2/252} = 0.089$, which is quite close to the standard error of the regression coefficient, 0.091.

In the previous estimates the parents are assumed to be a random sample of individuals from the population. If they were a group of selected parents, the regression estimates would not be affected, given that the bias produced would occur to the same extent in the numerator and the denominator of equation (6.1). However, the accuracy of the estimates would be affected, since a smaller variation in the parents would imply a greater error than that expressed by equation (6.5). It is possible, however, to reduce the standard error of the estimate without incurring bias, by selecting parents whose phenotypic values are found at the two extremes of the distribution of the character, since the regression depends greatly on these values (Hill, 1971). This can be seen in Figure 6.1, where it can be deduced that the presence of a few data in the lower right quadrant is responsible for the regression line leaning less than it would if these few data were ignored. If in Figure 6.1 only 5% of the parents of greater and shorter height were used (in total 26 pairs of data), the heritability estimate would be $h^2 = 0.60 \pm 0.16$, while the expected approximate standard error with that number of data would be $\sqrt{2/26} = 0.28$, almost double.

6.1.2 Estimation Based on the Degree of Resemblance between Sibs

When sibling groups are used, the estimation of genetic variance and heritability is obtained from the intraclass correlation coefficient (t) expressed in equation (6.2). The numerator, the variance between families σ_B^2, is also the covariance between sibs and, therefore, indicates their degree of resemblance. Suppose first the case in which a male mates with a certain number of females chosen at random from the population, having a single offspring with each of them. In this way we would obtain groups of half sibs that have in common their father but

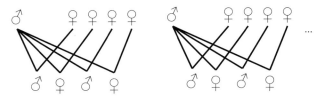

Figure 6.2 Design to create groups of half sibs for the estimation of heritability.

different mothers (Figure 6.2). The mean of the offspring is, as we had previously indicated, half the additive value A of the common father and, therefore, $\sigma_B^2 = V(A/2) = \frac{1}{4} V_A$, which is the covariance between half sibs. Substituting in (6.2), the intraclass correlation of half sibs is

$$t_{HS} = \frac{\frac{1}{4}V_A}{V_P} = \frac{1}{4}h^2. \tag{6.6}$$

An alternative demonstration of the covariance between half sibs ($\frac{1}{4}V_A$) is obtained from Table 3.2 (Chapter 3), averaging the square of half of the additive values for each genotype, that is, adding the products of the values of $(A/2)^2$ of each genotype by its genotypic frequency.

If the half sibs of each family had grown up in the same environment there could be a source of environmental bias that would contribute to increase their covariance, producing an overestimation of V_A and h^2. This will not occur, however, if the half sibs have grown in randomized environments, in which case $4t_{HS}$ is a good estimate of h^2.

Suppose now the case in which the families are full sibs, that is, the individuals of each group share mother and father. The deduction is simplified if we refer to the genotypes for a multiallelic locus. Suppose that the mother has genotype A_1A_2 and the father A_3A_4, that is, the parents are unrelated individuals. The four genotypes that can be found in the progeny with equal probability are A_1A_3, A_1A_4, A_2A_3 and A_2A_4 (Figure 6.3). If the additive value of the mother is A_m and that of the father A_f, that of any of the offspring is $\frac{1}{2}(A_m + A_f) + D_i$, where D_i is the dominance deviation corresponding to offspring i. Then, the mean genotypic value of the offspring is $\overline{O} = \frac{1}{2}(A_m + A_f) + \frac{1}{4}(D_1 + D_2 + D_3 + D_4)$ and the variance of this mean is $V(\overline{O}) = \frac{1}{4}V(A_m) + \frac{1}{4}V(A_f) + \frac{1}{16}V(D_1) + \frac{1}{16}V(D_2) + \frac{1}{16}V(D_3) + \frac{1}{16}V(D_4)$. Since these values are estimates of the relevant variance components (V_A or V_D), we have that $V(A_m) = V(A_f) = V_A$, $V(D_1) = V(D_2) = V(D_3) = V(D_4) = V_D$ and $V(\overline{O}) = \frac{1}{2}V_A + \frac{1}{4}V_D$.

Since full sibs usually share a common environment, at least the intrauterine environment or other maternal effects, this factor will be added to the observed variance among full-sib groups. Suppose that the total environmental variance V_E can be decomposed into variance due to the environment common to sib families (V_{EC}) and the intra-family environmental variance (V_{EW}), that is, $V_E = V_{EC} + V_{EW}$, and that we call $c^2 = V_{EC}/V_P$ the proportion of the phenotypic variance that is due to the common environment and, analogously, $d^2 = V_D/V_P$ the corresponding proportion for the dominance variance. Then, the phenotypic variance observed among full-sib groups would be $\frac{1}{2}V_A + \frac{1}{4}V_D + V_{EC}$, and the intraclass correlation between full sibs is

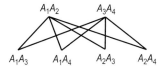

Figure 6.3 Mating between individuals to generate a group of full sibs where the possible genotypes in the offspring are indicated for a locus with four alleles ($A_1 \ldots A_4$).

$$t_{FS} = \frac{\frac{1}{2}V_A}{V_P} + \frac{\frac{1}{4}V_D}{V_P} + \frac{V_{EC}}{V_P} = \frac{1}{2}\,h^2 + \frac{1}{4}\,d^2 + c^2. \tag{6.7}$$

Note that $2t_{FS}$ estimates $h^2 + \frac{1}{2}d^2 + 2c^2$ and, therefore, the h^2 estimates obtained by this method will be inflated by dominance and common environment. The biases by c^2 have greater influence since this term is not weighted by a fraction. The common environmental variation usually implies an increase in the resemblance between sibs, that is, an increase in the variance between family means. However, in some cases it may imply a decrease in variance, as occurs when the individuals of the family must compete with each other for the same resources. For example, if they were fish and each sib family were raised in a container where food resources are limited, some individuals could be more effective than others in obtaining their food and would generate differences between sibs for a number of traits, reducing their covariance.

The error of heritability estimates by groups of full sibs is given by the following approximation of the sampling variance of the intraclass correlation coefficient

$$\sigma_t^2 \approx \frac{2[1 + (j - 1)t]^2(1 - t)^2}{j(j - 1)(n - 1)}, \tag{6.8}$$

where n is the number of families and j the number of individuals per family (see Visscher and Goddard, 2015 for a general formulation of the sampling error of the estimates of heritability). To simplify, it can be assumed that n and j are large enough to approximate $n - 1 \approx n$ and $j - 1 \approx j$, and that t is small enough so that $(1 - t)^2 \approx 1$. Equation (6.8) is then reduced to $\sigma_t^2 \approx 2(1 + jt)^2/nj^2$. Note that, for some n and j prefixed, σ_t^2 will be smaller the smaller t, that is, the smaller h^2. Therefore, the estimation method with sib groups will have less error when the heritability to be estimated is low. In contrast, for a prefixed volume of data (nj), the estimation of h^2 by regression will have less error than that from sibs for large values of t (Robertson, 1959). Thus, it can be recommended to estimate the heritability with groups of sibs when this is assumed low (it can be shown that, in general, when it is less than 0.25), and by regression when it is assumed high (greater than 0.25). Note, however, that the recommendation in the case of sib groups is based on the assumption that the number of individuals in each group (j) is high. In some cases, for example in humans, j is small, generally 2. In that case, equation (6.8) is reduced to $\sigma_t^2 \approx [(1 - t^2)^2]/n$, and the accuracy of the estimate is greater the higher the heritability.

Returning to the estimation of sib groups, if they are half sibs $h^2 = 4t$ and its standard error will be $SE[h^2] = 4SE[t] = 4\sigma_t = 4\sqrt{2}(1 + jt)/(j\sqrt{n})$ which, substituting $t = h^2/4$,

Table 6.1 *Example of estimation of genetic variance components in a design of sib groups*

Data						
$I = 3$		Father 1		Father 2		Father 3
$J = 2$	Mother 1	Mother 2	Mother 3	Mother 4	Mother 5	Mother 6
$K = 5$	26, 27, 29	25, 33, 25	30, 32, 34	29, 33, 27	32, 32, 34	28, 27, 25
	31, 25	30, 32	34, 30	34, 31	31, 35	29, 35

Sums						
X^2_{ijk}	27,601					
$X_{ij\cdot}$	138	145	160	154	164	144
$X_{i\cdot\cdot}$	283		314		308	
$X\cdots$	905					

Anova				
Factors	d.f.	SS	MS = SS/d.f.	E(MS)
Fathers	$I - 1 = 2$	$SSF = 54.07$	$MSF = 27.03$	$\sigma^2_e + K\sigma^2_m + JK\sigma^2_f$
Mothers	$I(J - 1) = 3$	$SSM = 48.5$	$MSM = 16.17$	$\sigma^2_e + K\sigma^2_m$
Error	$IJ(K - 1) = 24$	$SSE = 197.6$	$MSE = 8.23$	σ^2_e
Total	$IJK - 1 = 29$	300.17		

Calculations of the sums of squares

$$SSF = \sum_i \frac{X^2_{i\cdot\cdot}}{JK} - \frac{X^2_{\cdots}}{IJK} = [(283^2 + 314^2 + 308^2)/10] - [905^2/30]$$

$$SSM = \sum_{i,j} \frac{X^2_{ij\cdot}}{K} - \sum_i \frac{X^2_{i\cdot\cdot}}{JK} = [(138^2 + 145^2 + \ldots + 144^2)/5] - [(283^2 + 314^2 + 308^2)/10]$$

$$SSE = \sum_{i,j,k} X^2_{ijk} - \sum_{i,j} \frac{X^2_{ij\cdot}}{K} = 27{,}601 - [(138^2 + 145^2 + 160^2 + 154^2 + 164^2 + 144^2)/5]$$

$$TOTAL = \sum_{i,j,k} X^2_{ijk} - \frac{X^2_{\cdots}}{IJK} = 27{,}601 - [905^2/30]$$

can be approximated by $SE[h^2] = h^2\sqrt{2/n}$. The same approximate result is obtained for the estimation with full sibs. Therefore, the standard error of the estimate is inversely proportional to the square root of the number of families analysed. Equations (6.5) and (6.8) allow us to deduce what the volume of data analysed is needed in each case to obtain a standard error of the estimates appropriate to our interests. For example, if $SE[h^2] \approx h^2/4$ is desired, $n = 32$ families with numerous offspring are needed.

    Tables 6.1 and 6.2 show an example of the variance components estimation by means of an analysis of variance with data of full-sib and half-sib groups. In the example, $I = 3$ fathers are considered, each of which is crossed with $J = 2$ mothers having $K = 5$ descendants. The calculations allow us to obtain the observable components of the variance between fathers (σ^2_f), between mothers within father (σ^2_m), and of the error within families (σ^2_e), σ^2_T being the sum of the three components. With these observable components, the causal components (Table 6.2) of the variance can be calculated. The result of the example provides an estimate of the heritability of $h^2 = 0.40$. If there was no environment common to full-sib families $(c^2 = 0)$, the dominance variance component relative to the phenotypic variance would be $d^2 = 0.18$.

The estimation of heritability for the case of threshold traits can be consulted in Falconer and Mackay (1996, chapter 18).

6.1.3 Effect of Assortative Mating

The heritability estimation procedures presented involve random mating between parents for the evaluated trait. A relatively frequent case in nature, however, is that of assortative mating for the character, where individuals of similar phenotype tend to mate more frequently than they would by chance. This may be the case, for example, for human height, and is relatively common for body size in some animal species. Suppose that ρ is the phenotypic correlation of the character between the individuals of the pairs and h^2 the heritability of the character in the population, so that ρh^2 is the correlation between the additive values of the couple. The regression of the average value of the parents on that of their progeny is not affected by assortative mating, in the same way that it was not affected when extreme individuals were selected in the parental distribution (Section 6.1.1), that is, $b_{\overline{OP}} = h^2$. However, when only one parent is used, the corrected estimate for assortative mating is $b_{\overline{OP}} = \frac{1}{2}h^2(1+\rho)$. In the case of the estimation with full-sib groups, $t_{FS} = \frac{1}{2}h^2(1+\rho h^2)$ and with half sibs, $t_{HS} = \frac{1}{4}h^2(1 + 2\rho h^2 + \rho^2 h^2)$ (Nagylaki, 1978).

6.1.4 Estimation Based on the Degree of Similarity between Twins

A particular case of estimation of heritability with data of full sibs is that of twins. Monozygotic twins, which occur in very few species, including humans, are particularly useful because they allow the degree of genetic determination of a trait to be directly evaluated as they share 100% of their genes. The phenotypic differences between mono- and dizygotic twins are easily assessed by the so-called concordance between characters, which is the proportion of pairs of twins in which the same character is shown. For example, hair colour has a concordance of 89% between monozygotic twins, but only 22% between dizygotic twins. Likewise, diabetes has a concordance of 65% in the first case and 18% in the second (Cummings, 2014).

Table 6.2 *Estimation of the causal components of variance from the observable components in Table 6.1*

Observable components		Causal components
$\sigma_f^2 = \frac{MSF - MSM}{JK} = 1.087$	cov(HS)	$\frac{1}{4}V_A$
$\sigma_m^2 = \frac{MSM - MSE}{K} = 1.587$	cov(FS) − cov(HS)	$\frac{1}{4}V_A + \frac{1}{4}V_D + V_{EC}$
$\sigma_f^2 + \sigma_m^2 = 2.673$	cov(FS)	$\frac{1}{2}V_A + \frac{1}{4}V_D + V_{EC}$
$\sigma_e^2 = MSE = 8.233$	V_P − cov(FS)	$\frac{1}{2}V_A + \frac{3}{4}V_D + V_{EW}$
$\sigma_T^2 = \sigma_f^2 + \sigma_m^2 + \sigma_e^2 = 10.907$	V_P	$V_A + V_D + V_{EC} + V_{EW}$

Intraclass correlation coefficients		Estimates
$t_f = \sigma_f^2/\sigma_T^2 = 0.0996$	$\frac{1}{4}h^2$	$h^2 = 4t_f = 0.3985$
$t_m = \sigma_m^2/\sigma_T^2 = 0.1455$	$\frac{1}{4}h^2 + \frac{1}{4}d^2 + c^2$	
$t_{f+m} = \left(\sigma_f^2 + \sigma_m^2\right)/\sigma_T^2 = 0.2451$	$\frac{1}{2}h^2 + \frac{1}{4}d^2 + c^2$	
$t_m - t_f = 0.0459$	$\frac{1}{4}d^2 + c^2$	$d^2 + 4c^2 = 4(t_m - t_f) = 0.1836$

Table 6.3 *Observable and causal components of variance in the case of monozygotic (MZ) and dizygotic (DZ) twins, ignoring epistasis*

Observable components	σ_B^2	σ_W^2
Causal components		
Monozygotic twins (MZ)	$V_A + V_D + V_{EC}$	V_{EW}
Dizygotic twins (DZ)	$\frac{1}{2}V_A + \frac{1}{4}V_D + V_{EC}$	$\frac{1}{2}V_A + \frac{3}{4}V_D + V_{EW}$
Difference	$\frac{1}{2}V_A + \frac{3}{4}V_D$	$\frac{1}{2}V_A + \frac{3}{4}V_D$
Intraclass correlation coefficients		
$t_{MZ} = \sigma_B^2/\sigma_T^2$	$h^2 + d^2 + c^2$	
$t_{DZ} = \sigma_B^2/\sigma_T^2$	$\frac{1}{2}h^2 + \frac{1}{4}d^2 + c^2$	
$2t_{DZ} - t_{MZ}$	$c^2 - \frac{1}{2}d^2$	Approximate estimate of c^2
$2(t_{MZ} - t_{DZ})$	$h^2 + \frac{3}{2}d^2$	Approximate estimate of H^2

Twin studies allow $\frac{3}{2}$ for obtaining estimates of broad-sense heritability (H^2). Estimates are complicated, however, by the bias produced by the common environment (at least until birth). In the case of monozygotic twins, the observable component of variance between pairs (σ_B^2) includes all the genetic variance and the common environmental variance ($V_A + V_D + V_{EC}$), while the component of variance within pairs (σ_W^2) only includes the environmental variance not common to the sibs (V_{EW}) (Table 6.3). The components of the dizygotic twins are the same as those of full sibs (Tables 6.2 and 6.3). Assuming that the common and non-common environmental variances (V_{EC} and V_{EW}) are equal for both types of twins, an estimate very close to H^2 can be obtained. It is also possible to obtain an estimate close to c^2, the proportion of the variation due to the common environment (Table 6.3).

There are currently many estimates of heritability in humans based on twins data, with a huge variation depending on the traits and populations. For example, the heritability of the number of fingerprint crests is of the order of 0.9, that of height ranges between 0.6 and 0.8, that of intelligence (IQ scores) between 0.4 and 0.6 if estimated in children and 0.8 in adults, and that of infertility between 0.1 and 0.2. In a review of more than 2700 studies conducted between 1958 and 2012 for a total of about 18,000 characters that included data from more than 14 million pairs of twins, a global average heritability of $H^2 = 0.49$ was obtained (Polderman et al., 2015), as would be expected if there is a huge variation for the entire possible range of values between zero and one. An average value of c^2 of 0.174 was also found, subject to the assumption that the value of c^2 must be the same for mono- and dizygotic twins. In 69% of the cases, the results were compatible with the fact that the resemblance between twins is mainly due to additive effects, with relatively little influence of the common environment and non-additive genetic variation. Table 6.4 presents the averages of H^2 and c^2 averaged for different groups of traits, obtained by expressions $2(t_{MZ} - t_{DZ})$ and $2t_{DZ} - t_{MZ}$, respectively (Table 6.3), and ordered by increasing values of the heritability estimate. The characters related to cell activity, reproduction and social interactions are what have lower heritabilities and higher common environment, while skeletal, dermatological and ophthalmological characters have the highest average heritabilities.

Table 6.4 *Average estimates of broad-sense heritability (H^2) and proportion of the phenotypic variance due to common environment (c^2) for some types of human traits*

Type of character	H^2	c^2	Type of character	H^2	c^2
Cell activity	0.149	0.674	Immunological	0.494	0.147
Reproduction	0.313	0.320	Neurological	0.503	0.068
Social interaction	0.319	0.182	Haematological	0.505	0.324
Endocrine	0.395	0.322	Respiratory	0.545	0.094
Cardiovascular	0.436	0.149	Metabolic	0.584	0.191
Nutritional	0.436	0.115	Skeletal	0.591	0.265
Psychiatric	0.463	0.158	Dermatological	0.604	0.166
Cognitive	0.468	0.177	Ophthalmological	0.712	0.048

Note: Each estimate is the average of a variable number of studies. The standard error of estimates ranges between 0.006 and 0.054 for H^2 and between 0.005 and 0.090 for c^2.
Source: Data from Polderman et al. (2015).

6.1.5 Coefficients of Additive and Dominance Relationships

In the previous sections we have seen that the covariance between relatives represents a fraction of the additive variance and the dominance variance. This fraction is a function of the coefficient of additive relationships or theoretical correlation (r) in the case of the additive variance, and the coefficient of dominance relationships (u) in the case of the dominance variance. In general, the correlation of the additive values between two individuals X and Y is $r = f_{XY}/\sqrt{f_{XX}f_{YY}}$, where f_{XY} is the coancestry coefficient between X and Y (Chapter 4). Substituting the self-coancestries of X and Y of the denominator by the expression (4.4), we have $r = 2f_{XY}/\sqrt{(1+F_X)(1+F_Y)}$. The coefficient of additive relationships that weights the component of additive variance is the numerator of this expression, that is, the theoretical correlation under the assumption that the individuals are not inbred ($r = 2f_{XY}$), which is double the coancestry, or the expected proportion of genes that the individuals share if they are not inbred.

If the parents of X are A and B and the parents of Y are C and D, the coefficient of dominance relationships is $u = f_{AC}f_{BD} + f_{AD}f_{BC}$, that is, the probability that an allele of X is identical by descent to one of Y, and that the other allele of X is also identical by descent to the other of Y. This coefficient, therefore, only exists when X and Y share both parents (that is, when they are full sibs). Table 6.5 indicates the corresponding coefficients in the case of the relatives considered above. If we also consider the epistatic components of the variance (see Chapter 3), the corresponding coefficients are simply products of the previous ones, so that the covariance between relatives, in general, can be expressed as

$$\text{cov} = rV_A + uV_D + r^2V_{AA} + ruV_{AD} + u^2V_{DD} + r^3V_{AAA} + r^2uV_{AAD}$$
$$+ \ ru^2V_{ADD} + u^3V_{DDD} + \ldots \tag{6.9}$$

if the linkage between loci is ignored. Note that the contributions of the epistatic components present in equation (6.9) and described in Chapter 3 are increasingly smaller fractions of these. Therefore, it is expected that the contribution of epistasis to the covariance between relatives is generally small because the epistatic variance is, in general, smaller than the additive or

Table 6.5 *Coefficients of additive (r) and dominance (u) relationships for different types of relatives*

	r	u
Monozygotic twins	1	1
Full sibs	1/2	1/4
Half sibs	1/4	0
Parents and offspring	1/2	0

dominance ones. However, these terms of epistatic variance may bias to some extent the estimates obtained by the procedures described above.

The methods explained above for the estimation of heritability ignored epistasis. If this is considered, the estimates of heritability would be inflated by the epistatic components, as can be deduced from Table 6.5 and expression (6.9). Thus, parent-offspring regression estimates and half-sib estimates would be biased only by additive $\times$ additive components of variance. In contrast, estimates from twice the correlation of full sibs ($2t_{FS}$) would estimate $h^2 + (1/2)d^2 + (1/2)V_{AA}/V_P + (1/4)V_{AD}/V_P + (1/8)V_{DD}/V_P$, plus higher order epistatic components. Analogously, for twin studies, the value $2(t_{MZ} - t_{DZ})$ would estimate $h^2 + (3/2)d^2 + (3/2)V_{AA}/V_P + (7/4)V_{AD}/V_P + (15/8)V_{DD}/V_P$, plus higher order epistatic components (Lynch and Walsh, 1998, p. 583).

The coefficients of additive and dominance relationships applicable to populations with mixed inbreeding and random mating can be found in Walsh and Lynch (2018, p. 414).

6.2 Estimation of Genetic Correlation

The estimation of the genetic correlation can be done by the same procedures described in the previous sections. For two traits X and Y, the phenotypic (r_P), additive genetic (r_A) and environmental (r_E) correlations, the latter including all the environmental and non-additive genetic components, are

$$r_P = \frac{\text{cov}_P(X, Y)}{\sigma_{PX}\sigma_{PY}}, \quad r_A = \frac{\text{cov}_A(X, Y)}{\sigma_{AX}\sigma_{AY}} \quad \text{and} \quad r_E = \frac{\text{cov}_E(X, Y)}{\sigma_{EX}\sigma_{EY}}, \tag{6.10}$$

where $\sigma_{PX}^2 = \sigma_{AX}^2 + \sigma_{EX}^2$, and analogously for trait Y. Since $\text{cov}_P(X, Y) = \text{cov}_A(X, Y) + \text{cov}_E(X, Y)$, from equation (6.10) we deduce that $r_P = [\text{cov}_A(X, Y)] + \text{cov}_E(X, Y)]/(\sigma_{PX}\sigma_{PY})$ and, given that $h_X = \sigma_{AX}/\sigma_{PX}$ and $\sqrt{1 - h_X} = \sigma_{EX}/\sigma_{PX}$ (analogously for Y), substituting, we obtain the relation between the three correlations,

$$r_P = r_A h_X h_Y + r_E \sqrt{(1 - h_X^2)(1 - h_Y^2)}. \tag{6.11}$$

If the two traits have the same heritability, the formula (3.18), that was advanced in Chapter 3, is retrieved, and if $h_X^2 \approx h_Y^2 \approx 1/2$, $r_P \approx 1/2 (r_A + r_E)$.

Table 6.6 *Expected values of the mean of squares (MS) and the mean of products (MP) for traits X and Y in a variance–covariance analysis for a design with I half-sib groups with K individuals per group*

Factors	d.f.	$E[MS_X]$	$E[MS_Y]$	$E[MP_{XY}]$
Intergroup	$I - 1$	$\sigma^2_{WX} + K\sigma^2_{BX}$	$\sigma^2_{WY} + K\sigma^2_{BY}$	$cov_W + Kcov_B$
Error	$I(K - 1)$	σ^2_{WX}	σ^2_{WY}	cov_W

Relation with the causal components

$\sigma^2_{BX} = V_{AX}/4$

$\sigma^2_{BY} = V_{AY}/4$

$cov_B = cov_A/4$

Estimate of the correlation:

$$r_A = \frac{cov_B}{\sigma_{BX}\sigma_{BY}} = \frac{cov_A}{\sigma_{AX}\sigma_{AY}}$$

The estimation of the genetic correlation with data of parents and offspring is obtained by calculating the crossed covariances for each character. As in Section 6.1.1, the covariance between parents (P) and offspring (O) for trait X estimates half the additive variance for that trait, that is, cov(P_X, O_X) = 1/2 V_{AX}, and analogously for trait Y, cov(P_Y, O_Y) = 1/2 V_{AY}, while the crossed covariances between traits, cov(P_X, O_Y) and cov(P_Y, O_X) both estimate 1/2 cov$_A$(X,Y). Therefore, the estimate of the genetic correlation is obtained as

$$r_A = \frac{\frac{1}{2}\left[cov(P_X, O_Y) + cov(P_Y, O_X)\right]}{\sqrt{cov(P_X, O_X)cov(P_Y, O_Y)}} = \frac{cov_A(X, Y)}{\sqrt{V_{AX}V_{AY}}}. \tag{6.12}$$

In the case of sib families, an analysis of variances–covariances must be made. Table 6.6 presents the expected values of the means of squares in the case of a design of half-sib groups. With full-sib groups it would be the same, except that the components would be divided by 2, instead of 4.

The correlation estimates tend to have poor accuracy since they depend on the sampling errors of two characters and the corresponding estimates of heritability. The standard error can be approximated by

$$SE(r_A) \approx (1 - r^2_A)\sqrt{\frac{SE(h^2_X)SE(h^2_Y)}{2h^2_X h^2_Y}}. \tag{6.13}$$

Given that the standard errors of the heritabilities appear in the numerator, the designs that minimize the sampling variance of the heritability will also minimize the error of the correlation estimate.

6.3 Estimation of Variance Components and Prediction of Additive Values with Complex Structure of the Data

In the previous sections we have seen how to estimate the components of genetic variance and heritability with designs based on data from parents and offspring or groups of sibs, which is useful in experimental studies designed for this purpose. Once the heritability is known, the additive value of an individual can be estimated from the product of the heritability by the phenotypic value, as we saw in Chapter 3 (equation (3.17)), although the standard error of the estimate is very large unless the number of offspring or sibs is high. However, in many situations, the structure of the data is not so simple and it is possible to have individuals with diverse degrees of relatedness (sibs, parents, offspring, grandparents, cousins, etc.) determined by genealogical or molecular information. In addition, data are sometimes grouped into fixed-effect classes such as population, year, season, sex, and so on. The estimation of variance components and additive values is then made using all available information, both in the context of animal breeding (Henderson, 1984) and in natural populations (Kruuk et al., 2000; Kruuk, 2003). The most popular statistical technique for the estimation of variance components is maximum likelihood, although Bayesian inference and Monte Carlo Markov Chain methods are also often used (Sorensen and Gianola, 2002), and the usual method to estimate the additive values is BLUP. The basis of these methods will be explained in the next two sections.

6.3.1 Estimation of Mean and Variance by Maximum Likelihood

Maximum likelihood (ML; Fisher, 1921) is a very useful statistical method to estimate genetic parameters and perform tests of significance, presenting great advantages over alternative procedures when the data refer to arbitrary genealogies. Briefly, the likelihood of a hypothesis is the probability of observing the data given that hypothesis. To understand the procedure, consider a simple example that will be useful in later chapters. Suppose we want to estimate the rate of mutation per haploid genome and generation, and the appearance of a total of 1000 mutations in 1000 born individuals has been observed experimentally so the average number of mutations observed per individual in that generation is 1. To obtain the maximum likelihood estimation, a model or distribution of the data must be assumed. In the case of mutation, the most appropriate distribution is Poisson, so that if n is the number of mutations that appear and λ is the average number of mutations, the probability or likelihood (L) of this parameter is $L(\lambda) = (\lambda^n/n!)e^{-\lambda}$ and the natural logarithm of the likelihood is

$$\ln L(\lambda) = n\ln(\lambda) - \lambda. \tag{6.14}$$

Figure 6.4 illustrates the values of equation (6.14). The maximum likelihood estimate is the maximum value of the function, naturally $\lambda = 1$, for which $\ln L = -1$. To obtain the maximum of $\ln L$, the procedure is to derive the equation with respect to the parameter to be estimated and equalize to zero. In the case of equation (6.14), $d[\ln L(\lambda)]/d\lambda = (n/\lambda) - 1 = 0$, from where $\lambda = n$, the observed value, which would be equal to one in this example. The 95% confidence interval of the estimate is approximately $2\ln L$, as indicated by the dotted line in Figure 6.4, which corresponds to values of λ between 0.05 and 4.5. It is also possible to obtain how likely an estimate is in relation to another. For example, given that $n = 1$ mutations have been observed, the maximum likelihood estimate ($\lambda = 1$) would be $L(\lambda = 1)/L(\lambda = 3) = e^{-1}/(3e^{-3}) = 2.46$ times more likely than an estimate of $\lambda = 3$.

When it comes to estimating the components of variance of a quantitative trait, the procedure to follow is the same. Given a distribution of the observed data, the parameters of the distribution (means, variances, etc.) that maximize the likelihood of the observations are sought. Under the general assumption that quantitative traits follow a normal distribution, this is the one used for the likelihood. Thus, if data (y_i) of n independent individuals with normal

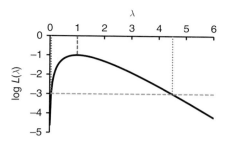

Figure 6.4 Illustration of the estimation of the rate of mutation per haploid genome and generation (λ) by maximum likelihood (L).

distribution and mean μ and variance σ^2 are available, the probability of the data vector (**y**) conditioned to the mean and variance is the product of normal densities of the data,

$$P(\mathbf{y}|\mu, \sigma^2) = \prod_{i=1}^{n} P(y_i|\mu, \sigma^2) = \prod_{i=1}^{n} \frac{1}{\sqrt{2\pi\sigma^2}} \exp\left(-\frac{(y_i - \mu)^2}{2\sigma^2}\right)$$

$$= \left(2\pi\sigma^2\right)^{-n/2} \exp\left(-\sum_{i=1}^{n} \frac{(y_i - \mu)^2}{2\sigma^2}\right), \tag{6.15}$$

and, taking logarithms,

$$\ln P(\mathbf{y}|\mu, \sigma^2) = -\frac{n}{2}\left[\ln(2\pi) + \ln(\sigma^2) + \frac{1}{n\sigma^2}\sum_{i=1}^{n}(y_i - \mu)^2\right]. \tag{6.16}$$

Calling $\bar{y} = (1/n)\Sigma y_i$ and $V = (1/n)\Sigma(y_i - \bar{y})^2$ to the observed mean and variance of the data, we have $\Sigma(y_i - \mu)^2 = \Sigma(y_i - \bar{y} + \bar{y} - \mu)^2 = \Sigma(y_i - \bar{y})^2 + \Sigma(\bar{y} - \mu)^2 + 2(\bar{y} - \mu)$ $\Sigma(y_i - \bar{y}) = nV + n(\bar{y} - \mu)^2 = n[V + (\bar{y} - \mu)^2]$, and noting that equation (6.16) is also the logarithm of the likelihood of the estimates of the mean and variance given the data set,

$$\ln L\left(\mu, \sigma^2|\mathbf{y}\right) = -\frac{n}{2}\left[\ln(2\pi) + \ln(\sigma^2) + \frac{V + (\bar{y} - \mu)^2}{\sigma^2}\right]. \tag{6.17}$$

Taking derivatives in equation (6.17) with respect to μ and σ^2 and equating them to zero,

$$\frac{d\ln\left(\mu, \sigma^2|\mathbf{y}\right)}{d\mu} = \frac{n(\bar{y} - \mu)}{\sigma^2} = 0 \tag{6.18}$$

$$\frac{d\ln\left(\mu, \sigma^2|\mathbf{y}\right)}{d\sigma^2} = -\frac{n}{2\sigma^2}\left[1 - \frac{V + (\bar{y} - \mu)^2}{\sigma^2}\right] = 0, \tag{6.19}$$

we obtain, respectively, that the maximum likelihood estimate of μ is $\hat{\mu} = \bar{y}$, that is, the observed mean, and the maximum likelihood estimate of the variance is $\hat{\sigma}^2 = V + (\bar{y} - \mu)^2$.

As can be noted, the variance estimate is an overestimation of V if the observed mean is different from the real population mean. Estimates of variance by maximum likelihood have the disadvantage of providing biased estimates because they do not take into account the degrees of freedom lost when estimating the fixed effects. This is corrected using restricted maximum likelihood (REML). The correction in the case in question would be to replace the term $(\bar{y} - \mu)^2$ with its expected value, which would be σ^2/n, the variance of the sample mean, that is, $\hat{\sigma}^2 = V + (\hat{\sigma}^2/n)$. Since σ^2 is unknown (it is what is intended to be estimated), it could be obtained through iterations, substituting the value for the one obtained in the previous iteration. That is, in iteration t, we would have $\hat{\sigma}_t^2 = V + (\hat{\sigma}_{t-1}^2/n)$. When the convergence is reached, $\hat{\sigma}_{t-1}^2 \approx \hat{\sigma}_t^2 = \hat{\sigma}^2$ and $n\hat{\sigma}^2 = nV + \hat{\sigma}^2$, resulting in

$$\hat{\sigma}^2 = \frac{n}{n-1}V = \frac{\sum_{i=1}^{n}(y_i - \mu)^2}{n-1}, \tag{6.20}$$

which is already an unbiased estimator of the variance.

6.3.2 REML Estimation with the Animal Model

The animal model, so called because it is generally applied to the phenotypic data of animals for a quantitative trait (obviously, in spite of the name, it is also applicable to plants), is usually expressed by a mixed model, that is, with fixed and random effects that, in its most common form would be

$$\mathbf{y} = \mathbf{Xb} + \mathbf{Za} + \mathbf{e} \tag{6.21}$$

(Henderson, 1984), where $\mathbf{y}$ is the vector of phenotypic values, $\mathbf{X}$ is the incidence matrix (of values 0 and 1) of the vector of fixed effects ($\mathbf{b}$), these being the general mean (μ) and the other fixed effects if any, $\mathbf{Z}$ the incidence matrix of the additive effects of individuals (vector $\mathbf{a}$), and $\mathbf{e}$ the vector of residual errors. Additive effects and residual errors follow a normal distribution $N(0, \sigma_A^2)$ and $N(0, \sigma_R^2)$, respectively.

Consider a small illustrative example corresponding to the data presented in Figure 6.5 that coincides with the first part of the genealogy of Figure 4.3 (Chapter 4). In the example we have the character data in individual A measured in two years, and there are no data for individual B.

In the mixed model of equation (6.21), the vector of fixed effects ($\mathbf{b}$) includes the general mean (μ) and the effect of years 1 and 2 (b_1 and b_2, respectively), the vector of additive effects ($\mathbf{a}$) contains the terms $a_A \ldots a_D$, and the vector of residual effects ($\mathbf{e}$) the terms $e_A \ldots e_D$, which we will assume are not correlated with each other. The expansion of equation (6.21) with the data in the example is then

$$\begin{pmatrix} 110 \\ 200 \\ 180 \\ 130 \end{pmatrix} = \begin{pmatrix} 1 & 1 & 0 \\ 1 & 0 & 1 \\ 1 & 0 & 1 \\ 1 & 1 & 0 \end{pmatrix} \begin{pmatrix} \mu \\ b_1 \\ b_2 \end{pmatrix} + \begin{pmatrix} 1 & 0 & 0 & 0 \\ 1 & 0 & 0 & 0 \\ 0 & 0 & 1 & 0 \\ 0 & 0 & 0 & 1 \end{pmatrix} \begin{pmatrix} a_A \\ a_B \\ a_C \\ a_D \end{pmatrix} + \begin{pmatrix} e_A \\ e_B \\ e_C \\ e_D \end{pmatrix}. \tag{6.22}$$

As we saw in Section 6.1.4, the additive genetic covariance between any two individuals is given by $r\sigma_A^2$ where r is the numerator of the coefficient of additive relationships

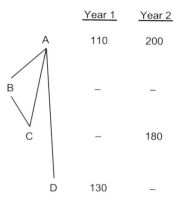

	Year 1	Year 2
A	110	200
B	–	–
C	–	180
D	130	–

Figure 6.5 Weight data in grams of four genealogically related individuals. From individual A, data are available in two successive years, while no data are available for individual B.

between individuals, that is, twice their coancestry, for example 1/2 if they were full sibs. Therefore, the matrix that includes all the additive covariances between individuals would be the variance–covariance matrix $\mathbf{G} = \mathbf{A}\sigma_A^2$, where $\mathbf{A}$ is the matrix with the numerator of the coefficient of additive relationships between individuals (Chapter 4) (we will later discuss the possibility of using the molecular coancestry matrix). This matrix is obtained by multiplying by two the coancestry values between individuals A to D that were calculated in Table 4.1 of Chapter 4. That is, matrix $\mathbf{A}$ is

$$\mathbf{A} = \begin{pmatrix} 1 & 0.5 & 0.75 & 0.5 \\ 0.5 & 1 & 0.75 & 0.25 \\ 0.75 & 0.75 & 1.25 & 0.375 \\ 0.5 & 0.25 & 0.375 & 1 \end{pmatrix}. \tag{6.23}$$

On the other hand, the residual variance–covariance matrix is $\mathbf{R}$, which would be reduced to $\mathbf{I}\sigma_R^2$ (where $\mathbf{I}$ is the identity matrix, with values of one in the diagonal and zero in the rest) if there are no interactions between the residual errors of individuals, as is usually assumed. The matrix of phenotypic variances–covariances is then $\mathbf{P} = \mathbf{Z}\mathbf{G}\mathbf{Z}' + \mathbf{I}\,\sigma_R^2$

$$\mathbf{P} = \begin{pmatrix} 1 & 0 & 0 & 0 \\ 1 & 0 & 0 & 0 \\ 0 & 0 & 1 & 0 \\ 0 & 0 & 0 & 1 \end{pmatrix} \begin{pmatrix} 1 & 0.5 & 0.75 & 0.5 \\ 0.5 & 1 & 0.75 & 0.25 \\ 0.75 & 0.75 & 1.25 & 0.375 \\ 0.5 & 0.25 & 0.375 & 1 \end{pmatrix} \sigma_A^2 \begin{pmatrix} 1 & 1 & 0 & 0 \\ 0 & 0 & 0 & 0 \\ 0 & 0 & 1 & 0 \\ 0 & 0 & 0 & 1 \end{pmatrix}$$

$$+ \begin{pmatrix} 1 & 0 & 0 & 0 \\ 0 & 1 & 0 & 0 \\ 0 & 0 & 1 & 0 \\ 0 & 0 & 0 & 1 \end{pmatrix} \sigma_R^2. \tag{6.24}$$

Then, the vector of phenotypic observations ($\mathbf{y}$) contains the vector of means $\mathbf{Xb}$ and the matrix of phenotypic variances–covariances $\mathbf{P}$. The likelihood function of the data is

$$L(\mathbf{y}|\mathbf{Xb}, \mathbf{P}) = (2\pi)^{-n/2}|\mathbf{P}|^{-\frac{1}{2}} \exp\left[-\frac{1}{2}(\mathbf{y} - \mathbf{Xb})'\mathbf{P}^{-1}(\mathbf{y} - \mathbf{Xb})\right],$$

and the logarithm of the likelihood of the estimates given the observations with the animal model, comparable to equation (6.17), is

$$\ln L(\mathbf{b}, \mathbf{P}|\mathbf{y}, \mathbf{Xb}) = c - \frac{1}{2}\log|\mathbf{P}| - \frac{1}{2}(\mathbf{y} - \mathbf{Xb})'\mathbf{P}^{-1}(\mathbf{y} - \mathbf{Xb}), \tag{6.25}$$

where c is a constant that can be ignored in obtaining the maximum, and $|\mathbf{P}|$ is the determinant of the matrix $\mathbf{P}$. As explained above, the estimates of $\mathbf{b}$ and $\mathbf{P}$ are obtained by taking derivatives in equation (6.25) with respect to these parameters, and equalling to zero. However, the solutions are not obtained in practice by this procedure, but by iterative processes simpler but costly from a computational point of view. As previously mentioned, the maximum likelihood estimates are biased by the existence of fixed effects and are corrected using restricted maximum likelihood (REML), which includes an adjustment in the vector of the observations ($\mathbf{y}$), so that they are made independent of the fixed effects (Patterson and Thompson, 1971; Meyer, 1989). In the case of the proposed example, due to the small number of data involved, it is not possible to obtain reliable estimates.

Model (6.21) can also be extended to other random effects such as those of common environment and indirect genetic effects (due to social interactions between individuals) that include maternal effects, which would imply adding new terms of random effects to the expression. Likewise, it can be extended to dominance effects, if it is assumed that there is no inbreeding, including the matrix of dominance relationships, $\mathbf{D}$, so that the matrix of dominance variances–covariances is $\mathbf{D}\sigma_D^2$ (for more details see chapter 27 of Lynch and Walsh, 1998). Finally, the animal model can be used in the detection of genotype–environment interactions (see Chapter 3), defining the character expressed in different environments as different traits, and determining if the genetic correlation between these is the unit, in which case there would be no $G \times E$ interaction, since, in this case, the allelic effects would be equal or proportional.

It is important to note that the variance component estimates provided by the animal model refer to the base population of the genealogy being considered, and are not affected by biases due to the finite census size, assortative mating, selection or inbreeding in later generations, since the matrix of additive relationships ($\mathbf{A}$) and, where appropriate, that of the dominance ones ($\mathbf{D}$), will take into account these factors. In general, the heritability estimates obtained with the animal model are usually lower than those obtained with the simpler procedures described above (parent-offspring or sib group analyses) probably because the latter methods are not able to eliminate as efficiently as the former the sources of bias of the estimates. In addition, the standard errors of the estimates obtained with the animal model are generally smaller than those obtained with simple methods.

6.3.3 Prediction of Additive Values by BLUP

When data of related individuals are available through complex genealogical relationships, the corresponding prediction of the additive values is made by BLUP (Best Lineal Unbiased Prediction), and the estimation of the fixed effects by BLUE (E for 'Estimation'), which is simply the generalized least squares estimator. Predictions and estimates are a linear function of the data, unbiased and optimal, hence their name.

BLUP allows for the estimation of additive values correcting for any type of fixed effects and using all available information, that is, that corresponding to all individuals related to the individual in question, enabling estimates of the additive value to be obtained even of individuals for which phenotypic data are lacking.

Consider the general model of equation (6.21) and remember that the expected value of the data is $E[\mathbf{y}] = \mathbf{Xb}$ and its phenotypic variance–covariance matrix is $\mathbf{P} = \mathbf{ZGZ'} + \mathbf{R}$, where $\mathbf{G} = \mathbf{A}\sigma_A^2$. In Chapter 3 we indicated that the best linear prediction of the additive value (A) of an individual is h^2P, where h^2 is the heritability of the trait and P the phenotypic value of the individual deviated from the population mean. Since the heritability is the regression of additive values on phenotypic ones, that is, $h^2 = b_{a \cdot y} = \text{cov}(y, a)/\sigma_P^2$, the prediction of the additive value is $\hat{a} = h^2P = \text{cov}(y, a)\sigma_P^{-2}(y - \mu)$ and, applied to the model of equation (6.21), we have that the BLUP prediction of additive values is

$$\hat{\mathbf{a}} = \text{cov}(\mathbf{y'}, \mathbf{a})\mathbf{P}^{-1}(\mathbf{y} - \mathbf{Xb}) = \mathbf{AZ'}\sigma_A^2[\mathbf{ZGZ'} + \mathbf{R}]^{-1}(\mathbf{y} - \mathbf{Xb}). \qquad (6.26)$$

The estimation of fixed effects by regression (generalized least squares estimation) or BLUE is

$$\hat{\mathbf{b}} = (\mathbf{X'P^{-1}X})^{-1}\mathbf{X'P^{-1}y}. \qquad (6.27)$$

As shown by the previous expressions, the prediction of the additive values requires the inversion of matrices $\mathbf{A}$ and $\mathbf{P}$, a complex task when the data to analyse are numerous. Fortunately, Henderson (1976) developed equations equivalent to (6.26) and (6.27) that do not require the inversion of $\mathbf{P}$ and also developed simple rules to obtain the inverse of $\mathbf{A}$ without inverting it, solving an important practical problem that prevented analysis with a large number of data. The mixed model equations of Henderson, which are used in practice for BLUP prediction and BLUE estimation, are

$$\begin{pmatrix} \mathbf{X'R^{-1}X} & \mathbf{X'R^{-1}Z} \\ \mathbf{Z'R^{-1}X} & \mathbf{Z'R^{-1}Z} + \mathbf{A^{-1}}\sigma_A^{-2} \end{pmatrix} \begin{pmatrix} \hat{\mathbf{b}} \\ \hat{\mathbf{a}} \end{pmatrix} = \begin{pmatrix} \mathbf{X'R^{-1}y} \\ \mathbf{Z'R^{-1}y} \end{pmatrix}. \qquad (6.28)$$

If there are no correlated errors, the residual variance–covariance matrix is simplified to $\mathbf{R} = \mathbf{I}\sigma_R^2$ and equation (6.28) is reduced to

$$\begin{pmatrix} \mathbf{X'X} & \mathbf{X'Z} \\ \mathbf{Z'X} & \mathbf{Z'Z} + \mathbf{A^{-1}}(\sigma_R^2/\sigma_A^2) \end{pmatrix} \begin{pmatrix} \hat{\mathbf{b}} \\ \hat{\mathbf{a}} \end{pmatrix} = \begin{pmatrix} \mathbf{X'y} \\ \mathbf{Z'y} \end{pmatrix}. \qquad (6.29)$$

The prediction of additive values requires prior knowledge of the additive and residual variances, which can be estimated, for example, by REML. These variances are, as mentioned above, those of the base population. Throughout the genealogy, the additive variance may change due to selection, census size reduction, and so on. However, these factors will not bias the predictions of the additive values as long as all individuals in the genealogy can be connected with those of generation 0 (that is, if there are no offspring of unknown parents), and when the individuals of the base population are a random sample of the population whose additive variance is used. This can be illustrated with a simple example. Consider the genealogy of Figure 6.5, which will be used in the next section. The additive value of, say individual C, would be half the additive value of its parents plus a deviation due to the Mendelian segregation of heterozygotes, $A_C = \frac{1}{2}A_A + \frac{1}{2}A_B + m_C$. But, in turn, the additive value of B would be half the additive value of its parents, one known (A) and the other unknown (?), supposedly also from the base population, plus the Mendelian deviation, that is, $A_B = \frac{1}{2}A_A + \frac{1}{2}A_? + m_B$. Therefore, substituting, $A_C = \frac{1}{2}A_A + \frac{1}{2}(\frac{1}{2}A_A + \frac{1}{2}A_? + m_B) + m_C = \frac{3}{4}A_A + \frac{1}{4}A_? + \frac{1}{2}m_B + m_C$, and the variance of A_C is then $V(A_C) = (9/16)V(A_A) + (1/16)V(A_?) + (1/4)V(m_B) + V(m_C)$. As we saw in Chapter 4, the Mendelian segregation term of an individual depends on the average coefficient of inbreeding $(\overline{F})$ of its parents $(\frac{1}{2}(1 - \overline{F})$; equation (4.5)) and, being σ_A^2, the additive variance of the base population,

$$
\begin{aligned}
V(A_C) &= \frac{9}{16}\sigma_A^2 + \frac{1}{16}\sigma_A^2 + \frac{1}{4}\sigma_A^2\left[\frac{1}{2}\left(1 - \frac{F_A}{2}\right)\right] + \sigma_A^2\left[\frac{1}{2}\left(1 - \frac{F_A + F_B}{2}\right)\right] \\
&= \frac{20}{16}\sigma_A^2 = 1.25\sigma_A^2 = (1 + F_C)\sigma_A^2,
\end{aligned}
\tag{6.30}
$$

since $F_A = F_B = 0$ in the example. That is, the variance of the additive value of individual C is equal to the additive variance of the base population modulated by the inbreeding coefficient of individual C (equation (6.23)). This property of BLUP is fundamental, given that it implies obtaining unbiased predictions even though the analysed population is subject to different processes such as selection, non-random mating, or changes in the population census size (Sorensen and Kennedy, 1984).

The prediction of dominance values can also be made by BLUP by introducing the corresponding dominance effects vector and the matrix of dominance relationships (Henderson, 1985; Lynch and Walsh, 1998, chapter 26).

6.3.4 Example of BLUP Prediction and BLUE Estimation

We will illustrate below the prediction of additive values by BLUP and the estimation of fixed effects by BLUE corresponding to the example of Figure 6.5. The mixed model is given by equations (6.21) and (6.22). Assume, for simplicity, that residual effects are only environmental in this case and follow a normal distribution $N(0, \sigma_E^2)$. Since we could not estimate the components of variance with such a small example, suppose that the base population from which individual A comes has variances $\sigma_A^2 = \sigma_E^2 = 0.5$. The terms of equations (6.29) of Henderson's mixed model using the matrices presented in equation (6.22) are

$$
\mathbf{X'X} = \begin{pmatrix} 1 & 1 & 1 & 1 \\ 1 & 0 & 0 & 1 \\ 0 & 1 & 1 & 0 \end{pmatrix} \begin{pmatrix} 1 & 1 & 0 \\ 1 & 0 & 1 \\ 1 & 0 & 1 \\ 1 & 1 & 0 \end{pmatrix} = \begin{pmatrix} 4 & 2 & 2 \\ 2 & 2 & 0 \\ 2 & 0 & 2 \end{pmatrix},
$$

$$
\mathbf{X'Z} = \begin{pmatrix} 1 & 1 & 1 & 1 \\ 1 & 0 & 0 & 1 \\ 0 & 1 & 1 & 0 \end{pmatrix} \begin{pmatrix} 1 & 0 & 0 & 0 \\ 1 & 0 & 0 & 0 \\ 0 & 0 & 1 & 0 \\ 0 & 0 & 0 & 1 \end{pmatrix} = \begin{pmatrix} 2 & 0 & 1 & 1 \\ 1 & 0 & 0 & 1 \\ 1 & 0 & 1 & 0 \end{pmatrix},
$$

$$
\mathbf{Z'X} = \begin{pmatrix} 1 & 1 & 0 & 0 \\ 0 & 0 & 0 & 0 \\ 0 & 0 & 1 & 0 \\ 0 & 0 & 0 & 1 \end{pmatrix} \begin{pmatrix} 1 & 1 & 0 \\ 1 & 0 & 1 \\ 1 & 0 & 1 \\ 1 & 1 & 0 \end{pmatrix} = \begin{pmatrix} 2 & 1 & 1 \\ 0 & 0 & 0 \\ 1 & 0 & 1 \\ 1 & 1 & 0 \end{pmatrix},
$$

$$
\mathbf{Z'Z} = \begin{pmatrix} 1 & 1 & 0 & 0 \\ 0 & 0 & 0 & 0 \\ 0 & 0 & 1 & 0 \\ 0 & 0 & 0 & 1 \end{pmatrix} \begin{pmatrix} 1 & 0 & 0 & 0 \\ 1 & 0 & 0 & 0 \\ 0 & 0 & 1 & 0 \\ 0 & 0 & 0 & 1 \end{pmatrix} = \begin{pmatrix} 2 & 0 & 0 & 0 \\ 0 & 0 & 0 & 0 \\ 0 & 0 & 1 & 0 \\ 0 & 0 & 0 & 1 \end{pmatrix},
$$

$$
\mathbf{X'y} = \begin{pmatrix} 1 & 1 & 1 & 1 \\ 1 & 0 & 0 & 1 \\ 0 & 1 & 1 & 0 \end{pmatrix} \begin{pmatrix} 110 \\ 200 \\ 180 \\ 130 \end{pmatrix} = \begin{pmatrix} 620 \\ 240 \\ 380 \end{pmatrix}
$$

$$
\mathbf{Z'y} = \begin{pmatrix} 1 & 1 & 0 & 0 \\ 0 & 0 & 0 & 0 \\ 0 & 0 & 1 & 0 \\ 0 & 0 & 0 & 1 \end{pmatrix} \begin{pmatrix} 110 \\ 200 \\ 180 \\ 130 \end{pmatrix} = \begin{pmatrix} 310 \\ 0 \\ 180 \\ 130 \end{pmatrix}. \tag{6.31}
$$

The numerator additive relationship matrix ($\mathbf{A}$) was presented in equation (6.23) and its inverse is

$$
\mathbf{A}^{-1} = \begin{pmatrix} 2.167 & -0.167 & -1 & -0.667 \\ -0.167 & 1.833 & -1 & 0 \\ -1 & -1 & 2 & 0 \\ -0.667 & 0 & 0 & 1.333 \end{pmatrix}. \tag{6.32}
$$

Therefore, the terms of equations (6.29) are

$$
\left(\begin{array}{ccc:cccc} 4 & 2 & 2 & 2 & 0 & 1 & 1 \\ 2 & 2 & 0 & 1 & 0 & 0 & 1 \\ 2 & 0 & 2 & 1 & 0 & 1 & 0 \\ \hdashline 2 & 1 & 1 & 2+2.167 & -0.167 & -1 & -0.667 \\ 0 & 0 & 0 & -0.167 & 1.833 & -1 & 0 \\ 1 & 0 & 1 & -1 & -1 & 1+2 & 0 \\ 1 & 1 & 0 & -0.667 & 0 & 0 & 1+1.333 \end{array} \right) \begin{pmatrix} \mu \\ b_1 \\ b_2 \\ \hdashline a_A \\ a_B \\ a_C \\ a_D \end{pmatrix} = \begin{pmatrix} 620 \\ 240 \\ 380 \\ \hdashline 310 \\ 0 \\ 180 \\ 130 \end{pmatrix}. \tag{6.33}
$$

Expression (6.33) leads to seven equations with seven unknowns:

$4\mu + 2b_1 + 2b_2 + 2a_A + a_C + a_D = 620;$

$2\mu + 2b_1 + a_A + a_D = 240;$

$2\mu + 2b_2 + a_A + a_C = 380;$

$2\mu + b_1 + b_2 + 4.167a_A - 0.167a_B - a_C - 0.667a_D = 310;$

$-0.167a_A + 1.833a_B - a_C = 0;$

$\mu + b_2 - a_A - a_B + 3a_C = 180;$

$\mu + b_1 - 0.667a_A + 2.333a_D = 130.$

Since the first is the sum of the second and third, there are six independent equations with seven unknowns. We can give a value of zero to the fixed effect of the first year ($b_1 = 0$), so that the resulting system of six equations and six unknowns is

$$\begin{aligned}
\mu &= (240 - a_A - a_D)/2;\\
b_2 &= (380 - 2\mu - a_A - a_C)/2;\\
a_A &= (310 - 2\mu - b_2 + 0.167a_B + a_C + 0.667a_D)/4.167;\\
a_B &= (0.167a_A + a_C)/1.833;\\
a_C &= (180 - \mu - b_2 + a_A + a_B)/3;\\
a_D &= (130 - \mu + 0.667a_A)/ 2.333.
\end{aligned}$$

(6.34)

The equations can be solved directly by solving the unknowns in (6.33), after eliminating the first equation and giving a value $b_1 = 0$. That is,

$$\begin{pmatrix} \mu \\ b_2 \\ a_A \\ a_B \\ a_C \\ a_D \end{pmatrix} = \begin{pmatrix} 2 & 0 & 1 & 0 & 0 & 1 \\ 2 & 2 & 1 & 0 & 1 & 0 \\ 2 & 1 & 4.167 & -0.167 & -1 & -0.667 \\ 0 & 0 & -0.167 & 1.833 & -1 & 0 \\ 1 & 1 & -1 & -1 & 3 & 0 \\ 1 & 0 & -0.667 & 0 & 0 & 2.333 \end{pmatrix}^{-1} \begin{pmatrix} 240 \\ 380 \\ 310 \\ 0 \\ 180 \\ 130 \end{pmatrix}.$$

(6.35)

The results are as follows:

$$\mu = 118.58;\ b_1 = 0;\ b_2 = 75.43;\ a_A = -1.59;\ a_B = -3.64;\ a_C = -6.41;\ a_D = 4.44.$$

However, obtaining the inverse required in (6.35) can be complicated if the size of the matrix is large. It is more common in practice to solve the system of equations (6.34) by an iterative process. For example, if we give zero value to all the unknowns in the first iteration, we obtain $\mu = 120$; $b_2 = 190$; $a_1 = 74.4$; $a_2 = 0$; $a_3 = 60$; and $a_4 = 55.72$. If we now introduce these values back into the equations we get a new list of values and so on, until an asymptotic solution is obtained. A convergence criterion must be set, for example that the change between two successive iterations is less than 0.01, as seen in Table 6.7 where the results of some iterations are shown. The asymptotic results coincide with the exact ones obtained previously. Thus, the general mean is $\mu = 118$ g, with a fixed effect of the second year of 75 g relative to the first, and the additive effects of the four individuals (A – D), deviated from the general mean and corrected by the year effect are, approximately, −1.6 g, −3.6 g, −6.4 g and 4.4 g, respectively.

6.3.5 Use of Molecular Coancestries with Genetic Markers

The availability of high-density genetic markers (using SNP chips or sequencing data, Pérez-Enciso, 2014) allows for the use of the molecular coancestry matrix (described in Chapter 4) as an alternative to the genealogical coancestry matrix, to carry out the estimation of variance–covariance components and predictions of additive and dominance values. We speak then of GREML and GBLUP (Meuwissen et al., 2001; Clark and van der Werf, 2013; see Chapter 11).

Table 6.7 *Result of iterations to obtain the solutions to equations (6.34)*

Unknown	Iter. 1	Iter. 2	Iter. 3	...	Iter. 40	...	Iter. 59	Iter. 60
μ	120	54.94	109.97		118.23		118.54	118.54
b_2	190	2.80	147.07		75.43		75.43	75.43
a_A	74.4	−5.49	48.57		−1.25		−1.56	−1.56
a_B	0	39.49	−10.61		−3.39		−3.62	−3.62
a_C	60	−18.53	52.09		−6.06		−6.37	−6.38
a_D	55.72	25.54	30.60		4.72		4.47	4.47

The molecular additive relationship matrix ($\mathbf{A_M}$) has the advantage of taking into account the Mendelian segregation, allowing us to distinguish, for example, between full sibs of the same family, so it can be used instead of the genealogical additive relationship matrix, $\mathbf{A}$, or combined with it (Legarra et al., 2009). A frequent situation in domestic animals and plants is that genealogical information of all individuals is available, but only a certain fraction of them has been genotyped for genetic markers. In this case, non-genotyped individuals can be distinguished from genotyped ones, and a matrix of mixed additive relationships can be used (Vitezica et al., 2011).

There are also other simpler methods for estimating genetic variance components with markers. Ritland (1996) proposed a heritability estimator when phenotypic data for a character and information on genetic markers are available in the absence of genealogies. The similarity between the phenotypic values (P) of a given trait between individuals i and j can be defined as $Z_{ij} = [(P_i − \bar{P})(P_j − \bar{P})]/\sigma_P^2$, where $\bar{P}$ and σ_P^2 are the mean and variance of the trait in the population. If the similarity between the phenotypes is due to the genes and the environment shared by the individuals, we can express $Z_{ij} = 2f_{ij}h^2 + r_e + e_{ij}$, where f_{ij} is the coancestry between individuals, which can be estimated with the markers (f_M), h^2 is the trait heritability, r_e is the possible correlation due to the environment common to individuals, and e_{ij} is the residual error. Therefore, the heritability can be obtained as the regression of the phenotypic similarities on the coancestries,

$$\hat{h}^2 = \frac{\text{cov}(Z, f_M)}{2\,\text{var}(f_M)}, \tag{6.36}$$

where var(f_M) is the variance of molecular coancestries. This method of estimation has the advantage of its great simplicity. However, it is not very accurate if the individuals used have little relatedness and the population is not structured in family groups (Rodríguez-Ramilo et al., 2007), as can occur in large natural populations, precisely the situation in which its use would be most recommended.

Another application of the animal model using genetic markers is the detection of loci that affect a quantitative trait and, for this, the effect of the locus in question is included as an additional random effect (George et al., 2000). Suppose that genotypic data of molecular markers are available, so that an animal model with traditional polygenic effects plus the effect of a specific marker can be considered, extending equation (6.21), we have

$$\mathbf{y} = \mathbf{Xb} + \mathbf{Za} + \mathbf{Zq} + \mathbf{e}, \tag{6.37}$$

where $\mathbf{q}$ is the vector of additive effects of the individuals associated with the marker. The polygenic additive effects are given, as before, by the variance–covariance matrix $\mathbf{A}\sigma_A^2$, where $\mathbf{A}$ is the genealogical additive relationship matrix, and the effects of the marker with matrix $\mathbf{A_M}\sigma_A^2$, where $\mathbf{A_M}$ is the additive relationship matrix for the marker (whose terms would be double the molecular coancestry between individuals). If the likelihood of model (6.21) is compared to that of model (6.37) by means of the likelihood ratio test, it can be determined if the considered marker has an associated effect on the trait, and this analysis can be repeated for all available markers. The discovery of a significant marker would suggest that this is associated in linkage disequilibrium with the locus that affects the trait and, knowing the genomic location of the marker, it is possible to deduce the location of the locus. In Chapter 11 we will study other methods for the detection of loci that affect quantitative traits using information from molecular markers.

6.3.6 Comparison between Estimates of Heritability with Genealogical and Molecular Data

The availability of SNPs for many species makes it possible to obtain estimates of heritability that can be compared with those obtained using designs of groups of relatives and their expected genealogical relationship. For example, chips with hundreds of thousands or millions of SNPs are available in humans, and numerous estimates of heritability for multiple characters have been derived from them. As we will see in Chapter 11, with SNP chips, genome-wide association studies (GWAS) are carried out aimed at trying to detect SNPs whose segregation is associated by linkage disequilibrium to some locus with effect on a particular character (causal locus). In this way it is possible to detect SNPs associated with the character and assign them a certain effect, so that, deducing its frequency in the population from the sample, it is possible to determine its contribution to heritability and compare it with that estimated by traditional methods. For example, as we mentioned earlier, the estimates of heritability for human height using twin data range between 0.6 and 0.8. Using GWAS and combining data from more than 250,000 individuals, 423 loci associated with the trait have been detected (Wood et al., 2014). However, the contribution of these variants only explains about 20% of the heritability obtained by twin studies. This difference, which is similar or even greater for other traits, has been called the 'missing heritability' problem and different arguments have been suggested to explain it (Maher, 2008), which will be discussed in Chapter 11. The most accepted is that many of the loci that control the character do not have an effect large enough so as to detect their association with a given SNP, or that the linkage disequilibrium between causal loci and SNPs is not strong enough for detection, perhaps because many of these loci are found at low frequencies in the population.

The estimation of the heritability can also be carried out using the information of all the SNPs of the chip, constructing the molecular coancestry matrix with them and making an estimation of the heritability by maximum likelihood. When this procedure has been carried out, estimates have been obtained that reach 50% of the heritability in the case of height and proportions similar or even lower for other traits (Vinkhuyzen et al., 2013; Zaitlen et al., 2013). The proportion of heritability that cannot yet be explained even using the molecular information of hundreds of thousands of SNPs is attributed to the incomplete linkage disequilibrium between SNPs and causal loci of the trait (Yang et al., 2010).

The accuracy of heritability estimates obtained from molecular marker data is much lower than that from designs that use genealogical information, and it is inversely proportional to the variance of the additive relationships between the individuals used in the estimation. Vinkhuyzen et al. (2013) describe the most relevant aspects. Consider first a design of full-sib pairs, as is usual in human data analyses. The expected coefficient of additive relationships is $r = 0.5$, which indicates, as already explained, the expected proportion of genes shared by two sibs. However, using molecular markers, a direct estimate of this proportion can be obtained. With this design of full-sib pairs in which additive relationships are estimated by markers, the variance of the heritability estimate is approximately $2/(n \times V(r_M))$, where n is the number of pairs and $V(r_M)$ is the variance of the additive relationships. This variance can be theoretically approximated by $V(r_M) \approx 1/(16L) - 1/(3L^2)$, where L is the total length of the genetic map in Morgans (M). In humans, the total map for the 22 autosomes is $L \approx 35$ M, so $V(r_M) \approx 0.04^2$. That is, the standard deviation of the expected proportion of genes shared by full sibs (0.5) is approximately 4%. As there are few cases of recombination per chromosome, the estimation can be made with few markers per chromosome. The estimate of heritability with this design has, therefore, an approximate standard error of $SE[h^2] = \sqrt{2/(n \times V(r_M))} \approx 35/\sqrt{n}$, so that a large number of pairs is needed to obtain accurate estimates of heritability with this method. Note that this error considerably exceeds that which would be obtained using genealogical estimates. In this case, we saw from equation (6.8) that, with groups of pairs of full sibs, $\sigma_t^2 \approx [(1 - t^2)^2]/n$ with which $SE[h^2] = 2(1 - t^2)/\sqrt{n}$ and, therefore, its maximum value is $SE[h^2] = 2/\sqrt{n}$. Obviously, genealogical estimates make a series of assumptions that are not necessary with molecular estimates. These reflect with greater reality the degree of coancestry between relatives, but are subject to a lower precision. For example, an analysis of 3375 full-sib pairs found an average molecular coancestry of 0.498 (range between 0.374 and 0.617) and a standard deviation of 0.036. The estimated heritability was 0.8, very similar to the genealogical estimates, but with a wide confidence interval (0.46 to 0.85).

If the design, instead of being made with groups of full sibs, was done with distantly related individuals, the variance of the heritability estimate would be much greater than in the case of a full-sib design, fundamentally because the value of $V(r_M)$ is lower. A great advantage of heritability estimates using genomic data is that it is possible to obtain estimates from a single full-sib family.

Problems

6.1 The aim is to carry out an experiment to estimate the heritability of a trait using pairs of parents and offspring. It is expected that the estimate of heritability will be around 0.6. It is intended to evaluate only one offspring per couple and a single parent or both. How many pairs of data would have to be evaluated, in each case, to obtain an estimate of heritability with a standard error equal to or less than 0.05?

6.2 An analysis of families has provided a value of the phenotypic correlation between full sibs of $t_{FS} = 0.12$. (a) What is the estimate of heritability that can be obtained with this data? (b) It is later discovered that the parents of the families did not mate randomly but with positive assortative mating for the character under study, having estimated that the phenotypic correlation between the individuals of the pairs is $\rho = 0.5$. How is the estimate of heritability modified? (c) If the phenotypic correlation between the pairs were maximal, what would be the value of the heritability?

6.3 In an analysis of mono- (MZ) and dizygotic (DZ) twins, the following components of the variance between pairs (B), within pairs (W) and total (T) have been obtained: $\sigma_B^2 = 0.41$, $\sigma_W^2 = 0.18$, $\sigma_T^2 = 0.59$ for monozygotic twins and $\sigma_B^2 = 0.25$, $\sigma_W^2 = 0.32$, $\sigma_T^2 = 0.57$ for dizygotic twins. What estimates of genetic components can these data provide?

6.4 The weights (P_i) of four mice are available: A (16.6 g), B (22.1 g), C (18 g) and D (12.4 g). Also, using genetic markers, we have their molecular coancestries ($f_{M,ij}$), the coancestry matrix being that indicated below. Estimate the heritability of weight.

$f_{M,ij}$	A	B	C	D
A	1	0.25	0.5	0.5
B	0.25	1	0.75	0.5
C	0.5	0.75	1.25	0.375
D	0.5	0.5	0.375	1

6.5 The aim is to estimate the heritability of a human trait using a design of full-sib pairs, estimating genomic relationships with molecular markers. How many pairs should be evaluated to obtain a standard error of the heritability estimate equal to or less than 0.1?

Self-Assessment Questions

1 The estimate of heritability obtained from the regression of offspring values on that of their parents will necessarily be biased if the latter are not a random sample of the population.

2 The variance of the means of full-sib families is equal to the covariance of the values of the members of the families.

3 The standard errors of heritability estimates obtained from full-sib families are always lower, the lower the value of the heritability.

4 For a given total number of evaluated individuals, the estimates of heritability by regression have less error if only one offspring per family is evaluated.

5 The coefficient of dominance relationship between two cousins is 1/8.

6 The phenotypic correlation is the sum of the genetic and environmental correlations.

7 The estimates of variance obtained by maximum likelihood are biased because they do not take into account the degrees of freedom lost when the fixed effects are estimated.

8 The prediction of additive values by BLUP is not biased by the fact that there is selection in the evaluated population.

9 Given that the estimates of heritability obtained by twin studies include dominance and epistatic components of variance, this may contribute to the fact that estimates provided by this method are higher than those obtained by means of molecular markers.

10 To obtain an estimate of heritability with a standard error of approximately 0.05 using molecular markers in pairs of full sibs in humans, of the order of 500,000 pairs would be needed.

7 Mutation

Concepts to Study

- Rate of mutation
- Mutational variance and heritability
- Probability of fixation of a mutation
- Quasi-neutral mutation
- Multiplicative model
- Synergistic and antagonistic epistasis
- Neutral mutation–drift and deleterious mutation–selection equilibria
- Mutation load
- Muller's Ratchet
- Hill–Robertson effect

Objectives for Learning

- To understand the importance of mutation in the evolutionary context and in the genetic improvement of plants and animals
- To learn how to estimate mutational parameters through mutation accumulation experiments, genomic estimates and fitness data in natural populations
- To understand the concept of mutational variance and its estimation, as well as the concepts of mutational heritability and mutational coefficient of variation
- To know the factors that determine the fixation of a mutation in a population of finite size
- To understand the concept of quasi-neutral mutation
- To distinguish between the different multilocus fitness models (additive, multiplicative and epistatic: synergistic and antagonistic)
- To know the predictions of genetic variation under the neutral mutation–drift model
- To understand the implications of the balance between deleterious mutation and selection
- To learn the concept of mutation load and its consequences on the population
- To understand the process called Muller's Ratchet
- To know the Hill–Robertson effect and its implications

7.1 Estimation and Analysis of Mutation in Quantitative Traits

Mutation is the source of population genetic variation on which natural or artificial selection acts to produce genetic changes leading to adaptive evolution or the economic improvement of plants and animals. In the case of single loci affecting qualitative traits or genes of major effect on quantitative traits the estimation of the frequency or rate at which mutations appear per generation is relatively simple for dominant mutations, since it is based directly on the count.

For example, if from 1 million births of phenotypically normal parents for the achondroplasia allele (a dominant mutation producing dwarfism) 10 individuals appear with the disease, the mutation rate per locus and generation will be $u = 10/(2 \times 10^6) = 0.5 \times 10^{-5}$, where the factor 2 of the denominator stems from the fact that each individual carries two alleles. For recessive alleles, calculations are not so straightforward and require certain assumptions about the frequency of the mutations hidden as heterozygotes in parents whose offspring are evaluated.

Typical mutation rates of individual loci with easily identifiable phenotypic effects in eukaryotes, such as colour mutations, allozyme variants or blood groups, are of the order of $u = 10^{-5} - 10^{-6}$ and the rate of nucleotide mutation per generation is about 10^{-8}, although it has a range of variation between 10^{-7} and 10^{-9} depending on the species (Kondrashov and Kondrashov, 2010). The mutation process in the germ line occurs during replication and mutation rates per nucleotide and round of replication are of the order of 5×10^{-11} (Drake et al., 1998). These data are compatible with the locus mutation rate mentioned above if multiplied by the number of nucleotides per gene (say 1000 as a typical value for coding genes) and by the number of rounds of replication from the appearance of the zygote to the formation of the gametes (between 10 and 100 in eukaryotes). The number of rounds of replication is precisely one of the reasons why the rate of mutation can vary between species. An interesting case in this respect, described by Crow (2000), is the disparity in the mutation rate between men and women. In the latter the number of rounds of replication from the zygote to birth, when all the eggs are already preformed, is 22. The female sex cells are then latent until fertilization, when the meiosis ends. In men, however, the production of spermatozoa is continuous throughout life. Thus, although the number of cell divisions of gametes is about 30 from the zygote to puberty, it increases continuously from that moment. For example, it is about 150 rounds at 20 years of age and about 610 at 40. Therefore, at this last age, the mutation rate in male gametes is $610/22 \approx 28$ times higher than in ova. In contrast, the rate of chromosomal mutations (such as breaks, inversions or translocations), which is in the order of 0.08 per haploid genome and generation in humans (Lynch, 2016), increases with mother's age, perhaps because of the cytoplasmic deterioration that occurs in oocytes over time.

The analysis of mutation in quantitative traits is complex, since the identification of the genotypes is not easy. Therefore, it is not possible to estimate the mutation rate of individual loci but of the whole set of loci that affect a trait and, with exceptions, only average estimates of the effect of mutations and their degree of dominance can be obtained. A parameter that is frequently used to quantify the impact of mutation on quantitative traits is the mutational variance (V_M), which is the increase in additive genetic variance per generation due to mutation. The ratio between this and the environmental variance constitutes the mutational heritability $h_M^2 = V_M/V_E$. In the case of the major fitness components, for which V_E is usually very high, it may be more appropriate to estimate the coefficient of mutational variation (CV_M; Houle et al., 1996), that is, the square root of the mutational variance divided by the mean of the trait. Knowledge of mutational parameters is fundamental, since the genetic constitution of a population and the consequences of inbreeding on the mean and variance of quantitative traits, which we will study in the next chapter, depend on them.

From an evolutionary perspective, the focus is primarily on mutations that affect fitness or its main components. In this respect we can distinguish beneficial, neutral and deleterious mutations. In general, by considering functional regions of the genome, it is assumed that most of the mutations occurring in such regions are deleterious, that is, they reduce the fitness of their carriers, basically because a random change in the sequence of a gene or expression regulatory element is much more likely to negatively disrupt its functionality than to cause a selective advantage to its carrier. Hence, this is the type of mutation that has the largest volume of empirical data. The estimation of mutational parameters is carried out by mutation accumulation experiments or by inferences from genomic data, which will be described later. However, in order to understand the basis of these methods we must first analyse the conditions that determine the removal or fixation of mutations in a population.

7.1.1 Probability of Fixation of a Mutation

As explained in Chapter 2, in a population of theoretical infinite size, mutations with a beneficial effect for fitness increase in frequency through the action of natural selection until they become fixed. Moreover, deleterious mutations decrease in frequency until their elimination. In finite size populations, however, where genetic drift comes into play, there is a certain probability that some advantageous mutations will be eventually lost and some deleterious ones will be fixed. Kimura (1957) applied diffusion theory (so called because it is applied in diffusion models of gases in Physics) to obtain various predictions about the evolution of allele frequencies in models governed by the forces of mutation, selection and drift. Among these predictions is that of the probability of fixation (P_f) of an allele with initial frequency q_0 in a stable size population. Assume the general fitness model in Table 2.3 (Chapter 2) by changing the signs of genotypic values only for the purposes of this section to emphasize that beneficial mutations increase fitness, so that the three genotypes AA, Aa and aa have fitnesses 1, $1 + sh$ and $1 + s$, respectively, where s is the selection coefficient associated with the allele under study (a), which may be advantageous ($s > 0$) or deleterious ($s < 0$), and whose initial frequency is q_0. If, for simplicity, we assume additive gene action ($h = 0.5$), the probability of allele fixation can be approximated by $P_f = (1 - e^{-2N_e s q_0})/(1 - e^{-2N_e s})$ which, in the case of a mutation, that is, an allele with initial frequency $q_0 = 1/2N$ in the population, is reduced to

$$P_f = \frac{1 - e^{-s(N_e/N)}}{1 - e^{-2N_e s}}, \qquad (7.1)$$

an expression also deduced by Malécot (1952).

When the mutational selection coefficient is small ($N_e s/N \ll 1$) and the product $N_e s$ is high ($N_e s \gg 1$), the numerator of equation (7.1) can be approximated by $s(N_e/N)$ and the denominator by 1, so that $P_f \approx s(N_e/N)$ and, if $N_e \approx N$, we arrive at $P_f \approx s$. That is, the probability of fixation of an additive mutation with selective advantage s is approximately equal to that advantage, a result obtained initially by Haldane (1927). Surprisingly, then, most beneficial mutations of small effect will be lost in a finite size population. The most general approximations for different dominance coefficients are presented in Table 7.1. If the population is not panmictic but has partial inbred matings, such as partial selfing, the general approximation of the probability of fixation for a favourable non-recessive mutation is $P_f \approx 2(N_e/N)s(\alpha + h - \alpha h)$ (Caballero and Hill, 1992b), where α is the Hardy–Weinberg disequilibrium factor. Note that in a population with partial selfing and Poisson distribution of offspring number, $N_e = N/(1 + \alpha)$ (Section 5.2.5 of Chapter 5), so that $P_f \approx 2s(\alpha + h - \alpha h)/(1 + \alpha)$ and, therefore, if the gene action is additive ($h = 0.5$), the probability of fixation is s, irrespective of the value of α, and the same is true if $\alpha = 1$, independently of the value of h.

Neutral mutations have a probability of fixation equal to their initial frequency ($1/2N$), an average fixation time of approximately $4N_e$ generations, and an average time until their elimination of $2(N_e/N) \ln (2N)$ generations. Therefore, their mean persistence time in the population is $[(1/2N) \times 4N_e] + [(1 - 1/2N) \times 2(N_e/N) \ln (2N)] = 2(N_e/N)[1 + \ln (2N)]$ generations.

When the mutation is deleterious and the product $N_e s \ll -1$, its destiny is almost invariably elimination by selection. If the mutation is not completely recessive, the number of copies contributed until its elimination is $-1/sh$ and its mean persistence time is given by $2[\ln(-1/sh) + 1 - \gamma]$ (Kimura and Ohta, 1969; García-Dorado et al., 2003), where $\gamma \approx 0.5772$ is Euler's constant.

For those mutations with an effect such that the absolute value of $N_e s$ is substantially less than 1, the numerator of equation (7.1) is approximated by $s(N_e/N)$ and the denominator by $2N_e s$, so that the probability of fixation is $P_f \approx 1/2N$, that is, the same as that for a strictly neutral mutation ($s = 0$). The reason is that, both when the mutation is advantageous ($s > 0$) or deleterious ($s < 0$), the action of genetic drift is more intense than that of selection and, therefore, the fate of the mutation depends mainly on chance. Under these conditions the mutations are called quasi-neutral. A more accurate approximation of the probability of fixation of this type of mutations due to Robertson (1960) is presented in Table 7.1.

To illustrate these predictions, consider a favourable additive mutation with effect $s = 0.01$ in a population with size $N_e = N = 1000$. Its probability of fixation is approximately equal to its selective value ($P_f \approx 0.01$). If the mutation were deleterious ($s = -0.01$; $sh = -0.005$) its probability of fixation would be negligible ($P_f \approx 10^{-11}$), the total number of copies before its elimination would be $-1/sh = 200$ as an average, and its mean persistence time

Table 7.1 *Prediction of the probability of fixation and mean persistence time measured in generations of a mutation with selection coefficient s and dominance coefficient h in a population of size N and effective size N_e*

Mutation	Condition	Approximate fixation probability	Mean persistence time
Favourable no recessive ($h > 0$)	$N_e sh \gg 2$	$P_f \approx \dfrac{2N_e sh}{N}$	
Favourable recessive ($h = 0$)	$N_e s \gg 1$	$P_f \approx \dfrac{\sqrt{2N_e s/\pi}}{N}$	
Neutral	$s = 0$	$P_f = \dfrac{1}{2N}$	$2(N_e/N)[1 + \ln(2N)]$
Quasi-neutral	$-1 < |N_e s| < 1$	$P_f \approx \dfrac{1}{2N} + \dfrac{N_e s}{3N}(1 + h)$	
Deleterious	$N_e s \ll -1$	$P_f \approx 0$	$2[\ln(-1/sh) + 1 - \gamma]$ $\gamma \approx 0.5772$

Note: The model assumes three genotypes *AA, Aa* and *aa* with fitnesses 1, $1 + sh$ and $1 + s$, respectively. Thus, beneficial mutations imply $s > 0$, and deleterious ones imply $s < 0$.

$2[\ln(-1/sh) + 1 - \gamma] = 11.44$ generations, which would have been about $2[1 + \ln(2000)] = 17.20$ generations if it were neutral instead of deleterious.

However, if the population has a much smaller size, for example $N_e = N = 10$, the probability of fixation of the same mutation would be 0.055 if it is beneficial and 0.045 if it is deleterious, becoming a quasi-neutral mutation with a probability of fixation close to the neutral one (0.05) and an average persistence of $2[1 + \ln(20)] = 8$ generations. Thus, a given mutation can behave as beneficial or harmful, or as quasi-neutral, depending on the effective size of the population in which it appears.

7.1.2 Estimation of the Rate of Mutation and Mutational Effects

As we have already indicated, mutational parameters, such as the mutation rate, the mean selection coefficient, the mean degree of dominance and the mutational variance, can be estimated by mutation accumulation experiments, and some of them through inferences obtained from genomic data. We will first describe the principles on which the former are based and the results obtained for different species and traits, particularly in the case of fitness components. Since the vast majority of mutations that affect fitness are deleterious, this is the type of mutation that is most frequently studied.

In the 1960s and 1970s, Mukai and coworkers carried out mutation accumulation experiments with *Drosophila melanogaster* using the balanced chromosome technique described in Chapter 5 (Figure 5.2) and initially proposed by Muller (1928). Basically, the method consists of extracting a single chromosome II from a given population by crossing an individual from that population with another from a strain with balanced chromosomes (such as *Cy/If* in Figure 5.2, or another analogous one with chromosomes carrying inversions that inhibit recombination, and visible markers for identification). From the cross we obtain a male of *Cy/+* genotype for chromosome II, where + refers to the wild chromosome, and this is crossed again with females of the *Cy/If* strain to obtain copies of the chromosome with which to found multiple independent lines. Each line is kept by crossing a single *Cy/+* male with *Cy/If* females, whereby the chromosome passes intact in heterozygosis from generation to generation except for the possible appearance of new mutations. In the different lines, mutations will accumulate and will only be lost if their deleterious heterozygous effects are large enough. After many generations, the chromosome fitness of each line can be evaluated in homozygosis by crossing males and females *Cy/+* from that line, as described in Figure 5.2a, or in heterozygosis, crossing *Cy/+* individuals of different lines, as illustrated in Figure 5.2b.

This method can only be carried out in species where chromosome manipulation is possible, but from the 1990s experiments were started using lines of size $N = 1$ (self-fertilization) or $N = 2$ (brother–sister mating), where the accumulation of mutations for the whole genome is evaluated, rather than for a single chromosome. Figure 7.1 illustrates a typical mutation accumulation experiment for a species with hermaphrodite reproduction, such as the nematode worm *Caenorhabditis elegans* (in this species males may also be present in a low proportion, but their occurrence can be avoided experimentally). In contrast to the balanced chromosome method, this procedure does not prevent recombination and mutations are expressed in homozygotes, increasing the efficiency of natural selection in their elimination. However, we recall that mutations behave as quasi-neutral if $N_e s \ll 1$ (again and hereafter we refer to the general

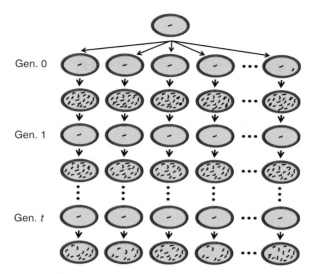

Figure 7.1 Mutation accumulation experiment design for a hermaphrodite species, such as *Caenorhabditis elegans*. From a single initial worm taken from a population that has gone through many generations of self-fertilization, so that it will be theoretically homozygous for the genome as a whole, descendants are taken to found the zero generation of numerous lines, each of which will be constituted by a single individual ($N = 1$) genetically identical for all of them. In each generation, the productivity (number of descendants) is evaluated, and the line is maintained from a single individual taken at random from the offspring. The accumulation of deleterious mutations will imply a reduction in the average productivity and an increase in its variance between lines with the course of generations, from whose magnitudes mutational parameters can be estimated.

model of deleterious mutations with fitnesses 1, $1 - sh$ and $1 - s$, respectively). Therefore, in a self-fertilization line ($N_e = 1$) all mutations with effect $s \ll 1/2$ behave approximately as neutral and would have a probability of fixation of $\sim 1/2$. Also, in brother–sister lines ($N_e \approx 2.5$) the same could be said of all mutations with $s \ll 2/5$, whose probability of fixation would be $\sim 1/4$. Mutations that are studied with these experiments may be spontaneous or induced with chemical or physical mutagens or through induced mobility of transposable elements (Mackay et al., 2001), but we will only focus on spontaneous mutations.

   If the trait evaluated is a fitness component, it can be directly measured (such as productivity, in the case of the example in Figure 7.1) or in competition with the starting population. In the case of species such as *Caenorhabditis*, individuals with the initial starting genotype may be frozen and resurrected with some success over time to be compared with individuals from the lines in a given generation. When evaluating fitness traits, the accumulation of deleterious mutations in each line is expected to reduce the mean value of the trait as a whole (ΔM), and as the number and effect of mutations accumulated over time in the different lines will differ between them, an increase in the variance of the trait values between lines (ΔV) will be expected over generations, as illustrated in Figure 7.1. The estimation of the values of ΔM and ΔV allows us estimates of certain mutational parameters to be obtained.

   Assume that a given mutation i contributes with genotypic values of 1, $1 - s_i h_i$ and $1 - s_i$ to the genotypes *AA*, *Aa* and *aa*, respectively, and that u_i is the probability with which mutation i appears. The total number of mutations appearing in the line per generation is $2Nu_i$ and, if these are quasi-neutral, the probability of fixation of each of them is $1/2N$, so the fixation rate per generation will be $2Nu_i \times 1/2N = u_i$, that is, equal to the mutation rate. If each mutation, once fixed in the line, reduces the mean of the trait in homozygosis by an amount s_i, the rate of reduction in mean per generation will be $\Delta M = \Sigma u_i s_i = \bar{s}\Sigma u_i = U\bar{s}$, where the summation is for all mutations, U is the haploid genomic mutation rate and $\bar{s} = (\Sigma u_i s_i)/\Sigma u_i$ is the average of the selective effects of mutations weighted by their mutation rate. Likewise, the increase in variance between lines per generation for the trait will be $\Delta V = \Sigma u_i s_i^2 = \bar{s^2}\Sigma u_i$, where $\bar{s^2} = (\Sigma u_i s_i^2)/\Sigma u_i$. Since the variance of mutational effects is, by definition of variance, $V(s) = \bar{s^2} - \bar{s}^2$, we have that $\bar{s^2} = \bar{s}^2 + V(s)$ and, substituting, $\Delta V = [\bar{s}^2 + V(s)]\Sigma u_i$. Considering now the square of the coefficient of variation of the effects, $C^2 = V(s)/\bar{s}^2$, we arrive at that $\Delta V = \bar{s}^2(1 + C^2)\Sigma u_i = U\bar{s}^2(1 + C^2)$. Thus, if one obtains empirical estimates of ΔM and ΔV, maximum estimates of $\bar{s}$ and minimum estimates of U can be obtained from

$$\frac{\Delta V}{\Delta M} = \bar{s}(1 + C^2); \quad \frac{\Delta M^2}{\Delta V} = \frac{U}{(1 + C^2)}. \tag{7.2}$$

This estimation procedure is called the Bateman–Mukai method (Bateman, 1959; Mukai, 1964) and, as can be seen in equations (7.2), the estimates obtained depend on the coefficient of variation of the mutational effects (C).

    With this method it is also possible to estimate the mutational variance, V_M, the increase in additive genetic variance per generation due to new mutation, and the corresponding mutational heritability. Recall from Chapter 3 (Table 3.2 and equation (3.9)) that, by replacing a by s_i, the average effect of a mutation is $\alpha_i = s_i h_i - 2d_i q_i$, or approximately $\alpha_i \approx s_i h_i$ if $d_i q_i \approx 0$, and the additive variance is $V_A = 2\Sigma \alpha_i^2 q_i(1 - q_i) \approx \Sigma(s_i h_i)^2/N$. As $2NU$ mutations are produced per generation,

$$V_M = 2NU\bar{s^2 h^2}/N = 2U\bar{s^2 h^2}. \tag{7.3}$$

If mutations were strictly additive ($h_i = 0.5$; $d_i = 0$), the expression is simplified to $V_M = U\bar{s^2}/2$, which can be estimated by $\Delta V/2$. The variance between lines for the trait ($V_{B,t}$) will increase with generations (t) at a rate of $V_{B,t} = t\Delta V$, such that

$$V_{B,t} = tU\bar{s^2} = 2tV_M. \tag{7.4}$$

    Although the distribution of effects of deleterious mutations is generally unknown, accumulated information indicates that there are more mutations of small effects than of large effects. A distribution with great versatility that is often used to investigate and model the mutational effects is the gamma distribution, which is illustrated in Figure 7.2 for different values of the shape parameter of the distribution (β). The distribution statistics are $\bar{s} = \theta\beta$, $\bar{s^2} = \theta^2\beta(1 + \beta)$, $V(s) = \theta^2\beta$ and $C^2 = V(s)/\bar{s}^2 = 1/\beta$, where θ is the scale parameter of the distribution. Note that small values of β indicate very leptokurtic distributions, where the vast majority of mutations have small effects and a few very large effects. The distribution asymmetry is given by $2/\sqrt{\beta}$ and the relative kurtosis by $3(1 + 2/\beta)$, which tends to 3 if the effects are equal ($\beta \to \infty$) and grows indefinitely if $\beta \to 0$. Suppose, for example, that the distribution of effects has a shape parameter of $\beta = 1$, which corresponds to the exponential

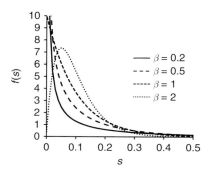

Figure 7.2 Gamma distribution of mutational effects, with density function $f(s) = [s^{\beta-1}e^{-s/\theta}]/[\theta^{\beta}\Gamma(\beta)]$, where θ is the scale parameter and β the shape parameter of the distribution and $\Gamma(\beta)$ is the gamma function, which equals $(\beta - 1)!$ if β is an integer. The mean value of mutational effects in the example shown is $\bar{s} = 0.1$.

distribution, $f(s) = e^{-s/\theta}/\theta$. In that case $C^2 = 1/\beta = 1$ and the estimates of U and $\bar{s}$ with the Bateman–Mukai method (equations (7.2)) would produce 50% overestimations of $\bar{s}$ and 50% underestimations of U. Nevertheless, it should be noted that, although the gamma distribution is useful, the distribution of mutational effects cannot always be approximated by it, and in some cases it may be more complex (Eyre-Walker and Keightley, 2007; Bataillon and Bailey, 2014).

Figure 7.3 presents a summary of the estimates of the deleterious mutation genomic rate U and the mean homozygous effect of mutations $\bar{s}$, obtained with the Bateman–Mukai method for a variety of species. In multicellular eukaryotes, the averages are $U = 0.08$ (median 0.04) and $\bar{s} = 0.22$ (median 0.16) (panels a and b). These averages include 31 estimates, approximately half of which correspond to viability or other fitness components in *Drosophila melanogaster* and the remainder to fitness traits in *Caenorhabditis*, *Amsinckia* and *Arabidopsis*. The possible biases of some of the estimates, particularly the highest ones for U, have been widely debated and could be overestimates. The reasons given are varied, such as the possibility of non-mutational changes in trait means occurring in experiments and produced by contaminations or adaptive changes.

Virus estimates denote a much larger mutational rate, around unity (panel c), consistent with the idea that viruses often have much higher mutation rates than bacteria or eukaryotes. In contrast, U estimates for unicellular organisms (bacteria and yeast) are two orders of magnitude lower than those of multicellular eukaryotes, in accordance with the fact of carrying out a single round of replication per generation.

Estimates of the lethal mutation rate are relatively easy to obtain with the *Drosophila* balanced chromosome design, and a widely accepted value is $U = 0.015$ per haploid genome and generation (Simmons and Crow, 1977), although later estimates have suggested values twice as high.

The mean mutational coefficient of variation for fitness traits is $CV_M = 0.026$ with a standard deviation of 0.027 (Halligan and Keightley, 2009). Moreover, the average mutational heritability ($h_M^2 = V_M/V_E$) of eukaryotes and yeast for fitness components is 1.26×10^{-3} (panel d), that is, V_M is in the order of 1/000th of V_E, although smaller estimates have

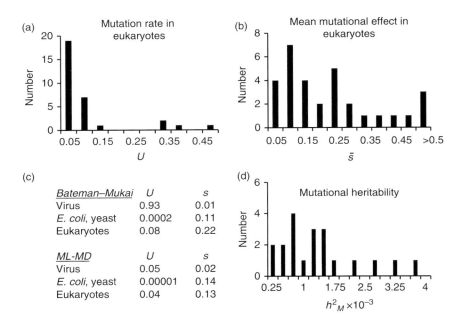

Figure 7.3 Estimates of (a) the mutation rate per haploid genome and generation (U), (b) mean homozygous effect of mutations ($\bar{s}$) and (d) mutational heritability (h_M^2), obtained from mutation accumulation experiments and compiled by Halligan and Keightley (2009) and García-Dorado et al. (2004). All estimates refer to the Bateman–Mukai method, summarized in panel (c), except for a summary of those obtained with other methods, also indicated in panel (c). Estimates of eukaryotes exclude yeast.

been obtained from populations of wild plants ($1-2 \times 10^{-4}$; Rutter et al., 2010). Nevertheless, the assumption of the infinitesimal model in obtaining the estimates imply that these may be underestimations by a factor of 3 (Keightley et al., 1993).

The lack of knowledge of the distribution of mutational effects causes the estimation method of Bateman–Mukai to produce estimates with bias of unknown magnitude, since the estimates depend on the square of the coefficient of variation of the mutational effects (C^2; equations (7.2)). However, the results of mutation accumulation experiments can also be analysed using more advanced statistical techniques, such as maximum likelihood (ML; Keightley, 1994, Shaw et al. 2002), and minimum distance (MD; García-Dorado, 1997), which overcome this obstacle and allow the estimation of the shape parameter of the distribution (β) to be obtained. The estimates of this parameter are very variable but many of them are higher than unity (Halligan and Keightley, 2009), with the average being $\beta = 1.9$ (excluding those indistinguishable from zero or infinity), indicating a distribution with a bell shape (see Figure 7.2). This high value of β is consistent with the relative similarity between the estimates of U and $\bar{s}$ for eukaryotes achieved by the Bateman–Mukai method and those obtained with the other methods (Figure 7.3c), since the bias in the estimates of the former depends on the variation of mutational effects, that is, the value of C^2 in equations (7.2), and this is smaller the greater is β.

The rate of deleterious mutation can also be estimated from molecular data. The comparison between nucleotide divergences of evolutionarily closed species allows the inference of the fraction of mutations with deleterious effects (Kondrashov and Crow, 1993). Thus, an estimate of the genomic rate of deleterious mutation can be obtained as

$$U = U_n \Phi, \tag{7.5}$$

where U_n is the rate of nucleotide mutation per generation for the genome as a whole, and Φ is the fraction of mutations whose deleterious effect is large enough for selection to determine its elimination ($s \gg 1/[2N_e]$). The magnitude of Φ can be estimated by comparing the substitution rates of mutations that are assumed to be exclusively neutral (π_s) (e.g. those in pseudogenes or non-functional archaic transposable elements) with the total divergence observed between two twin species (π_g). The fraction of non-fixed mutations is, therefore, $\Phi = 1 - \pi_g/\pi_s$. For example, the value of Φ comparing the sequences of *Drosophila melanogaster* and *D. simulans* is 0.58, providing an estimate of $U \approx 0.6$ (Haag-Liutard et al., 2007). Similar analyses with human genome data compared to the chimpanzee genome indicate that they occur on the order of $U_n = 35$ new nucleotide mutations per haploid genome, most of which have paternal origin, and $U \approx 1.1$ deleterious mutations (Keightley, 2012). On the other hand, the analysis of non-synonymous mutations that produce amino acid polymorphisms in humans suggests that the mean effect of the mutations in homozygosis is $s \approx 0.03$ and that the distribution of effects is very leptokurtic ($\beta \approx 0.2$) (Boyko et al., 2008).

The great difference found between molecular mutation rates and those obtained in mutation accumulation experiments, as well as in kurtosis in both cases (compare β values of 0.2 and 2 in Figure 7.2) suggests that mutational effects detected in mutation accumulation experiments are only those of substantial effect, there being no sufficient power to detect many small-effect mutations ($s < 5 \times 10^{-4}$; García-Dorado et al., 2004) in these types of experiments (see also Charlesworth, 2015). These mutations can be of great importance from an evolutionary point of view, considering large populations and also considerable periods of time. However, from the perspective of reduced population sizes in selection or conservation programmes, it is the mutations of substantial effect detected with mutation accumulation experiments that are most significant. Therefore, both types of estimates have interest, whether evolutionary or applied.

7.1.3 Estimation of the Dominance Coefficient

Mutation accumulation experiments for fitness traits also allow estimates of the average coefficient of dominance to be obtained using the values of the trait in the lines and in crosses between pairs of them. A first type of estimate is obtained as the ratio between the value of the trait from the cross between two lines and the sum of the values of the trait in those lines. Since different mutations are expected to be fixed in each line, these will be found in heterozygosis in the cross. Thus, the decline in the mean of the trait that occurs in the cross between two lines, divided by the sum of the declines that occur in those lines, constitutes the so-called *ratio estimate*, which gives the average of the dominance coefficients weighted by the selection coefficients $\hat{h}_R = \Sigma s_i h_i / \Sigma s_i$ (Mukai, 1969b). This method, however, has the drawback of producing biased estimates if part of the decline observed over generations in the mean of the trait is of environmental origin, something that is difficult to disregard in most cases.

A method that does not present the above-mentioned bias consists in obtaining the regression of the heterozygotes (trait values at crosses between lines) on the sum of the

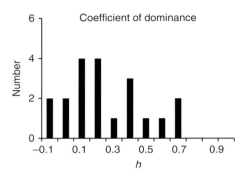

Figure 7.4 Estimates of the average coefficient of dominance of mutations obtained from mutation accumulation experiments by the regression of mutational effects of heterozygotes on that of homozygotes. The estimates correspond to fitness components in *Drosophila*, *Caenorhabditis* and yeast compiled by García-Dorado et al. (2004) and Caballero (2006). The mean value of the estimates of h is 0.21, and the median is 0.12.

homozygotes (sum of trait values in the parental lines), which provides an estimate of the average dominance coefficient weighted by the square of the selection coefficients $\hat{h}_b = \Sigma s_i^2 h_i / \Sigma s_i^2$ (Mukai and Yamazaki, 1968). Figure 7.4 presents a summary of regression estimates for fitness components in *Drosophila*, *Caenorhabditis* and yeast. Although with great variation, the average dominance coefficient is 0.2. This average is in agreement with that obtained from the reanalysis of the estimates obtained with the ratio estimate, when the possibility of environmental changes in fitness is taken into account (García-Dorado and Caballero, 2000).

Obtaining information on the joint distribution of selection and dominance coefficients is an arduous task, since it is necessary to estimate the effects of each mutation individually, being particularly difficult for small-effect mutations. The analysis of data for fitness traits mutations in five experiments with *Drosophila*, yeast, *Caenorhabditis* and *Arabidopsis* indicates a variance of h values between 0.12 and 0.28 and a negative correlation between s and h values of the order of -0.3 (Caballero, 2006). On the other hand, the dominance coefficient generally accepted for lethal genes is $h \approx 0.02$ (Simmons and Crow, 1977), indicating almost total recessivity. In general, it is assumed that large-effect mutations in homozygosis tend to be recessive, whereas small-effect mutations tend to be additive, although with a large variation, real or due to estimation error (Caballero and Keightley, 1994). These observations can be plotted in a model for which the values of h are obtained with equal probability in the interval $[0, e^{-ks}]$, where k is a constant that can be calculated for the gamma distribution by the expression $k = \theta^{-1}[(2\bar{h})^{-1/\beta} - 1]$, with $\bar{h}$ being the desired average value of h. An example of this model is presented in Figure 7.5.

Since a large fraction of the observed mutations are recessive, Fisher (1928) postulated the hypothesis that they were initially additive and that dominance modifiers appeared at other loci that would cause the heterozygote expression to become similar to that of the wild homozygous genotype. If this type of evolution of dominance occurred, it would be very slow, since it would involve the appearance of mutations in the modifier loci and their subsequent increase in frequency. A more generally accepted explanation, which is furthermore compatible with the observation that small-effect mutations tend to be

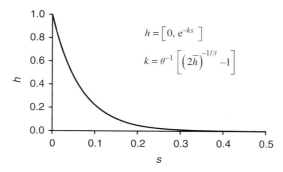

Figure 7.5 Model to describe the relationship between the values of the selection coefficient of the mutations (s) and the dominance coefficient (h). The values of h are obtained with uniform probability in the interval between 0 and $\exp(-ks)$ (values below the line), where k is a constant that allows the calculation of the desired mean h value given a gamma distribution of mutational effects with shape parameter β and scale parameter θ. In the example shown, $\bar{s} = 0.1, \bar{h} = 0.2, \beta = 1, \theta = 0.1$, and $k = 6.667$.

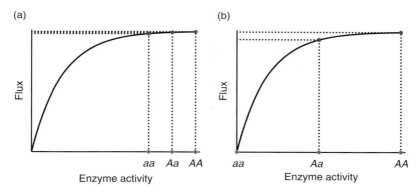

Figure 7.6 Metabolic pathway flux model based on the enzymatic activity of an enzyme of the pathway. The wild genotype (AA) implies the maximum activity and the maximum flux of the pathway. Mutation a reduces the enzymatic activity with additive gene action (the heterozygote has intermediate activity between those of the homozygotes). When the mutation effect is small (panel a), the pathway flux for the heterozygote is approximately intermediate between those of the two homozygotes, so the locus gene action is approximately additive or only partially recessive ($h = 0.32$ in the example of the figure). However, if the mutation has a high activity-reducing effect (panel b), the heterozygote flux is more similar to that of the wild homozygote, so the mutation has recessive gene action on the phenotype ($h = 0.06$ in the example of the figure).

additive (at least on average) and large-effect ones tend to be recessive, is the physiological theory of dominance of Wright (1934). This can be explained in terms of the kinetics of metabolic pathways (Kacser and Burns, 1981). It is thought that many loci with effect on quantitative traits can be genes that encode enzymes or regulate their expression. It has been observed that the enzymatic activity of the heterozygotes is intermediate (or practically so) between those for the corresponding homozygotes. However, the flux of a metabolic pathway involving several enzymes generally has a non-linear relationship with the enzymatic activity of each enzyme. The argument is illustrated in Figure 7.6 where the amount of flux and, therefore, the amount of end-product, is plotted from a typical multienzyme metabolic pathway as a function of the enzymatic activity of one of the enzymes involved in the same pathway. When the enzymatic activity is zero, there is no flux in the pathway. If the activity increases slightly, there is a large increase in the flux of the pathway, but when the activity reaches saturation levels the flux increases very little, since this will depend on the activity of the other enzymes involved in the pathway.

Suppose now, as illustrated in the figure, that the activity of the wild type *AA* genotype implies a large enzymatic activity corresponding to the maximum flux of the pathway. If we consider a mutation *a* whose effect is a slight reduction in activity (panel a), and that of the heterozygote *Aa* is intermediate between the two homozygotes, the heterozygote flux will be approximately intermediate between the two homozygotes and, therefore, the gene action for the trait will be additive. However, if the mutation is of larger effect (panel b) so that the activity is zero, and the heterozygote has again intermediate activity between the two homozygotes, the flux will no longer be intermediate this time, but very similar to that of the wild homozygote, resulting in recessive gene action of the mutation. Obviously, loci with effects on quantitative traits not only encode enzymes, but with this model it is possible to explain the tendency for large-effect mutations to be recessive without the need for modifiers of dominance.

7.1.4 Beneficial Mutations and Summary of Mutational Parameters for Fitness

So far we have only considered the parameters corresponding to deleterious mutations. The rate and effects of beneficial mutation can only be estimated in organisms that reproduce sufficiently fast, such as viruses, bacteria and yeast, where mutation accumulation experiments can be performed with very high population sizes and durations of thousands of generations, as well as with clonal analyses. The results suggest beneficial mutation rates in *E. coli* and yeast on the order of $U \approx 2 \times 10^{-5}$, with mean effects on the order of $s \approx 0.01$ or lower (Sniegowski and Gerrish, 2010). Since the overall genomic mutation rate in *E. coli* and yeast has been estimated to be 3×10^{-3} (Drake et al., 1998), this would indicate that beneficial mutations make up about 1% of all mutations. On the other hand, considering the deleterious mutation rates in these species obtained with the Bateman–Mukai method (Figure 7.3c), the deleterious mutation fraction would constitute 7% of all mutations. Considering humans, it has been estimated that between 1% and 10% of mutations can be deleterious (Keightley, 2012; Rands et al., 2014). It is difficult to draw conclusions about the distribution of beneficial mutation effects although some evidence suggests an approximately exponential distribution (Eyre-Walker and Keightley, 2007; Bataillon and Bailey, 2014).

Figure 7.7 presents a general summary of the mutational parameters for fitness and its major components. The average values and typical intervals of mutation rates, effects and dominance coefficients for each type of mutation are included only as a guideline since, as seen above, there is a large variation between estimates for different traits, species and estimation methods.

The mutation rates mentioned in this chapter refer to the germ line. However, it is also important to note the relevance of somatic mutations, particularly deleterious ones, because of their impact on the development of many diseases, such as cancer. It has been estimated that the somatic mutation rate may be of the order of 50 times higher than the germ

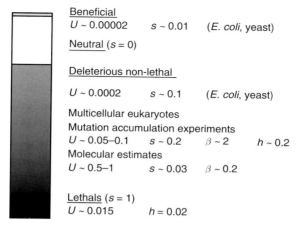

Beneficial
$U \sim 0.00002$ $s \sim 0.01$ (*E. coli*, yeast)

Neutral ($s = 0$)

Deleterious non-lethal
$U \sim 0.0002$ $s \sim 0.1$ (*E. coli*, yeast)

Multicellular eukaryotes
Mutation accumulation experiments
$U \sim 0.05–0.1$ $s \sim 0.2$ $\beta \sim 2$ $h \sim 0.2$
Molecular estimates
$U \sim 0.5–1$ $s \sim 0.03$ $\beta \sim 0.2$

Lethals ($s = 1$)
$U \sim 0.015$ $h = 0.02$

Figure 7.7 Summary of types of mutations in functional regions of the genome according to their impact on fitness. The size of the bars illustrates the relative magnitude of each class of mutation (beneficial, neutral, deleterious and lethal) as a rough guide, since it depends on the species and the type of estimation, as indicated at the right of the figure, where mean values and typical indicative intervals are given for the mutation rate per haploid genome and generation (U), the average effect of mutations on homozygosis (s), the dominance coefficient (h) and the shape parameter of the distribution of effects (β).

line mutation rate, which means that an adult cell may contain thousands of new mutations (Lynch, 2016). Even though only a fraction of 1% were deleterious, the result is a growing accumulation of harmful mutations with age.

7.1.5 Combined Effect of Mutations and Environmental Factors

The additive combination of the mutational effects accumulated on the genome would imply a linear decline in the mean fitness of the population (Figure 7.8), this being $W = 1 - \Sigma s_i$, where the summation is for all mutations i. However, a widely accepted model for modelling multilocus effects for fitness is the multiplicative one (independent effects), with mean fitness $W = \Pi(1 - s_i)$, which is also often used to combine the effects of different components of fitness. For example, if an individual carries mutations that reduce its mating success by 30%, its fecundity by 10% and its viability by 20%, the individual's fitness is assumed to be reduced by $(1 - 0.3) \times (1 - 0.1) \times (1 - 0.2) = 0.5$, that is, 50%, instead of 60% if the effects were added together. If the effects are small and not very numerous, the difference between the additive and multiplicative models is negligible. If fitness is measured on a logarithmic scale, the slope of the mean decline with the multiplicative model is linear, so it is common to analyse changes in fitness on this scale.

In reality, however, it may be that the combined effect of mutations is greater or lower than it would have been by multiplying or adding their effects separately. In the former case we speak of synergistic (also called narrowing, negative or reinforcing) epistasis and in the latter of antagonistic (also called diminishing return, positive or attenuating) epistasis, both represented in Figure 7.8. There are many possible ways of modelling epistasis and one possibility was explained in Chapter 3. There is some experimental evidence that the decline

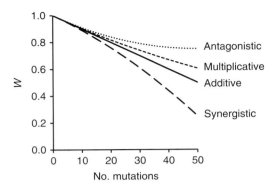

Figure 7.8 Effect of different multilocus mutational models on the mean fitness of the population (W) assuming, for simplicity, that all mutations (n) have the same deleterious effect in homozygosis ($s = 0.01$). In the additive model, the effects of mutations are summed, that is, $W = 1 - ns$. In the multiplicative model, the fitnesses contributed by each locus are multiplied, $W = (1 - s)^n$. For the synergistic epistatic model, the effect of the mutations is greater than the sum of their effects separately, which can be represented by the quadratic model $W = 1 - ns - (ns)^2$. Finally, in the antagonistic epistatic model, the combined effect of the mutations is less than the sum of their effects separately, which can be represented by the model $W = 1 - ns + (ns)^2$.

in fitness in eukaryotic mutation accumulation experiments fits the quadratic model of synergistic epistasis in Figure 7.8 (Mukai, 1969a; Whitlock and Bourguet, 2000), although other data indicate absence of epistasis (Halligan and Keightley, 2009). It has been suggested (Sanjuán and Elena, 2006) that species with simpler genomes, such as RNA viruses, tend to show antagonistic epistasis, whereas bacteria and yeast absence of epistasis, and multicellular eukaryotes synergistic epistasis.

Another question of interest is whether the deleterious effects of mutations increase in adverse environments. It is well known that some mutations are only deleterious in certain environments, for example there are alleles in *Drosophila* that are only lethal above a certain temperature, which are called conditional lethals (Vermeulen and Bijlsma, 2004). The data on the impact of the environment on mutational effects, reviewed by Martin and Lenormand (2006), indicate that the mutational variance (ΔV or V_M) tends to be higher in adverse environments but the decline in mean (ΔM) is not. This would imply that the U estimates obtained by the Bateman–Mukai method would be underestimations, and the estimates of $\bar{s}$ overestimations, if they were evaluated in adverse environments (see expressions (7.2)). However, most of the estimates obtained with this method have been carried out in benign environments, so they are not expected to be heavily biased for this reason.

Finally, the rate of mutation may also depend on the genetic background in which it is produced. For example, the number of mutations could increase if the population is not well adapted to its environment. Fisher (1930) suggested that if a population is close to its optimal constitution, only beneficial mutations of small effect could appear, but if the population is far from its optimum, advantageous mutations of much greater magnitude may appear. In fact, it has been observed that in populations with a very low average fitness, the rate of beneficial mutation is much higher than in populations with a normal average fitness. Many of these mutations are compensatory, meaning that they restore part of the decline in fitness due to mutations occurring in other loci (Poon and Chao, 2005).

7.1.6 Estimation of Mutational Parameters for Neutral or Quasi-Neutral Traits

The analysis of mutational parameters for quantitative traits other than fitness can also be carried out by mutation accumulation experiments, as those described above. It can also be performed by artificial selection experiments, starting from a population lacking genetic variability, which may have been obtained by crossing techniques with balanced chromosomes inhibiting recombination or by multiple generations of rapid inbreeding. The main theoretical foundations for these estimations are explained next.

Consider the situation in which the trait under study is neutral, that is, it is uncorrelated with fitness. In this scenario, genetic variability is determined exclusively by its increase due to new mutation and its reduction by genetic drift. Considering first individual loci, let us recall that in Chapter 4 (Section 4.3.1) it was deduced that, in the neutral mutation–drift equilibrium, an inbreeding coefficient $\hat{F} = 1/(4Nu + 1)$ and a heterozygosity $\hat{H} = 1 - \hat{F}$ are reached (Malécot, 1948), where u is the per locus mutation rate. Substituting N for N_e (Chapter 5) we have, therefore, that the heterozygosity in the neutral mutation–drift equilibrium is

$$\hat{H} = \frac{4N_e u}{4N_e u + 1} \approx 4N_e u,$$ (7.6)

holding the approximation for $4N_e u \ll 1$. That is, under the neutral model, the heterozygosity in the mutation–drift equilibrium is a function of the effective size and the mutation rate per locus. The term $\theta = 4N_e u$ is often referred to as the scaled mutation rate, and can be thought of as a measure of the rate of mutation for the whole population (Charlesworth and Charlesworth, 2010, p. 206). The denominator of equation (7.6) is also the so-called effective number of alleles (Kimura and Crow, 1964), mentioned in Chapter 4 (Section 4.4.4), that is, the number of alleles of the locus in the population if they all would segregate at the same frequency. Note that equation (7.6) allows the estimation of the effective population size to be inferred from estimates of mutation rate and heterozygosity. For example, using nucleotide diversity data (equivalent to H) in human populations it has been estimated that $H \approx 10^{-4}$ (Li and Sadler, 1991). With mutation rates of the order of $u \approx 10^{-9}$ per nucleotide, N_e estimates of between 10,000 and 20,000 are obtained, which is the historical effective size range assumed in humans (Charlesworth, 2009).

Let us now consider the same neutral argument for a quantitative trait in which the contribution of mutation each generation is quantified by the mutational variance (V_M). Recall that the genetic variance is a function of heterozygosity (equation (3.9), Chapter 3) and that this is reduced by drift in a fraction $(1 - 1/2N)$ each generation (equation (4.19), Chapter 4) where, in general, N must be replaced by the effective population size N_e (Chapter 5). Therefore, the additive genetic variance in each generation t is the result of its decrease by drift and its increase by new mutation, $V_{A(t)} = V_{A(t-1)}(1 - 1/2N_e) + V_M$. After a certain number of generations ($t \gg N_e$) both processes will be compensated for by reaching an equilibrium in which $V_{A(t-1)} = V_{A(t)} = \hat{V}_A$. Substituting,

$$\hat{V}_A = 2N_e V_M \qquad (7.7)$$

(Lynch and Hill, 1986), which determines the magnitude of the additive variance in the neutral mutation–drift equilibrium. An example of this process of regeneration of genetic variance by mutation in populations of moderate effective size ($N_e \approx 500$) was empirically demonstrated in *Drosophila melanogaster* by Amador et al. (2010). Starting from a population initially devoid of genetic variation, this increased over time to reach a genetic variance of the order of that commonly seen in laboratory populations in a period of time predicted by mutations subject to weak selection (400–500 generations).

Strictly, equation (7.7) applies only to the neutral case. Quantitative traits are not generally neutral, but if the effective population size is small, the equation can be roughly used. For example, if lines of pairs of individuals are founded with brother–sister mating each generation, the effective size is $N_e \approx 2.5$ (Chapter 5), so the additive variance of equilibrium will be $\hat{V}_A \approx 5V_M$. If the mutational variance is, as we saw earlier, approximately equal to $10^{-3}V_E$, we have $\hat{V}_A \approx 5 \times 10^{-3}V_E$, so the equilibrium heritability will be $\hat{h}^2 = \hat{V}_A/V_E = 0.005$. This shows that a line of fast inbreeding maintains very little genetic variation, as expected.

When several mutation accumulation lines with effective size N_e are available initially without genetic variation, the genetic variance between lines increases with generations at a rate $2\Delta F V_A$, where ΔF is the increase in inbreeding in one generation (equation (4.12), Chapter 4), equal to $1/2N_e$ if N is replaced by N_e. Therefore, the rate of increase in variance between lines is V_A/N_e. After t generations the variance between lines will be approximated by $V_{B,t} = tV_A/N_e$. Substituting V_A for its equilibrium value (equation (7.7)) we have $V_{B,t} = t(2N_e V_M)/N_e$, so that $V_{B,t} = 2tV_M$, as expressed in equation (7.4). Returning to the example of brother–sister lines, the ratio between the variance between lines ($V_{B,t}$) and the phenotypic variance within lines ($V_{W,t}$) would be, using expression (7.7), $V_{B,t}/V_{W,t} = (2tV_M)/(2N_e V_M + V_E)$. Substituting $N_e \approx 2.5$ and $V_M \approx 10^{-3}V_E$, we have $V_{B,t}/V_{W,t} \approx 0.002t$. Thus, at $t = 20$ generations, for example, the variance between lines is expected to be 4% of the phenotypic variance within lines.

If the experiment carried out is one of artificial selection on a population initially lacking genetic variation (Caballero et al., 1991), we will see in Chapter 9 that the response to selection (the change in the population mean) in generation t due to new mutation can be approximated by

$$R_t \approx 2tN_e V_M i/\sqrt{V_P} = t\hat{V}_A i/\sqrt{V_P} \qquad (7.8)$$

(Hill, 1982), where i is the intensity with which the selection is applied and V_P is the phenotypic variance of the trait, making it possible to estimate V_M from the observed response.

The values of the square of the coefficient of mutational variation and of the heritability for morphological traits in *Drosophila melanogaster* obtained with the previous methods are somewhat lower than for fitness ($CV_M = 0.001$–0.012 (Houle et al., 1996) and $h_M^2 = 0.5$–0.85×10^{-3} (García-Dorado et al., 1999)). If the traits under study are not strictly neutral and additive, estimates of V_M will be underestimates. In fact, the analysis of individual mutations for these traits suggests that genes with low effect have gene action indistinguishable from additivity, whereas those of large effect are usually recessive and have a deleterious pleiotropic effect on fitness, it being common to find lethal genes with a substantial effect on morphological traits (López and López-Fanjul, 1993b; Mackay, 2001). Nevertheless, pleiotropy, although supposedly frequent in quantitative traits, may be difficult to detect on many occasions (Wagner and Zhang, 2011).

7.2 Implications of Deleterious Mutations in Populations of Large Size

7.2.1 Equilibrium between Deleterious Mutation and Selection

In Chapter 2 we saw that in the absence of mutation and drift, natural selection reduces the frequency of deleterious alleles until their elimination. However, the continuous appearance of these by mutation implies that a balance will be reached between the reduction in frequency by selection and its increase by mutation. Consider the general model of a locus with genotypes AA, Aa and aa, genotype frequencies p^2, $2pq$ and q^2, and fitnesses 1, $1 - sh$ and $1 - s$, respectively. As we saw in Section 2.8.3 (Chapter 2), the frequency of allele A after selection is $p' = [p^2 + pq(1 - sh)]/[1 - 2pqsh - q^2s]$. If one considers the possibility that the allele A mutates to the form a with probability u (the probability of back-mutation can be ignored because it is much smaller), the expression of p' must be multiplied by $(1 - u)$, the probability that the allele remains as A in the next generation. When the equilibrium between the reduction in frequency by selection and its increase by mutation is reached, we will have that $p' = p = \hat{p}$, $\hat{p}$ being the equilibrium frequency. Substituting we get $\hat{p} = [\hat{p}^2 + \hat{p}\hat{q}(1 - sh)](1 - u)/[1 - 2\hat{p}\hat{q}sh - \hat{q}^2s]$ and, simplifying, we arrive at the second degree equation $s\hat{q}^2(1 - 2h) + sh\hat{q}(1 + u) - u = 0$. If the deleterious allele is recessive ($h = 0$), the solution is

$$\hat{q}^2 = \frac{u}{s}, \tag{7.9}$$

and if there is partial recessivity ($h > 0$) and it is held that $h^2s \gg u$ (so that q^2 and qu are negligible), the solution approaches

$$\hat{q} \approx \frac{u}{sh}. \tag{7.10}$$

Equation (7.9), corresponding to a recessive allele, gives the maximum possible frequency of a deleterious allele at the deleterious mutation–selection balance for a certain value of s. As we saw in Chapter 2, recessivity implies the masking of an allele effect in heterozygotes, making the action of natural selection inefficient in its elimination. A small effect of the allele on heterozygosis implies a substantial reduction of the equilibrium frequency. For example, a recessive lethal allele ($s = 1$, $h = 0$) that appears by mutation with frequency $u = 10^{-6}$ per generation will be maintained in an infinite population with an equilibrium frequency $\hat{q} = 0.001$ (equation (7.9)). However, if its dominance coefficient value were $h = 0.02$ (the dominance value generally associated with lethals), the equilibrium frequency would be two orders of magnitude lower ($\hat{q} \approx 0.00005$; equation (7.10)). If the population is not panmictic but has partial inbred matings so that α is the Hardy–Weinberg disequilibrium factor, expression (7.10) becomes $\hat{q} \approx u/[s(\alpha + h - \alpha h)]$ (Charlesworth and Charlesworth, 2010, p. 161).

7.2.2 Mutation Load

The presence of segregating deleterious mutations implies a reduction in the mean fitness of the population compared to the one it would have if no such mutations existed. This decrease in mean fitness is the so-called mutation load, formulated by Haldane (1937). Let us consider again the general model of genotype frequencies and fitnesses for a locus of the previous section. In the deleterious mutation–selection equilibrium, the mean fitness of the population is $W = 1 - 2\hat{p}\hat{q}sh - \hat{q}^2 s$ and the mutation load is defined as $L = (W_{max} - W)/W_{max}$, where W_{max} is the maximum possible fitness which, in the model considered, is equal to unity, so that $L = 1 - W = 2\hat{p}\hat{q}sh + \hat{q}^2 s$. We will keep henceforth the assignment of $W_{max} = 1$ for simplicity.

If we consider a recessive allele ($h = 0$), substituting its equilibrium frequency (equation (7.9)) we obtain $L = \hat{q}^2 s = (u/s)s = u$. This is a surprising result at first glance, since it implies that in the deleterious mutation–selection equilibrium, the load only depends on the mutation rate, regardless of the frequency or effect of the allele that produces it. This is because the mutations with large effect have a great impact on the fitness of the individual but are found at low frequencies in the population, whereas mutations of low effect have less impact on the fitness of the individual but are more frequent, so that both factors compensate for each other and, therefore, the load will only depend on the mutation rate.

If the deleterious allele is not completely recessive ($h > 0$), substituting equation (7.10) and assuming that $q^2 s$ is negligible since $p \approx 1$, we have $L \approx 2\hat{q}sh = 2(u/sh)sh = 2u$. Again, the load only depends on the mutation rate, although in this case it is twice that caused by the recessive mutations. The explanation provided by Haldane is that upon the elimination by selection of a recessive mutation, this is carried out by eliminating a homozygous genotype, which involves removing two copies of the allele. In contrast, non-recessive mutations are found primarily in heterozygotes, and their elimination involves only one copy. Therefore, the latter produce twice as much load.

If we consider the joint effect of all genomic loci with presumably deleterious alleles, assuming a multiplicative model and taking into account that full recessivity should be infrequent, the mean fitness of the population is $W = 1 - L = \Pi(1 - 2u)$, where the product involves all mutations of the genome. If the locus mutation rate is small, we can approximate $1 - 2u$ by e^{-2u}. Thus,

$$L = 1 - W \approx 1 - e^{-2U}, \tag{7.11}$$

where $U = \Sigma u$. If U is not large, $L \approx 2U$.

A very illustrative way of understanding the concept of load is due to King (1966), with a correction of Kondrashov and Crow (1988) (see Agrawal and Whitlock, 2012). Considering the trait viability, selection produces the mortality of a given fraction of the population. If the mean number of deleterious mutations in the individuals of this fraction is z, and the mean number of mutations carried by the survivors is y, the mean number of mutations removed by the mortality produced by selection is $z - y$. Since the probability of mortality is equal to the load (L), the mean number of mutations eliminated per individual by selection is

$L(z - y)$. This number must be, in the equilibrium, equal to that of mutations appearing $(2U)$, hence $L(z - y) = 2U$, and

$$L = \frac{2U}{z - y}. \tag{7.12}$$

If a deleterious mutation is considered with mutation rate u and an effect in the heterozygote, it can be assumed that $z = 1$ and $y = 0$, from which Haldane's result of $L = 2u$ is obtained. If the mutation is recessive and mortality occurs, therefore, in the homozygote, $z = 2, y = 0$ and $L = u$, as expected. But equation (7.12) applies not only to the case of independent (multiplicative) effects of mutations, but is general. Suppose that $U = 0.5$. With a multiplicative model of mutational effects, the load indicated by equation (7.11) would be $L = 0.63$, which would imply, using equation (7.12), that the mean number of mutations removed by selection is $z - y = 1.59$. But if the mean number of mutations removed per individual were, for example, double $(z - y = 3.18)$, then the load would be about half, $L = 0.31$. One of the mechanisms by which the number of mutations eliminated by selection may be greater than that expected with a model of independent mutations is synergistic epistasis (Figure 7.8) observed in some experiments. This would mean a much more efficient removal of the mutation load by natural selection (Crow, 2000), particularly if selection acts by truncation so that a fraction of the population carrying a number of mutations greater than a given value is eliminated. Agrawal and Whitlock (2012) discuss other mechanisms that would reduce the mutation load in natural populations.

For a non-panmictic population with partial inbred matings the load is approximately $L \approx 1 - \exp[-U(2 - a)/(a + h - ah)]$ (Charlesworth and Charlesworth, 2010, p. 167), which shows that the equilibrium load decreases as the inbreeding coefficient increases, reflecting the purging of deleterious mutations by inbreeding.

The expected genetic variance for fitness in an infinite population under the deleterious mutation–selection balance can be obtained for a locus with equilibrium frequency $2\hat{p}\hat{q}$ in the heterozygotes and effect sh, as $V_{G,W} = 2\hat{p}\hat{q}(sh)^2$. Substituting $\hat{p} \approx 1$ and $\hat{q}$ by equation (7.10) we have $V_{G,W} = 2(u/sh)(sh)^2 = 2ush$. Assuming additivity between loci, $V_{G,W} = 2\Sigma ush$, or

$$V_{G,W} \approx 2U\overline{sh}. \tag{7.13}$$

Substituting equation (7.3), and assuming constant values of s and h, (7.13) is reduced to $V_{G,W} \approx V_M/sh$.

A recurrent debate on mutation load is its plausibility given the deleterious mutation rates, particularly those estimated with genomic methods. If we consider, as an extreme example, the human genomic mutation rate indicated above ($U = 1.1$), the mean fitness would be $W = e^{-2U} = 0.11$. If fitness directly translates into survival probability at face value, what is termed hard selection (Wallace, 1975), this would mean that a pair of individuals would need to have 20 children so that an average of two could survive simply to maintain the population size as invariable, which seems unrealistic. However, note that the load refers to the fitness of a population relative to the mutation-free genotype (W_{max}), probably non-existent in any natural population. The problem, therefore, disappears almost completely if differences in fitness are considered in relative terms for a given population or species, or according to gene frequencies or population density (the so-called soft selection), in which case the load would manifest in the form of genetic variation for fitness among the individuals of the population. For example, if the absolute mean fitness of a population of constant size is 0.11 (relative to that of a mutation-free genotype), an individual with an absolute fitness of 0.10 would have a relative fitness of $0.10/0.11 = 0.91$. A theoretical analysis of this model was carried out by Lesecque et al. (2012).

Another interesting debate, applicable mainly to human populations and already discussed by Muller (1950), refers to the possible accumulation of deleterious mutations caused by the relaxation of selection in human populations, due to the improvement of environmental conditions, the advances of medicine and the development of well-being (Crow, 1997; Lynch, 2016). It seems logical to think that these factors would reduce the deleterious effect of many mutations, which would lead them to higher equilibrium frequencies, although the load would not be affected as it depends only on the mutation rate. During the first half of the twentieth century, eugenic currents came to be applied in the form of sterilizations even in the 1970s in several countries. The result of the application of these measures would necessarily be a failure since, as we saw in Chapter 2, the elimination of recessive mutations is very ineffective since they are masked in heterozygotes and appear continuously by mutation (see Kevles, 1998).

In addition to the mutation load, there are other types of load related to the reduction in the average fitness of the population for various causes. For example, there is a load due to the segregation of heterozygotes in the case of overdominant fitness models, another due to recombination, when certain allelic associations are advantageous, or the loads due to environmental heterogeneity or inbreeding, among others. The most relevant types of load are, however, the mutation load already described and the inbreeding load, which we will study in the next chapter, and which refers to the decline in average fitness when inbreeding exposes deleterious recessive mutations in homozygosis. Crow (1970) presents a lucid description of the consequences of different types of load, as well as the concept of cost of selection, also due to Haldane, which refers to the dynamic process that must be carried out by selection to spread a favourable mutation towards fixation.

7.2.3 Estimation of the Average Dominance Coefficient in the Mutation–Selection Equilibrium

We saw in Section 7.1.3 that the average dominance coefficient can be estimated in mutation accumulation experiments from the mean of the trait in homozygous lines and their crosses. This estimation method can also be applied in populations with standing genetic variation using the balanced chromosome technique, which, under the assumption of deleterious mutation–selection equilibrium in the population, allows estimates of the dominance coefficient to be obtained (Mukai et al., 1972; Fernández et al., 2004). Using the balanced chromosome strains described above, homozygous individuals can be obtained for a given chromosome, and also individual carriers of different chromosomes that would be heterozygous for all loci in which the two chromosomes carry different alleles. As we have already indicated, at mutation–selection equilibrium the total number of individuals carrying a copy of a deleterious allele with effect s_i and dominance h_i is $q_i = u_i/s_ih_i$ (equation (7.10)), so that the variance of the trait for heterozygous individuals would be $\sigma_y^2 = \Sigma q_i s_i^2 h_i^2 = \Sigma(u_i/s_ih_i)s_i^2 h_i^2 = \Sigma u_i s_i h_i$, where the summation involves all mutations on the chromosome. Likewise, the variance of the trait for the sum of the corresponding homozygous individuals would be $\sigma_x^2 = \Sigma q_i s_i^2 = \Sigma(u_i/s_ih_i)s_i^2 = \Sigma u_i s_i/h_i$, and the covariance between them, $\sigma_{x,y} = \Sigma q_i s_i^2 h_i = \Sigma(u_i/s_ih_i)s_i^2 h_i = \Sigma u_i s_i$. Therefore, the regression of the heterozygotes on the homozygotes is

$$\hat{h}_b = b_{y.x} = \frac{\sigma_{x,y}}{\sigma_x^2} = \frac{\Sigma u_i s_i}{\Sigma u_i s_i/h_i} = \frac{\bar{s}}{\overline{s/h}}, \tag{7.14}$$

which estimates the harmonic mean of h weighted by s.

On the other hand, the inverse of the regression of homozygote values over that of heterozygotes,

$$\hat{h}_{1/b} = \frac{1}{b_{x.y}} = \frac{\sigma_y^2}{\sigma_{x,y}} = \frac{\Sigma u_i s_i h_i}{\Sigma u_i s_i} = \frac{\overline{sh}}{\bar{s}}, \tag{7.15}$$

estimates the arithmetic mean of h weighted by s. With this method, Mukai and coworkers obtained estimates of $\hat{h}_b$ between 0.2 and 0.4, and of $\hat{h}_{1/b}$ between 0.6 and 1.0, in various natural populations of *Drosophila melanogaster*. The disadvantage of this method is that it assumes the deleterious mutation–selection equilibrium and violations thereof may involve considerable biases, particularly in the estimates of $\hat{h}_{1/b}$ (Fernández et al., 2005).

7.2.4 Estimation of Mutational Parameters in the Mutation–Selection Equilibrium from Data of Panmictic and Inbred Populations

If data on means and variances of fitness are available for panmictic and inbred populations, the mutation rate, the mean selection coefficient of the mutations and their dominance coefficient can be obtained (Morton et al., 1956; Deng and Lynch, 1996; Caballero, 2006). The necessary calculations are presented in Table 7.2. Suppose that estimates of the mean (W_P) and genetic variance (V_{WP}) for a fitness component are obtained experimentally in a large size population, and that self-fertilizations of the individuals of that population are carried out so that estimates of the mean (W_S) and variance (V_{WS}) of individuals from a generation of self-fertilization are obtained. The fitnesses of the genotypes in each case are shown in Table 7.2. A joint variable (T) can also be obtained by combining the fitness values of the parents and their inbred offspring (Deng, 1998).

The mean fitness of the panmictic population is, from equation (7.11), $W_P = e^{-2U}$ which, for simplicity can be expressed by the additive approximation, $W_P \approx 1 - 2U$. The genetic variance expected for fitness in an infinite population with random mating in the deleterious mutation–selection equilibrium is given by expression (7.13). Likewise, from the genotype frequencies and fitnesses of Table 7.2, it is possible to obtain the expected mean of the inbred populations, the variance of the T variable and its covariance with P, and the difference between the mean fitness of the panmictic and inbred populations (δ), which is called inbreeding depression, and which we will study in detail in Chapter 8. Other variances and covariances are not necessary.

Table 7.2 *A method for estimating mutational parameters with data from natural populations subjected to inbreeding assuming deleterious mutation–selection balance*

Genotype	AA	Aa	aa
Frequency	p^2	$2pq$	q^2
$W_{\text{panmixia}} = P$	1	$1 - sh$	$1 - s$
$W_{\text{selfing}} = S$	1	$\dfrac{1}{4} + \dfrac{1 - sh}{2} + \dfrac{1 - s}{4}$	$1 - s$
$T = 4S - 2P$	2	$2 - s$	$2 - 2s$

Means	Variances and covariances
$W_P = 1 - 2U$	$V_{WP} = 2U\overline{sh}$
$W_S = 1 - U\left[1 + \frac{1}{2}\left(\overline{1/h}\right)\right]$	$V_{WT} = 2U\left(\overline{s/h}\right)$
$\delta = W_S - W_P = U\left[1 - \frac{1}{2}\left(\overline{1/h}\right)\right]$	$\text{cov}_{WP,T} = 2U\bar{s}$

Estimations of h	Estimations of U and s
$\hat{h}_b = \dfrac{\text{cov}_{WP,T}}{V_{WT}} = \dfrac{\bar{s}}{\left(\overline{s/h}\right)}$	$\hat{U} \approx \dfrac{2\hat{h}_b\delta}{2\hat{h}_b - 1}$
$\hat{h}_{1/b} = \dfrac{V_{WP}}{\text{cov}_{WP,T}} = \dfrac{\overline{sh}}{\bar{s}}$	$\hat{s} \approx \dfrac{V_{WP}}{2U\hat{h}_{1/b}}$

Source: Deng and Lynch (1996), Deng (1998) and Caballero (2006).

The estimates of the dominance coefficient can be obtained in a similar way to that presented in equations (7.14) and (7.15) and, from these approximations and the previous variances and covariances, we can obtain estimates of U and $\bar{s}$ that would be unbiased if the values of h and s were not correlated and did not violate the supposition of deleterious mutation–selection equilibrium. A negative correlation between s and h produces overestimations of U and underestimations of $\bar{s}$. Using this method with slight modifications, estimates of $\hat{U} = 0.74$, $\hat{s} = 0.14$, $\hat{h} = 0.30$ and $V_M/V_E = 4.6 \times 10^{-4}$ were obtained in *Daphnia* (Deng and Lynch, 1997). Likewise, using an estimate of $\hat{U} = 0.36$ for viability in *Drosophila*, Lynch et al. (1995) estimated an average value of $\hat{h} = 0.39$. If the most typical value in eukaryotes (0.08; Figure 7.3) is used, the estimated average is $\hat{h} = 0.22$, close to the consensus value in eukaryotes.

7.3 Mutation and Recombination

7.3.1 Evolutionary Advantage of Recombination

One of the most important evolutionary aspects for which the fixation of beneficial or deleterious mutations is relevant is the evolution of sex. It is well known that sexual reproduction carries the so-called two-fold cost of sex (Maynard Smith, 1978), because at an equal number of descendants per individual, the size of an asexual population doubles in each generation while that of a sexual population is maintained constant (Figure 7.9). This cost must be added to those inherent in the search for a mate and the competition for mating in the sexual species.

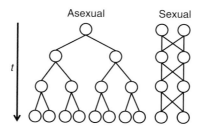

Figure 7.9 Illustration of the two-fold cost of sex. If the number of offspring per individual is two, an asexual population doubles its size each generation (t), while that of a sexual population remains constant.

A factor that can balance the two-fold cost of sex is the evolutionary advantage of recombination, a hypothesis elaborated basically by Fisher (1930) and Muller (1932). If beneficial mutations occur in a population at different loci, the probability of fixing each one will be approximately independent of the others if there is recombination, but this is not the case in the absence of recombination. In sexual species, mutations that arise in different individuals can be combined in the same genome by recombination, as illustrated in Figure 7.10a. However, in asexual species two beneficial mutations can only be incorporated into the same genome if the second one occurs in an individual carrying the first (Figure 7.10b). This implies that the fixation of beneficial mutations is much faster in populations with recombination, providing a net selective advantage to sexual reproduction in this respect, although this is probably not enough to overcome the two-fold cost of sex (Hartfield et al., 2010).

Muller (1964) developed an analogous argument for the case of deleterious mutations. In sexual species, deleterious mutations can be assembled in the same genome by recombination, which facilitates their simultaneous elimination and, in theory, may increase the possibility of recomposing a mutation-free genome by crossbreeding and recombination (Figure 7.10c). In asexual species, in contrast, mutations accumulate in the genomes without it being possible to reconstruct a genome with a mutation number lower than the minimum present at a given generation. Thus, although selection eliminates those genomes with the highest number of mutations, the minimum number of mutations carried by a given genome grows irreversibly with the passage of generations in a process called Muller's Ratchet, by analogy with the toothed mechanism of certain machines which only allow the advance in one direction.

This selective disadvantage of the absence of recombination is illustrated in Figure 7.11, which shows the decline in the mean fitness of a population by mutation when different sizes of the genome (L; in Morgans) are considered. In the absence of recombination ($L = 0$), the fitness decline is much more pronounced than in genomes with recombination. The rate at which mutations are fixed in the absence of recombination depends to a large extent on the magnitude of the genomic class with the fewest mutations, which can be approximated by $N\exp[-U/s]$ if the deleterious mutation rate is U and the selection coefficient of the mutation is s (Haigh, 1978). The fixation probability of mutations (both deleterious or beneficial) for any rate of recombination (including absence of recombination) can be obtained from an extension of the formula to predict the effective population size under linkage (Chapter 5, expression (5.17)) and Kimura's expression (7.1) for the probability of fixation (Santiago and Caballero, 2016).

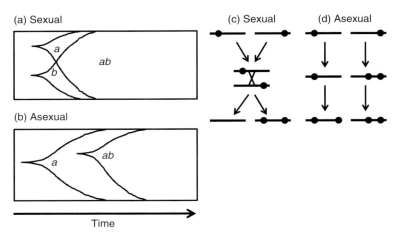

Figure 7.10 (a, b) Adaptations of figures taken from Muller (1932) and Crow and Kimura (1965). (a) Illustrates the advantage of recombination of sexual species in the fixation of beneficial mutations (*a* and *b*) carried by different individuals, which can be incorporated into the same genome by recombination, thus facilitating their fixation. (b) In asexual species, on the other hand, for two beneficial mutations to be incorporated in the same genome, the second one must occur in a carrier of the first. (c) Illustrates the advantage of recombination in reducing the number of deleterious mutations (circles) in a genome (lines) by recombination, whereas in the absence of this, (d) the number of mutations per genome increases irreversibly, which is called Muller's Ratchet.

7.3.2 The Hill–Robertson Effect

Felsenstein (1974) showed that Fisher-Muller's hypothesis of the advantage of recombination in the fixation of beneficial mutations and the elimination of deleterious ones is a phenomenon closely related to the so-called Hill–Robertson effect, described in the context of the response to artificial selection (Hill and Robertson, 1966). When two alleles of different loci segregate, interference can occur between them, so that their probability of fixation is smaller than that if they did not segregate simultaneously. This can occur even in the absence of linkage disequilibrium and interaction between loci. As we saw in Chapter 5, Robertson (1961) pointed out that selection on a locus increases the magnitude of genetic drift on a second unselected (neutral) one, even if this is not linked to the first, due to the random association that can occur between them and is maintained for several generations before gradually disappearing. The result is a reduction of the effective population size ascribed to the neutral locus, expressed by equation (5.13). The Hill–Robertson effect is an extension of this argument when the two loci are linked. In this case, the association between the two loci will persist over a larger number of generations and will be slowly diluted by recombination, involving a progressive reduction of the effective population size associated with the neutral locus as illustrated in Figure 5.1. When two linked genes increase in frequency due to selection, the relevant effective size decreases, reducing the effectiveness of selection. If there is recombination, the effect is reduced but, in the absence of recombination, it is magnified.

Applied to deleterious mutations, the Hill–Robertson effect also predicts an increase in the rate of fixation of deleterious mutations in the absence of recombination. It is, therefore,

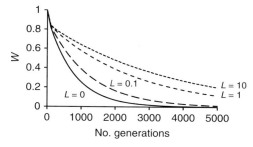

Figure 7.11 Decrease in the mean fitness of a haploid population of size $N = 100$ in which deleterious mutations are generated with rate $U = 0.2$ per genome and generation with mutational effects (s) exponentially distributed with mean 0.03, considering different sizes of the genome (L; in Morgans). $L = 0$ implies complete absence of recombination, while the case with free recombination is very similar to that of $L = 10$.

evident that this effect represents the same phenomenon as that described by Fisher and Muller. The ultimate reason for interference between loci induced by the Hill–Robertson effect is the development of negative linkage disequilibrium as a consequence of selection. Let us assume two mutations occur at two linked loci. If they are beneficial and are in coupling phase (on the same chromosome) they will tend to be fixed quickly and, if they are deleterious, will tend to be lost easily. But if, on the contrary, they are in repulsion phase, they will remain in the population for longer producing negative disequilibrium. Although this effect was postulated as a consequence of selection, it can also be generated by drift of independent loci.

Problems

7.1 In a population with size $N = 100$ individuals and effective size $N_e = 50$ an additive mutation occurs with favourable effect $s = 0.1$ in the homozygote. (a) What is the probability that the mutation will end up being fixed in the population? (b) What would be that probability if the mutation were recessive? (c) If the mutation were deleterious and additive, what would be its mean persistence time and the average number of copies present in the population prior to its elimination?

7.2 In a mutation accumulation experiment where 100 crossbred brother–sister lines of *Drosophila melanogaster* were used, the mean viabilities initially and in generation 100 were 0.8 and 0.35, and the corresponding variances between lines 0.05 and 0.14, respectively. If the distribution of effects of mutations is assumed to be exponential, what are the estimates that can be obtained for the genomic mutation rate, U, the mean effect of mutations, $\bar{s}$, and the mutational variance, V_M?

7.3 A strain of *Drosophila melanogaster* devoid of genetic variation was created by means of crosses using balanced chromosomes. With this strain, lines were founded with effective size $N_e = 20$ and maintained for 100 generations in panmixia. We then evaluate a neutral quantitative trait whose mutational variance has been estimated to be $V_M = 0.0005$. What would be the approximate expected value of the variance between lines and the additive variance of the trait in the last generation?

7.4 The rate of mutation of lethal recessives is 2.6×10^{-6} per gene and generation and the frequency of such alleles in a population is 1.4×10^{-4}. Are these estimates compatible with complete recessivity of the lethal alleles? If not, what would be the most likely value of the dominance coefficient, h?

7.5 The following data correspond to the fitness value of six individuals of a population of large size and the mean fitness of their progeny by self-fertilization. Is it possible to estimate the deleterious genomic mutation rate, U, the mean effect of the mutations, $\bar{s}$, and its average dominance coefficient, $\bar{h}$?

Fitness of individuals	0.78	0.75	0.81	0.79	0.83	0.85
Fitness of selfed progeny	0.63	0.61	0.70	0.69	0.72	0.74

Self-Assessment Questions

1 Multicellular eukaryotes generally have a higher rate of mutation per round of replication than prokaryotes or unicellular eukaryotes.

2 The probability of fixation of a mutation with selective advantage $s = 0.01$ in a population with size $N = 10$ is approximately equal to that of a neutral mutation.

3 The Bateman–Mukai method of estimation of the mutation rate and the mean effect of mutations in mutation accumulation experiments will give overestimations of U and underestimates of the mean s if the effect of the mutations is highly variable.

4 Although an enzymatic mutation had additive gene action for enzyme activity, its effect on the flux of the metabolic pathway in which it participates may be recessive if the effect of the mutation is large.

5 It has been estimated that in *E. coli* and yeast approximately 1% of the mutations can be beneficial.

6 The additive model implies independence of effects between loci for fitness.

7 For a purely neutral trait, the genetic variance of equilibrium increases indefinitely by mutation as the effective population size increases.

8 Welfare and medicine are expected to increase the mutation load in human populations.

9 The absence of recombination is the reason why faster fixation of deleterious mutations occurs in species with asexual reproduction.

10 Muller's Ratchet also occurs in autogamous species.

8 Consequences of Inbreeding

Concepts to Study

- Inbreeding depression
- Dominance and overdominance hypotheses of inbreeding depression
- Inbreeding load and number of lethal equivalents
- Q_{ST} index of population differentiation for quantitative traits
- Outbreeding depression
- Mutational meltdown
- Self-incompatibility and heterostyly in plants
- Automatic selective advantage of self-fertilization
- Heterosis
- General and specific combining ability
- Minimum effective population size of a viable population

Objectives for Learning

- To learn how to calculate the components of the genotypic variance in a non-panmictic population
- To understand the concepts of inbreeding depression and inbreeding load, and their interpretation under the dominance and overdominance hypotheses
- To understand the redistribution of genetic variance in sets of lines of small census size
- To learn how to estimate the index of genetic differentiation for quantitative traits, Q_{ST}, and understand its possible biases
- To comprehend the concept of outbreeding depression and its implications
- To learn the meaning of mutational meltdown and its consequences
- To learn how to predict the magnitude of the inbreeding load that is purged by natural selection
- To know the functioning of self-incompatibility systems in plants
- To recognize the hypothesis of the automatic selective advantage of self-fertilization
- To understand the concept of heterosis and its use in the genetic improvement of plants and animals
- To know the meaning of the minimum effective size of a viable population
- To know the procedures used for the conservation of diversity in *ex situ* conservation programmes and its consequences
- To understand how molecular information provided by neutral markers can be used as a complement to the estimates of quantitative genetic variation

8.1 Effects of Inbreeding on the Mean and Variance of Quantitative Traits

As we saw in Chapter 4, inbreeding produces changes in the genotype frequencies that imply an increase in the frequency of homozygotes and a reduction in that of heterozygotes (equations (4.15)–(4.17)). These changes usually alter the mean and variance of the quantitative traits, sometimes with important consequences for the population. Inbreeding depression, that is, the change generated by inbreeding in the mean of quantitative traits, is one of those consequences, and it is manifested as a deterioration of fitness of consanguineous individuals relative to non-consanguineous ones (Charlesworth and Charlesworth, 1999; Charlesworth and Willis, 2009). Inbreeding depression is a phenomenon well known by plant and animal breeders and conservation managers, who generally try to prevent matings between related individuals in order to avoid an increase in inbreeding. The inverse phenomenon is also common, heterosis or generation of vigorous offspring in crosses between unrelated inbred individuals, a method used for a long time in the genetic improvement of plants and animals (Darwin, 1876). The negative consequences of inbreeding are evident in our own species, that is, it is well known that the crossing of related individuals usually leads to the appearance of malformations and hereditary diseases in their offspring. It is perhaps because of this that most human societies have prohibited incestuous unions and, also, many wild species have innate mechanisms that impede or minimize the likelihood of such crossings. Finally, the changes generated by inbreeding on the variance of quantitative traits have an impact on the response to selection and the adaptation of populations to the environment where they live.

8.1.1 Decomposition of the Genotypic Value and Variance in a Non-Panmictic Population

In Chapter 3 we studied the partition of the genotypic value and variance for a locus in its additive and dominance components in the case of a panmictic population. In this section we will analyse this partition for the case of a non-panmictic population in which matings occur that generate a certain inbreeding F. The corresponding expressions are presented in Table 8.1 and a numerical example in Table 8.2.

As already indicated, inbreeding produces an increase in the frequency of homozygotes and a reduction in that of heterozygotes (Table 8.1). From the sum of products of genotypic values by their frequencies, the contribution of the locus to the population mean is obtained, $M = aq + 2dpq - 2dpqF$. The difference between this and the corresponding (M_0) for a panmictic population ($F = 0$) (Table 3.2, Chapter 3) is the inbreeding depression (δ),

$$\delta = M_0 - M = 2dpqF = F\sqrt{V_D}, \tag{8.1}$$

that is, it depends on the root square of the dominance variance (equation (3.10), Chapter 3).

Note, first, that for inbreeding depression to occur, it is a necessary condition that there is allelic segregation in the locus ($p, q \neq 0$) and inbreeding ($F \neq 0$), and that the depression is maximum with intermediate allele frequencies. Second, inbreeding depression requires recessiveness, that is, that the dominance effect (d) is different from zero. With additive gene action ($h = 0.5; d = 0$), the mean of the character is not affected by inbreeding and, therefore, there is no inbreeding depression. Note also that, despite what the word depression may suggest, the change in the mean with inbreeding is not necessarily a reduction, since this change occurs in the direction of the recessive allele value. If this reduces the mean, as in the example of Table 8.2, the mean of the trait will decrease with inbreeding, but if it increases it, the opposite will occur. This is precisely the reason for which inbreeding depression generally implies a reduction in the mean when it comes to reproductive traits since, as we saw in Chapter 7, most deleterious mutations that segregate in populations are recessive. In essence, by increasing the frequency of the homozygotes, inbreeding reveals the deleterious nature of alleles whose action, in the absence of inbreeding, would be masked in the heterozygotes due to their low frequency. When a mutation is rare, it is unlikely that two unrelated individuals of the population are carriers of it. However, two relatives will have a greater chance of each carrying a copy of any deleterious mutation from a common ancestor. These mutations could, therefore, appear in homozygotes in their descendants, manifesting their detrimental effect.

This explanation of inbreeding depression is the so-called dominance hypothesis. An alternative hypothesis widely debated is the overdominance one, which attributes this type of gene action to inbreeding depression for fitness. This hypothesis, currently considered a minority view given the scarcity of empirical data that support it (Hedrick, 2012), has been, however, suggested by the apparent overdominance often found for genetic markers, which can be generated by linkage disequilibrium under a model of deleterious mutations. Let's illustrate the situation with the following example. Assume a neutral marker (M) with two alleles (M_1 and M_2) and that each of them is in complete linkage disequilibrium with a deleterious allele at another locus, M_1 with allele a of locus A (with alleles A and a) and M_2 with allele b of locus B (with alleles B and b). Therefore, there are two gametic types in repulsion phase, aM_1B and AM_2b, and three genotypes, aM_1B/aM_1B, aM_1B/AM_2b and AM_2b/AM_2b. Suppose, to simplify, that the fitnesses are 1, $1 - sh$ and $1 - s$ for the two loci subject to selection. We have then that the genotypic fitnesses are $(1 - s)^2$ for each of the homozygotes and $(1 - sh)^2$ for the heterozygote. If $h < (1 + s)/2$ there will be apparent overdominance for marker M, even though none of the loci show overdominant gene action. It is possible, however, that real overdominance for some loci is responsible for some of the observed inbreeding depression in certain situations.

The consequences of inbreeding depression are evident in many domestic species that are highly selected and subject to high inbreeding rates. For example, dog breeds are maintained with high inbreeding and many of them present characters that would reduce their survival in the wild, such as predisposition to arthritis of the Labrador Retriever or respiratory problems of bulldogs and Pekingese. In the next section we will study the procedure for estimating inbreeding depression from data on inbred populations.

As said, expression (8.1) refers to a single locus. When considering all the loci that affect the trait, the inbreeding depression would be the sum (Σ) of this expression for all of them, $2F\Sigma d_i p_i q_i$, if an interlocus additive model is assumed. If for most of the loci that control the trait, the recessive allele is the one that reduces its value, $\Sigma d_i p_i q_i > 0$, it is said that there is directional dominance and there will be net depression. However, if the recessive allele reduces the expression of the character in some loci ($d_i > 0$), but increases it in others ($d_i < 0$), the positive and negative values can be cancelled out ($\Sigma d_i p_i q_i \approx 0$) with net absence of depression. In this situation it is said that there is non-directional dominance, that is, that there is no depression despite the existence of dominance for the loci that affect the trait. In the characters related with fitness there is usually, as already indicated, directional dominance

Table 8.1 *Decomposition of the genotypic value of a locus and calculation of the mean and additive variance in a population with inbreeding coefficient* F

Genotype	A_1A_1	A_1A_2	A_2A_2
Genotype frequencies	$p^2 + pqF$	$2pq(1-F)$	$q^2 + pqF$
Genotypic value (G)	0	ah	a
G value deviated from M	$0 - M$	$ah - M$	$a - M$
Additive value (A)	$-2q\alpha$	$(p-q)\alpha$	$2p\alpha$
Dominance deviation (D)	$-2dq(q+pF)\frac{1-F}{1+F}$	$\frac{2d[F+pq(1-F)^2]}{1+F}$	$-2dp(p+qF)\frac{1-F}{1+F}$
Dominance effect (d)	$d = a(h-1/2)$		
Genotypic mean (M)	$M = aq + 2dpq - 2dpqF = aq + 2dpq(1-F)$		
Average effect of allelic substitution (α)	$\alpha = \frac{e}{1+F} = ah - \frac{2d(q+pF)}{1+F}$		
Average allelic excess (e)	$e = \alpha(1+F) = ah(1+F) - 2d(q+pF)$		
Additive variance (V_A)	$V_A = 2\alpha epq = 2\alpha^2 pq(1+F)$		

Note: Compare with Tables 3.2 and 3.3 (Chapter 3), where the case of a non-inbred population is presented.

Table 8.2 *Example of the contribution of a locus to the mean and variance of a trait in a panmictic or inbred population*

Inbreeding (F)	$F = 0$			$F = 0.5$		
Genotype	A_1A_1	A_1A_2	A_2A_2	A_1A_1	A_1A_2	A_2A_2
Genotype frequencies	0.36	0.48	0.16	0.48	0.24	0.28
Genotypic value (G)	0	8	10	0	8	10
G value deviated from M	-5.44	2.56	4.56	-4.72	3.28	5.28
Additive value (A)	-4.48	1.12	6.72	-4.16	1.04	6.24
Dominance deviation (D)	-0.96	1.44	-2.16	0.75	-2.99	1.28
Dominance effect (d)	3			3		
Genotypic mean (M)	5.44			4.72		
Average effect of allelic substitution (α)	5.60			5.20		
Average allelic excess (e)	5.60			7.80		
Additive variance (V_A)	15.05			19.47		
Dominance variance (V_D)	2.07			1.61		
Genotypic variance (V_G)	17.13			21.08		
Inbreeding depression ($2dpqF$)	0			0.72		

Note: Allele frequencies $p = 0.6$ and $q = 0.4$ for alleles A_1 and A_2, respectively, homozygous effect $a = 10$ and coefficient of dominance $h = 0.8$. Parameters from Table 3.2.

and, therefore, inbreeding depression. However, for many other characters, for example some morphological ones, usually dominance is not directional and, therefore, there is little or no depression. We saw in Figure 3.4 of Chapter 3 that reproductive traits have a greater dominance variance than morphological ones. This observation is parallel to the magnitude of the corresponding inbreeding depression, greater in the reproductive traits than in the morphological ones (DeRose and Roff, 1999), in line with the dependence of inbreeding depression on the dominance variance (equation (8.1)).

Table 8.1 also shows the partition of the genotypic value of a locus (G) in additive (A) and dominance (D) values in an inbred population. Recall that this partition, detailed in Chapter 3, involved obtaining the average effect of allelic substitution (α) defined as the difference between the average effects associated with each allele that depend on allele frequencies and the population mean, M; and that the additive values of the different genotypes are the sum of the average effects of the alleles carried by the relevant individual. As we saw in Chapter 3, the average effects can be obtained in two ways. One is to calculate them as the mean of the trait for the individuals that carry a copy of the allele in question, giving rise to expressions (3.4) and (3.5). The other, which corresponds to the original definition of Fisher (1918), is obtained as the regression coefficient of the genotypic values (G) on the allelic dose (zero, one and two alleles A_2 for the three genotypes), such as illustrated in Figure 3.1 (Chapter 3). In the case of panmictic populations, the two procedures lead to the same result: the average effect of allelic substitution, $\alpha = ah - 2dq$ (equation (3.6)). However, in non-panmictic populations, each procedure produces a different solution (Fisher, 1941; Kempthorne, 1957). The first procedure gives rise to the so-called average allelic excess (e), whose general expression is $e = ah(1 + F) - 2d(q + pF)$ (Table 8.1), while the second results in the average effect of allelic substitution, $\alpha = e/(1 + F)$, equivalent to that obtained in Chapter 3. When the population is panmictic ($F = 0$), the average effect and average excess coincide (Table 8.2). The additive variance can then be obtained in a general way as

$$V_A = 2e\alpha pq = 2\alpha^2 pq(1 + F).\tag{8.2}$$

Expression (8.2) seems to indicate that inbreeding always increases the additive variance of a panmictic population in a fraction F. However, this is not the case, since the average effects may be different in panmictic and inbred populations. Figure 8.1 illustrates some examples of additive and dominance variance for a locus using the genotypic values of Table 8.2 and different frequencies and dominance coefficients. When the locus is additive ($h = 0.5$), V_A increases linearly with F, and the same occurs if the allele frequency is intermediate ($q = 0.5$) regardless of the value of h. But if $q \neq 0.5$ and there is dominance, V_A can increase or decrease with respect to that of the panmictic population. The dominance variance, V_D, which expression can also be obtained as the product of the squared dominance deviations by the genotype frequencies from Table 8.1, can initially increase with F to decline to zero when $F = 1$ (Figure 8.1).

The above models assume absence of epistatic interactions between loci. Providing the population is in gametic phase equilibrium only epistatic interactions involving dominance contribute to inbreeding depression (Walsh and Lynch, 2018, p. 257).

8.1.2 Estimation of Inbreeding Depression and Inbreeding Load

Given that the characters that show the greatest inbreeding depression are reproductive ones, we will consider the model of fitness values indicated in Chapters 2 and 7, which is presented in Table 8.3 for a non-panmictic population. In Chapter 7 we defined the mutation load as the reduction in fitness due to the segregation of deleterious mutations in the population. If the population is panmictic ($F = 0$) and we consider the model of Table 8.3, the mean fitness of the population for a locus is $W_0 = 1 - 2pqsh - q^2s$ and, therefore, the mutation load is $L = 1 - W_0 = 2pqsh + q^2s$. In an

Table 8.3 *General model of fitness for a locus in an inbred population*

Genotype	A_1A_1	A_1A_2	A_2A_2
Genotype frequency	$p^2 + pqF$	$2pq(1-F)$	$q^2 + pqF$
Fitness	1	$1 - sh$	$1 - s$
Mean fitness	$W_F = 1 - qFs - q^2s(1-F) - 2pqsh(1-F)$		

Note: The first of the three terms that are subtracted from the unit in the mean fitness (W_F) indicates the probability of mortality of homozygotes by inbreeding and the last two that of the homozygotes and heterozygotes in the non-inbred population.

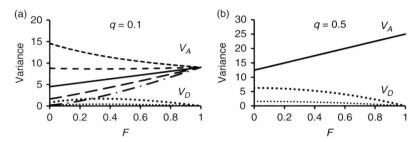

Figure 8.1 Additive (V_A) and dominance (V_D) variances for a locus with allele frequencies p and q, homozygous effect $a = 10$ and coefficient of dominance h (see Table 8.1) in a population with inbreeding coefficient F and frequencies (a) $q = 0.1$ and (b) 0.5. The continuous or broken lines correspond to V_A with values (from top to bottom) of $h = 1, 0.75, 0.5, 0.25$ and 0 (panel a) or any value of h (panel b). The dotted lines of both panels correspond to V_D with values of $h = 0$ or 1 (thick points) and $h = 0.25$ or 0.75 (thin points).

inbred population the mean fitness (W_F) can be obtained analogously by the product of the frequencies and fitnesses in Table 8.3 and the resulting load is $L = 1 - W_F = Fqs + (1-F)$ $(2pqsh + q^2s)$ (Morton et al., 1956), where the first term is the load in consanguineous individuals and the second in non-consanguineous ones. Reordering, we get

$$L = \left(2pqsh + q^2s\right) + F\left(qs - q^2s - 2pqsh\right) = A + FB. \tag{8.3}$$

The first term (A) is the load expressed in the panmictic population, to which the possible decline in the average fitness due to environmental factors could be added. The second contains the B factor, the load masked in the heterozygotes, which is expressed with inbreeding, for which it is called inbreeding load. Substituting the dominance effect, which for the model of Table 8.3 is $d = s(1/2 - h)$, in B we obtain $B = 2dpq$, and $BF = \delta = 2dpqF$, that is, the magnitude of the inbreeding depression (equation (8.1)).

If we now consider a multiplicative multilocus model, that is, with independent mutational effects, the total load would be expressed as $L = 1 - W_F = e^{-(A + BF)}$ and the logarithm of mean fitness is $\ln(W_F) \approx -A - BF$, where

$$A = -\ln(W_0); \qquad\qquad BF = -\ln(W_F/W_0). \tag{8.4}$$

Therefore, the mean fitness of the population is

$$W_F = W_0 e^{-BF}, \tag{8.5}$$

and the inbreeding depression is

$$\delta = 1 - e^{-BF}. \tag{8.6}$$

If a single W_F estimate is available, the inbreeding load is obtained from equation (8.4). For example, the estimated average proportion of mortality in human populations in the offspring of unrelated individuals is 0.069, while the same proportion is 0.177 in children of cousins (Stern, 1973, p. 494). Subtracting these values from the unit we have $W_0 = 0.931$ and $W_F = 0.883$. Since the inbreeding coefficient of inbred individuals is 1/16, we have that $B = -\ln(0.883/0.931)/(1/16) = 0.84$, which can also be approximated as the difference in mortality rates divided by F, that is, $B \approx (0.117 - 0.069)/(1/16) = 0.77$.

When the mean fitness of the population is available for different levels of inbreeding, the rate of inbreeding depression or inbreeding load can be estimated from the regression of the logarithm of mean fitness on inbreeding. Figure 8.2 shows some examples of inbreeding depression in brother–sister mated lines of *Drosophila melanogaster*. Figure 8.2a presents the decline in a competitive index between inbred and non-inbred individuals obtained experimentally for different values of F. The decline pattern is similar to what would be obtained with a multilocus multiplicative fitness model (Figure 7.8, Chapter 7) and, in fact, the corresponding decline in logarithmic scale is linear, as expected for that model (Figure 8.2b). The regression slope is −3.13. The depression rate is also usually expressed as a percentage of the mean of the character in the non-inbred population (ln $(W_{F=0}) = 1.37$ in the example of Figure 8.2b), therefore $\delta = 3.13/1.37 = 2.3$, that is, the estimated depression rate is 2.3% of the mean of the trait for each 1% increase in inbreeding. In a meta-analysis in animals, DeRose and Roff (1999) found an average inbreeding depression rate of 0.47% of the mean per 1% increase in inbreeding for life-history traits in animals while only 0.09% for morphological traits. Likewise, Leroy (2014) found an average of 0.14% for livestock species.

On some occasions, like that in Figure 8.2c, the depression is non-linear but accelerates for high levels of inbreeding. This result may suggest the existence of synergistic epistasis, as described in Figure 7.8 (Chapter 7). In the case of Figure 8.2c the fit of the dots to a quadratic regression is better than to a linear one. Nevertheless, one way to prove statistically the non-linearity of inbreeding depression, avoiding the problems inherent to the regression, is to compare the change in the mean of the character between two low levels of inbreeding and two high levels that do not overlap with the first ones, and verify the statistical significance of the difference between both changes through a Student's t test (Lynch and Walsh, 1998, p. 267).

When inbreeding depression is estimated experimentally throughout generations in inbred lines and, in addition, there are available measures of the trait in one or more non-inbred lines contemporaneous with the inbred ones, it is possible to correct the changes observed in the mean due to purely environmental factors (Lynch and Walsh, 1998). The method consists of performing the partial regression of the phenotypic mean of the inbred lines in generation t ($Z_{F,t}$) on the average of the control lines ($Z_{0,t}$) and the coefficient of inbreeding (F_t),

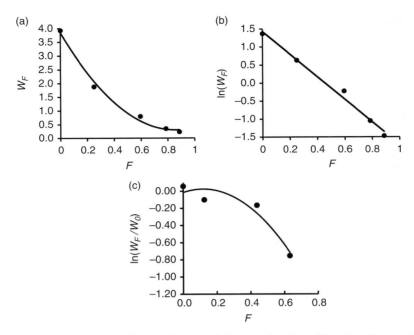

Figure 8.2 Changes in the average fitness of the population as a function of the inbreeding coefficient in brother–sister mating lines in *Drosophila melanogaster*. (a) Data of Latter and Robertson (1962) for a competitive index between flies of inbred lines and of a control strain with visible markers. (b) Data of panel (a) when the fitness is represented in logarithmic scale. (c) Productivity data (number of pupae per female; data from Domínguez-García et al., 2019), expressed as the log of the ratio between the average productivity for lines with different levels of inbreeding (W_F) and that of a simultaneous outbred control (W_0). The curved or straight lines in each case indicate the quadratic or linear fitting to the data.

$$Z_{F,t} = a + bZ_{0,t} + \delta F_t + e_t, \tag{8.7}$$

where a is a constant, b is a partial regression coefficient, δ is the inbreeding depression corrected by environmental trends and e_t is the deviation from the mean to the regression in generation t. The mean of the inbred lines corrected by the environmental effects is $Z^*_{F,t} = Z_{F,t} - b(Z_{0,t} - Z_0)$, where Z_0 is the mean of the control lines in all generations. The inbreeding depression δ can also be obtained as the regression of $Z^*_{F,t}$ on F_t. Lynch and Walsh (1998) provide more details about the methods for estimating the inbreeding depression, its sampling variance and statistical tests of hypotheses. Note, however, that equation (8.7) applies strictly to characters with additive gene action between loci and in the absence of selection.

In the case of fitness components, the inbreeding load (B) is often expressed as the number of lethal recessive equivalents, that is, a group of deleterious mutations that, combined in an additive manner, would produce the same inbreeding depression as a lethal recessive allele. For example, a lethal recessive equivalent could be constituted by 10 recessive mutations whose individual effects in homozygosis are a reduction of 10% in fitness, or 100 mutations with effects of 1% and so on.

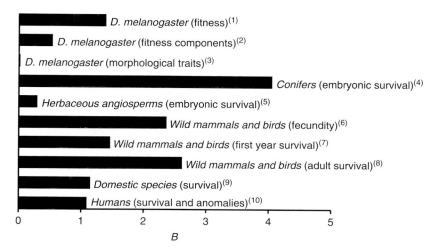

Figure 8.3 Summary of averages of estimates of inbreeding load (B) expressed in number of lethal equivalents for various species and traits (estimates compiled by Lynch and Walsh (1998) and O'Grady et al. (2006), and estimates in *Drosophila* from Bersabé and García-Dorado (2013) and López-Cortegano et al. (2016)). [1] Measurement of competitive fitness against a marker strain and productivity. [2] Egg-to-pupa viability, egg-to-adult viability, fecundity in males and females, longevity in males and mating success. [3] Body weight, wing and thorax length and number of sternopleural and abdominal bristles. [4–5] Embryonic survival for different species. [6–8] Estimates of mean fertility and early and late survival in various species of mammals and wild birds. [9] Averages of survival estimates in sheep, pigs, cows, quail and chickens. [10] Average survival estimates, congenital anomalies, mental retardation and congenital heart disease. Standard deviations of the estimates are [1] 0.86, [2] 0.28, [3] 0.05, [4] 1.24, [5] 0.33, [6] 1.33, [7] 0.79, [8] 2.55, [9] 1.24 and [10] 0.81.

Figure 8.3 presents a summary of average estimates of the inbreeding load expressed in number of lethal equivalents for different groups of species and traits. Estimates in *Drosophila melanogaster* indicate $B \approx 1$–2 for fitness measures. For morphological traits, however, there is practically no inbreeding depression, as previously indicated. In this species, the techniques of chromosomal analysis by strains with balanced chromosomes (Figure 5.2, Chapter 5) have also allowed us to deduce that approximately half of the inbreeding load for viability is attributable to lethal genes (Simmons and Crow, 1977), although this does not necessarily apply to other species.

The embryonic survival in conifers leads to very high B values, around four lethal equivalents, that contrasts with the much lower values corresponding to annual herbaceous angiosperms. The explanation for the greatest mutational load in the former may be their higher rate of mutation per generation, perhaps caused by their long generation interval (Lynch and Walsh, 1998).

The estimates of B for mammals and wild birds have average values between 1.5 and 2.5 lethal equivalents. Given that the traits analysed for these species (fecundity and early and late survival) are partial components of fitness, this indicates that the overall inbreeding load for fitness can be of the order of six lethal equivalents (O'Grady et al., 2006). Most recent estimates confirm these values or others even higher (Hedrick and García-Dorado, 2016), particularly for estimates obtained with genomic data. This high load can have a large impact on the viability of populations and, for example, with $B = 6$, a 5% increase in inbreeding implies an expected inbreeding depression of $\delta = 1 - e^{-BF} = 1 - e^{-6 \times 0.05} = 0.26$, that is, a decline in fitness of 26%, although this prediction is an overestimate because it does not consider the effect of the genetic purging that we will see later.

Finally, the average B values for viability traits in domestic animals and survival characters and congenital anomalies in humans are somewhat higher than one lethal equivalent. The estimates of B are higher for wild species than for domestic ones (Leroy, 2014), a result compatible with the hypothesis that inbreeding depression is greater in natural environments than in benign ones provided to domestic or captive species (Armbruster and Reed, 2005). The interaction between inbreeding depression and environmental conditions is a widely debated topic. Wright (1977) already suggested that more consanguineous individuals would be more sensitive to environmental stress and, therefore, that inbreeding depression would be greater in adverse environments. Experimental results generally confirm this hypothesis (Fox and Reed, 2011), which is explained partly because in adverse environments the phenotypic variation increases due to the interaction between inbreeding depression and stress (Reed et al., 2012). However, in 24% of the studies the opposite result was found, that is, there was less depression in more stressful environments, which suggests that the stress–inbreeding depression relationship cannot be generalized.

The magnitude of the inbreeding load can be predicted for a population at deleterious mutation–selection balance as a function of the mutation rate and the mutational effects. Given that the load is $B = \sum 2d_i p_i q_i$, where the sum is for all mutations, substituting $d_i = s_i(\frac{1}{2} - h_i)$ we obtain $B = \sum s_i p_i q_i (1 - 2h_i)$. Recalling that at the deleterious mutation–selection balance $q_i = u_i/(s_i h_i)$ (equation (7.10), Chapter 7) and $p_i \approx 1$ for deleterious mutations in large populations, we obtain

$$B \approx \sum_i u_i \left(\frac{1}{h_i} - 2 \right) = U \left(\overline{1/h} - 2 \right), \tag{8.8}$$

that is, B is a function of the genomic mutation rate U and the harmonic mean of h. For example, if we assume $U = 0.2$ and a harmonic mean of h of 0.1, that is, $\overline{1/h} = 10$, we obtain a prediction of $B = 1.6$, which is a reasonable value for fitness components in many species (Figure 8.3). Assuming constant values of s and h we have $B \approx 2Ud/sh$.

The estimates of A (equation (8.4)), which include the load of a panmictic population and sources of mortality due to environmental causes, are usually lower than those of B. For example, for survival characters and congenital anomalies in humans, the value of A is approximately 10% that of B, on average, reaching 30% for domestic species, but it is expected that the values of A are even larger in wild species in their natural environment due to the greater impact of mortality for non-genetic natural causes in this environment.

Notice, finally, that with the values of A and B it is possible to estimate the dominance coefficient of mutations under the assumption of deleterious mutation–selection balance. Since $B = \sum 2d_i p_i q_i = \sum s_i p_i q_i (1 - 2h_i)$ and $A + B = \sum s_i q_i$, the ratio $B/(A + B) \leq 1 - [(\sum 2s_i h_i q_i)/(\sum s_i q_i)]$ and, substituting $q_i = u_i/(s_i h_i)$, we obtain that $A/[2(A + B)] \geq (\sum u_i)/(\sum u_i/h_i)$, that is, the ratio $A/[2(A + B)]$ is an estimate of the harmonic mean of h for new mutations, which will be an overestimation if there are causes of environmental mortality that inflate the value of A. The results obtained with the values of A and B mentioned above indicate a harmonic mean of h between 0.02 and 0.15, compatible with the arithmetic means of h presented in Chapter 7 (Figure 7.4), since harmonic means are expected to be lower than arithmetic means.

8.2 Inbreeding in Panmictic Populations of Reduced Census Size

There are notable differences between the inbreeding resulting in panmictic populations of reduced population size and that corresponding to a population of high census size and non-panmictic mating. In a theoretically infinite population with non-panmictic mating the changes in the mean and genetic variance are reversible in the sense that they obey changes in the genotype frequencies but not in the allele frequencies. Thus, if the population repro-duces in panmixia, the population mean and its variance would revert to the state prior to the inbreeding process. In populations of reduced size, on the contrary, the changes in mean and variance are irreversible because they involve changes in the allele frequencies, with chance loss and fixation of alleles, which can have important evolutionary consequences and for the conservation of genetic resources.

8.2.1 Redistribution of Within- and Between-Line Genetic Variance

Suppose that a large size population in which a biallelic locus A with alleles A and a segregates with frequencies p_0 and q_0, respectively, is subdivided into multiple lines of size N that are maintained individually in panmixia in the absence of selection. The situation is similar to that illustrated in Figure 2.4 (Chapter 2). The allele frequencies in the different lines will change erratically due to genetic drift, with its expected variance in generation t expressed by equation (2.9), $\sigma_{q,t}^2 = p_0 q_0 F_t$. The genotype frequencies in the set of lines are given by equations (2.10) to (2.12), which are also those of Table 8.1, that is, $p_0^2 + p_0 q_0 F_t$, $2p_0 q_0(1 - F_t)$ and $q_0^2 + p_0 q_0 F_t$ for the genotypes AA, Aa and aa, respectively, where F_t is the inbreeding coefficient reached in generation t due to the finite census size of the lines. Suppose now that the locus has an effect on a quantitative trait with genotypic value a in homozygosis and additive gene action. The genetic variance in the starting population will be $V_{G0} = 2a^2 p_0 q_0 = (a^2/2)p_0 q_0$, since the average effect of allelic substitution is $\alpha = a/2$ in this case. In generation t, the variance in a line i will be $V_{Gi,t} = (a^2/2)p_{i,t}q_{i,t}$, where $p_{i,t}$ and $q_{i,t}$ are the corresponding allele frequencies, which will be in Hardy–Weinberg equilibrium since mating in each line is panmictic. The average variance of the trait in the lines will then be $V_{GW} = E[V_{Gi,t}] = (a^2/2)E[p_{i,t}q_{i,t}]$, where E indicates the expected value or average between lines. Since the expected value of the heterozygote frequency in generation t is $2p_0 q_0(1 - F_t)$, we have that the expected genetic variance in the lines is $E[V_{Gi,t}] = (a^2/2)p_0 q_0(1 - F_t)$ and that

$$V_{GW,t} = V_{G0}(1 - F_t) \qquad (8.9)$$

(Wright, 1951). That is, in the case of an additive and neutral locus, the additive variance is reduced linearly with inbreeding in the same fraction as the expected heterozygosity and, after many generations, uniformity within the line ($V_{GW} = 0$) will be reached.

Now consider the variance of the line means for the trait. In generation t the expected mean of the lines is given by $M = aq + 2dpq$ (Table 3.2), which, with additive gene action, is reduced to $M = aq$. The variance of the mean, in the absence of selection, is then $V(M) = V(aq) = a^2 V(q) = a^2 \sigma_{q,t}^2 = a^2 p_0 q_0 F_t$ and, substituting the expression of the variance in the starting population, V_{G0}, we have

$$V_{GB,t} = 2F_t V_{G0} \tag{8.10}$$

(Wright, 1951). That is, in the case of an additive locus, the variance of the means of the lines for the trait increases linearly with inbreeding. The total genetic variance would be, therefore,

$$V_{GT,t} = V_{GW,t} + V_{GB,t} = V_{G0}(1 + F_t), \tag{8.11}$$

which coincides with that of a large non-panmictic population (equation (8.2)). The corresponding observable components of phenotypic variance would approximately be $V_{PW} = V_{GW} + V_E - (V_{PW}/n) = n(V_{GW} + V_E)/(n+1)$, and $V_{PB} = V_{GB} + (V_{PW}/n)$, respectively, where V_E is the environmental variance, which is assumed to be constant, and the term V_{PW}/n is the variance of the sampling error of the line means corresponding to estimates based on n data per line. The within- and between-line components of the genetic variance must be obtained empirically through an analysis of variance.

The evolution of the within- and between-line variances with the inbreeding coefficient expressed in equations (8.9) and (8.10) is illustrated with an example in Figure 8.4a. In this case, the initial genetic (additive) variance is 0.125 which is reduced to zero in the within-line component and increases to $2V_{G0} = V_{GT} = 0.25$ in the between-line component.

The above expressions apply exclusively to an additive model. When there is dominance, the within- and between-line components can evolve differently, as demonstrated by Robertson (1952). Some examples of this evolution for a recessive locus with different initial frequencies are presented in panels (b–d) of Figure 8.4. When the recessive allele is initially at a frequency of 0.5 (panel b), the component of within-line variance decreases with the increase in inbreeding and the between-line component increases to $2V_{G0}$, as in the additive model (panel a). However, as expected in the case of fitness, if the frequency of the recessive allele is low (panel c), the within-line variance (as well as the additive and dominance variances) increases to a certain value of F and then declines. The reason for this temporary increase in the within-line variance is the following. As we already know, genetic drift produces a random change in the allele frequencies with the same expected magnitude both for increased or decreased frequency. However, in the case of a recessive allele, the corresponding change in the additive variance is not the same, as can be seen in Figure 3.2 (Chapter 3). If the frequency of the allele is low, a small increase in its frequency increases the additive variance to a greater extent than the reduction in variance corresponding to the same decrease in frequency. Therefore, changes in allele frequency by genetic drift slightly increase the additive variance up to a certain level of inbreeding. When the population is highly consanguineous, the genetic variance is obviously reduced to zero due to the fixation of one of the alleles. Finally, if the recessive allele was initially at high frequency (panel d), the

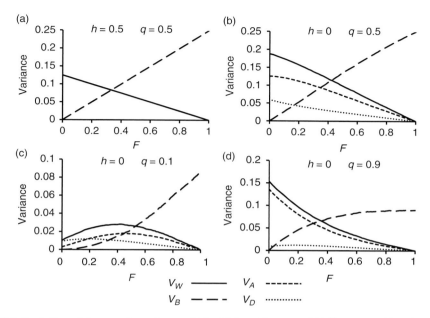

Figure 8.4 Evolution of the genetic variance within (V_W) and between (V_B) lines, as well as that of the additive (V_A) and dominance (V_D) variance, where appropriate, for increasing values of the inbreeding coefficient (F) for a locus with effect on a quantitative trait $a = 1$ in homozygosis and ah in heterozygosis and initial frequency q. The results have been obtained by simulation of a set of 100 lines of size $N = 20$ maintained independently for 150 generations. The V_W and V_B components are obtained by analysis of variance. (a) Additive model, where the evolution of V_W ($= V_A$) and V_B is given by equation (8.9) and (8.10), respectively. (b–d) Model of a recessive locus with different initial allele frequencies.

within-line variance is reduced and the between-line one increases with the increase in F, but reaches its maximum value asymptotically.

Epistasis can also increase the additive variance with inbreeding. Wade and Goodnight (1998) suggested the importance of a possible 'conversion' of the epistatic variance into additive variance after a population bottleneck and several theoretical studies show this result with neutral models of variation (Barton and Turelli, 2004; López-Fanjul et al., 2006).

The increase in epistatic variance could confer some evolutionary importance to the processes that occur when a population is founded with few individuals (the so-called founder effect), since the excess of additive variance generated could be used to adapt the population to its new environment. However, studies that predict a significant increase in the additive variance after a population bottleneck refer to strictly neutral models. When components of fitness are considered, the predicted increase in variance is substantially reduced due to the purging of deleterious mutations by selection. The result is illustrated in Figure 8.5 where it is observed that the increase in additive variance due to dominance and epistasis is substantially reduced when selection comes into play, almost completely removing the effect of epistasis on the variance (Pérez-Figueroa et al., 2009). Thus, inasmuch as the genetic variance is due to unconditional deleterious alleles, the increase in genetic variance attributable to inbreeding will allow an

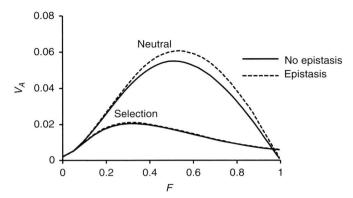

Figure 8.5 Change in the additive genetic variance (V_A) with different values of the inbreeding coefficient (F) (simulation data of Pérez-Figueroa et al., 2009b). A base population of high census size at mutation–selection equilibrium was simulated whose genetic variation for fitness was due to deleterious mutations with mutation rate $U = 0.05$, effects in homozygotes distributed exponentially with mean 0.2 in homozygosis and average dominance coefficients $h = 0.2$, distributed according to the model of Figure 7.5 (Chapter 7). From this base population, lines of size $N = 10$ were established that were maintained for 100 generations, whose additive variance is shown in the figure considering a neutral or selection model. The continuous lines assume absence of epistasis, while the discontinuous ones consider strong synergistic epistasis between pairs of loci.

increase in purifying selection, known as purging, that tends to reduce inbreeding depression, but an increase in the rate of evolution of a population after a bottleneck is not expected.

Empirical evidence for laboratory data for morphological traits indicates, in general, an absence of increases in additive variance after population bottlenecks, but substantial increases are often observed for fitness components (Van Buskirk and Willi, 2006; Taft and Roff, 2012). The results are compatible with the general assumption that in the former, which show little or no inbreeding depression, there is a general absence of non-additive genetic variation, while in the latter there is greater non-additive genetic variation and they show a high inbreeding depression. It is difficult, however, to discern experimentally whether part of the increment observed in the additive variance after a population bottleneck is due to epistasis or can be explained only by dominance.

8.2.2 Genetic Differentiation in Quantitative Traits

The genetic variation between different populations for a quantitative trait that, in the case of an additive neutral model, is given by equation (8.10), can be expressed by a dimensionless parameter named Q_{ST} by Spitze (1993) and that is analogous to the fixation index F_{ST} studied in Chapter 4. The definition of Q_{ST} arises from expressions (8.10) and (8.11) obtained by Wright (1951) and its relation to F_{ST} deduced by Lande (1992). In the same way that F_{ST} constitutes a measure of the allele frequency differentiation between populations, Q_{ST} provides an average of the genetic divergence for the trait as a function of total variation, and is expressed as

$$QST = \frac{V_{GB}}{V_{GB} + 2V_{GW}},$$ (8.12)

where the '2' of the denominator comes from the fact that the genetic variance between populations is proportional to $2F_{ST}$ (equation (8.10); Wright, 1951), and should be omitted in the case of a haploid species. Substituting expressions (8.9) and (8.10) in (8.12) and noticing that $F = F_{ST}$ if there is random mating within lines, we have that $Q_{ST} = 2F_{ST}V_{G0}/[2F_{ST}V_{G0} + 2V_{G0}(1 - F_{ST})] = F_{ST}$. That is, the expected value of Q_{ST} for an additive and neutral quantitative trait is equal to the value of F_{ST} for neutral loci. The comparison between Q_{ST} and F_{ST} values can be used as a tool to detect the action of selection on the trait. The panmictic groups under study may be isolated lines, such as those discussed in the previous section, or sub-populations of the same population, among which there may be migratory flow.

   The level of neutral differentiation between sub-populations could be obtained by estimating F_{ST} with neutral molecular markers (Chapter 4). On the other hand, the genetic differentiation for the trait must be estimated in an experimental design in which individuals of different families and sub-populations grow and are evaluated in the same experimental environment and the data are analysed by an analysis of variance. For example, if individuals from sib families from different sub-populations are evaluated, the analysis would provide estimates of the genetic variance between sub-populations (σ^2_{GB}), between families within sub-population (σ^2_f) and within families (σ^2_w). If the trait is additive and there are no common environmental effects, the heritability would be estimated as $h^2 = 2\sigma^2_f/(\sigma^2_f + \sigma^2_w)$, as deduced from equation (6.7), and $Q_{ST} = \sigma^2_{GB}/(\sigma^2_{GB} + 2\sigma^2_{GW})$ (equation (8.12)), the genetic variance within sub-populations being $\sigma^2_{GW} = 2\sigma^2_f$. A result in which $Q_{ST} > F_{ST}$ could indicate that the selection favours different average values of the character in each sub-population, which is called divergent or diversifying selection (LeCorre and Kremer, 2003). Conversely, $Q_{ST} < F_{ST}$ would indicate that the population is subjected to a selective force that tends to maintain the same average trait value in all sub-populations (convergent selection). The different types of selection will be studied in more detail in Chapter 10.

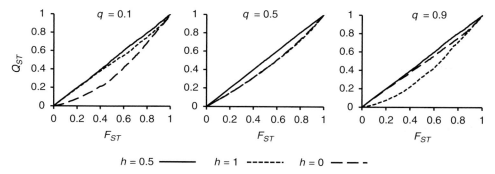

Figure 8.6 Comparison between Q_{ST} and F_{ST} indices for a locus with initial frequency q and effect $a = 1$ in homozygosis and ah in heterozygosis on a quantitative trait. The results have been obtained by simulation, as explained in the footnote of Figure 8.4.

If the individuals of the different sub-populations have not grown in the same experimental environment, Q_{ST} estimates could be biased upwards if they include differences in the expression of the trait due to the environment associated with each sub-population. Finally, on some occasions it is not even possible to obtain genetic components of within- and between-sub-population variance, and only phenotypic components are available, in which case the statistic is called P_{ST} and has a merely descriptive application.

The equality $Q_{ST} = F_{ST}$ is strictly predicted for a neutral and additive trait. The presence of dominance generally implies that $Q_{ST} < F_{ST}$ in the absence of selection (López-Fanjul et al., 2003; Goudet and Büchi, 2006), as illustrated in Figure 8.6, so that dominance should not hide the footprint of divergent selection. When there is selection, dominance generally accentuates the increase of Q_{ST} relative to that of F_{ST}, facilitating its detection (Santure and Wang, 2009). Conversely, dominance may hide the footprint of convergent selection since, in this case, a result is expected in which $Q_{ST} < F_{ST}$ with or without selection.

Epistasis may also imply a value of Q_{ST} lower than F_{ST}. The effect of additive by additive epistatic variance, for example, is to increase both the within- and between-sub-population variance. In this case, the first can be approximated by $V_{GW} = V_A(1 - F_{ST}) + 4F_{ST}(1 - F_{ST})V_{AA}$ and the second by $V_{GB} = 2F_{ST}V_A + 4F_{ST}^2 V_{AA}$ (Whitlock, 1999), where V_A and V_{AA} are the additive and additive $\times$ additive epistatic variances in the base population. Substituting in equation (8.12) we obtain $Q_{ST} = [F_{ST}V_A + 2F_{ST}^2 V_{AA}] / [V_A + 2F_{ST}(2 - F_{ST})V_{AA}] \leq F_{ST}$, that is, the additive $\times$ additive epistasis implies a value of Q_{ST} lower than F_{ST} in the absence of selection. Synergistic epistasis also implies $Q_{ST} < F_{ST}$ in the absence of selection (López-Fanjul et al., 2003). Therefore, again, the presence of epistasis does not hide the footprint of divergent selection but it can do so with convergent selection.

The Q_{ST} value of a quantitative trait contains information on the evolutionary history of the population similar to that of F_{ST} of neutral loci (Rogers and Harpending, 1983). However, the distribution of F_{ST} values is very heterogeneous even for neutral loci. Therefore, when investigating the variation for a single quantitative trait, the Q_{ST} value should be compared to the distribution of F_{ST} values for multiple markers, rather than their average value (Whitlock, 2008). When we want to analyse the Q_{ST} value of multiple traits we can use the genetic covariance matrix within populations (**G**) and between populations (**D**). Note that, from equations (8.9) and (8.10), it follows that $V_{GB} = [2F_{ST}/(1 - F_{ST})]V_{GW}$, so, similarly, under a neutral model of additive traits, it is expected that $\mathbf{D} = [2F_{ST}/(1 - F_{ST})]\mathbf{G}$ (Martin et al., 2008).

The experimental results mostly indicate that $Q_{ST} > F_{ST}$ (Leinonen et al., 2013), which suggests that a considerable part of the divergence observed between populations for quantitative traits must be attributed to the selective pressure imposed by the different particular environmental conditions. The genetic adaptation of populations to their local conditions has an important implication in the field of conservation. When individuals from different populations are crossed, it is possible that the hybrids show a reduced fitness compared to their parents, which is called outbreeding depression (Edmands, 2007), a concept closely related to the genotype–environment interaction discussed in Chapter 3.

8.2.3 Mutational Meltdown and Purging of the Inbreeding Load

In previous sections we have seen that inbreeding is responsible for a decline in the mean fitness of the population due to the expression of recessive deleterious mutations in homozygosis. If inbreeding increases because the population census size has been reduced, the consequences depend on the new effective size. If this is relatively high, natural selection purges, at least partially, the inbreeding load that remained initially hidden in the large population, avoiding most of the inbreeding depression and leading to a new equilibrium with less inbreeding load. If the census size is small, purging is less effective and, in addition to the process of inbreeding depression, the fixation of new deleterious mutations may occur due to the prevalence of genetic drift over selection. Recall (Chapter 7) that quasi-neutral deleterious mutations can be fixed with a probability close to that of a neutral allele ($1/2N$), if the product of the effective population size (N_e) by the selection coefficient of the mutation (s) is much less than unity, that is, $N_e s \ll 1$. Therefore, the smaller the effective size of a population, the faster the average fitness declines due to the fixation of the deleterious mutations that occur continuously in the population, what can be called drift load. This deterioration can produce a reduction of the effective size which, in turn, implies a greater probability of fixing other mutations. This recurrent cycle, which can lead, together with inbreeding depression, to the extinction of the population, is called mutational meltdown (Lynch and Gabriel, 1990; Lynch et al., 1995). Note that although inbreeding depression and mutational meltdown are distinct processes, both are a consequence of reduced census sizes and consanguineous matings. Inbreeding depression is due to the expression of the inbreeding load present in the ancestral population by means of inbreeding, while mutational meltdown is due to the fitness deterioration attributable to new mutations and not only to that ascribed to recessive mutations. The impact of mutational meltdown depends on the rate of deleterious mutation and the effects and dominance of mutations, among other factors (Whitlock and Bürger, 2004).

     The accumulation of deleterious mutations and the inbreeding depression that occurs in population of reduced size can be partially restricted by natural selection. Except in populations of very low census size, selection continuously eliminates deleterious mutations, especially those with substantial effect in heterozygotes, although it acts less efficiently against fully or partially recessive mutations. These, however, can also be purged to some extent by inbreeding. This requires a moderate effective size (large enough for selection to be intense but small enough to generate appreciable inbreeding) or consanguineous matings that occur deliberately or spontaneously. In general, purging will be effective for loci with value $N_e d > 1$ (García-Dorado, 2012), where N_e is the effective population size and d the purging coefficient (studied in Chapter 4), which equals the dominance effect defined in Table 8.1. For example, simulations that considered a model of partially recessive mutations that occurred at a rate $U = 0.5$ per haploid genome and generation and an average homozygous effect of 0.04 showed a 10% decline in mean fitness in populations of size $N = 50$ during a period of 100 generations, instead of the 53% expected in the absence of purging (Pérez-Figueroa et al., 2009b). Purging has important consequences for conservation genetics (Leberg and Firmin, 2008), although it is difficult to detect, particularly in conservation programmes (Crnokrak and Barrett, 2002; Boakes et al., 2007), unless designed experiments are carried out for this purpose (López-Cortegano et al., 2016).

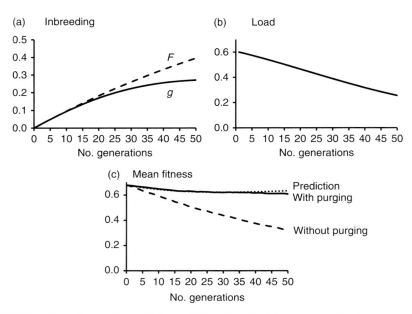

Figure 8.7 Effect of purging on the coefficient of inbreeding, the inbreeding load and the mean fitness of the population. A high census size population ($N = 1000$) was simulated during 10,000 generations in which deleterious mutations were produced with genomic rate $U = 0.2$, mutational effects with $s = 0.1$ and coefficient of dominance $h = 0.2$. The dominance effect and purging coefficient are, therefore, $d = s(1/2 - h) = 0.1 (0.5 - 0.2) = 0.03$. The initial inbreeding load ($B_0 \approx 0.6$) is predicted by equation (8.8). A population of size $N = 50$ was established from the starting population, which was maintained for 50 generations and whose parameters are described in the figure averaged over 100 replicates. (a) Expected inbreeding coefficient without purging (F) and purged inbreeding coefficient (g) calculated by equation (4.26) using $N = 50$ and $d = 0.03$. (b) Predicted reduction of the inbreeding load (B_t) approximated by equation (8.14). (c) Average fitness of the simulated population (W_t) in the absence of purging (dashed line), with purging (solid line) and predicted by equation (8.15) (dotted line).

The consequences of purging the inbreeding load are a reduction in the amount of inbreeding depression, a decrease in the purged inbreeding coefficient relative to the genealogical one, and a decline in the average fitness of the population that is lower than expected without purging. The joint effects of inbreeding depression and purging can be predicted as a function of the purged inbreeding coefficient (g), which depends on the purging coefficient (d), both defined in Chapter 4 (García-Dorado, 2012). In a population of effective size N_e, the purged inbreeding coefficient in generation t is given by equation (4.26), $g_t = [(1/2N) + (1-1/2N)g_{t-1}](1-2dF_{t-1})$, where $F_t = 1-[1-(1/2N)]^t$ is the expected inbreeding coefficient without purging. With the passage of generations ($t \rightarrow \infty$), $F_t \rightarrow 1$ and g_t tends towards an equilibrium value that can be approximated by

$$\hat{g} = \frac{1 - 2d}{1 + 2d(2N - 1)}. \tag{8.13}$$

Due to the effect of purging, the inbreeding load is reduced over generations, so that it can be approximated in generation t by

$$B_t = B_0 g_t (1 - F_t)/F_t, \tag{8.14}$$

where B_0 is the initial inbreeding load, which can be predicted by equation (8.8) if the ancestral population had a very high census size. Likewise, the fitness of the purged population can be predicted by equation (8.5) by substituting F for g, that is,

$$W_t = W_0 e^{-B_0 g_t}. \tag{8.15}$$

Figure 8.7 shows an example of simulations and predictions of the main parameters in the evolution of a population under a model of deleterious mutations, considering or not purging. The purged inbreeding coefficient (g_t) is substantially reduced with the generations of inbreeding (panel a) producing a gradual reduction of inbreeding load (panel b) obtained with equation (8.14). The result is a much lower decline in the average fitness of the population than would be expected without purging (panel c). The prediction of the average fitness by means of equation (8.15) (dotted line in panel c) approximates the simulations results quite well.

8.3 Evolution of Inbreeding in Natural Populations

One of the evolutionary consequences of inbreeding depression is the development of genetic systems to prevent inbreeding. Sexual reproduction, for example, in addition to its advantages over asexual reproduction that were studied in Chapter 7, is a mechanism that prevents selfing in species with separate sexes. Many animal species have also developed behavioural patterns that prevent mating between relatives. Likewise, the majority of human societies prohibit consanguineous unions, probably due to the continued observation by our ancestors of diseases and developmental anomalies in the corresponding children. Curiously, crossings between half sibs are usually forbidden while those of uncle/niece or aunt/nephew are not, although both produce the same coefficient of inbreeding in their offspring. But it is particularly in the plant kingdom that the mechanisms to avoid self-fertilization stand out, many of them already studied by Darwin (1877). In many species there are self-incompatibility genetic systems determined by the genotype of the pollen or the plant that produces it. These systems are controlled by one or more self-sterility loci (S) with multiple alleles (S_1, S_2, S_3, ...). An example of one of these systems is illustrated in Figure 8.8. The pollen grains carrying some of the alleles of the self-sterility locus present in the genotype of the recipient plant will not be able to form pollen tubes or the fertilization will be aborted. Only those grains that carry different alleles will be able to fertilize the recipient female gamete. Note that this system not only prevents self-fertilization but also reduces the probability of mating between relatives if they share an allele. For example, the first and second plants of the figure starting from the left share the S_1 allele, so their cross-fertilization will be restricted.

In addition to the systems of incompatibility mentioned above, there are, in many plants, morphological or physiological systems in which the male and female sexual organs are spatially separated (herkogamy) or functional at different times (dichogamy), thus reducing the probability of self-pollination. In some hermaphroditic species individuals are classified into two or three types according to the relative length of the stamens and pistil,

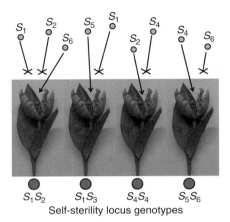

Figure 8.8 Illustration of a self-incompatibility system using a self-sterility locus (S, with multiple alleles $S_1, S_2, S_3, \ldots$). Only pollen grains with an allele different from those of the recipient plant can fertilize it, preventing self-fertilization and reducing the probability of biparental inbreeding.

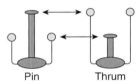

Figure 8.9 Illustration of heterostyly (distily) or reciprocal herkogamy, in which there are two floral forms in the population characterized by different lengths of stamens and pistils. Pollination is more frequent between different forms (arrows) than between equal forms, reducing the likelihood of self-fertilization or crosses between related plants.

which is called heterostyly. Figure 8.9 illustrates a case of reciprocal heterostyly or herkogamy in which there are two floral forms (in this case, distily) in the population, individuals with long style and short stamens (Pin form) or with short style and long stamens (Thrum form). Pollinators specialized in feeding in the internal or external part of the flower will facilitate pollination between the two forms and reduce the likelihood of self-pollination. In the genus *Primula*, for example, the genetic determination of this floral character is determined by three closely linked loci, with dominance relationships $A > a$, $B > b$ and $C > c$. The genotypes *ABC/ABC* or *ABC/abc* determine the Pin form while the triple recessive genotype *abc/abc* determines the Thrum form. When recombination breaks haplotypic combinations and homostylic plants are generated where self-fertilization is not impeded, the offspring suffer from inbreeding depression and are removed by natural selection, which maintains the genetic system. Most species that present heterostyly also have self-sterility systems due to diallelic incompatibility (Figure 8.8). One hypothesis maintains that incompatibility systems evolve first as a system to avoid inbreeding, with inbreeding depression being the selective agent, and heterostyly subsequently emerging as a mechanism to promote effective pollen transfer between the two types, reducing their waste in incompatible systems (Charlesworth and Charlesworth, 1979). Alternatively, heterostyly could arise prior to the appearance of

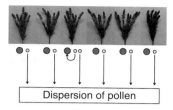

Figure 8.10 Illustrative diagram of the automatic selective advantage of self-fertilization hypothesized by Fisher (1941). Large circles indicate ovules and small ones, pollen grains. The third plant from the left, able to self-fertilize, will have a 50% fitness advantage with respect to the others.

incompatibility systems, which would then appear as a mechanism to prevent occasional failures of the imperfect heterostylic system (Lloyd and Webb, 1992).

Despite the harmful effects of inbreeding and the natural mechanisms to avoid it, many species of plants and some animals reproduce by self-fertilization, either totally or partially. The most accepted hypothesis is that autogamous species have evolved from allogamous ones. In the laboratory it is also possible to generate highly inbred populations through self-fertilization or mating between sibs, although multiple inbred lines must be established since many will become extinct by inbreeding depression and mutational melt-down. After several generations, a large part of the inbreeding load will have been purged in the surviving lines, so that they can remain indefinitely. Something similar is what may occur in nature for the formation of autogamous species. This will require that in the first evolu-tionary steps inbred individuals are able to resist the consequences of inbreeding depression. The model that explains the evolution of self-fertilization is due to Fisher (1941), and it is the so-called automatic selective advantage hypothesis of self-fertilization that is illustrated in Figure 8.10.

Let us assume a population of allogamous plants of constant census size where each plant contributes on average a grain of pollen and an ovule for cross-fertilization that will give rise to the next generation. If a mutation occurred in a plant that allowed its self-fertilization (the case of the third plant in Figure 8.10) and still could provide pollen to compete with the others by cross-fertilization, the average contribution of this plant would be 50% higher than the others. Their offspring by self-fertilization, however, would be compromised by inbreed-ing depression, and if the plant were a carrier of a recessive lethal, one would expect that 1/4 of the offspring would not survive. In this case, the 50% advantage would overcome the 25% mortality and the mutation that enables self-fertilization would increase in frequency, leading the population to a majority or completely autogamous state. If it were a carrier of two recessive lethals, one would expect a mortality of 7/16 in the inbred progeny, still less than 1/2, so that the advantage of the mutation that makes self-fertilization possible would still persist, with the same outcome. With three recessive lethals, however, the mortality would be greater than the advantage produced by self-fertilization, so the mutation would not thrive. It is evident, therefore, that only in populations with an inbreeding load approximately equal to or less than two lethal equivalents could this evolutionary process occur successfully. It will be more likely, therefore, that this process will be successful in populations of reduced size, where it is expected that the number of lethal equivalents will be lower, due to previous purging. In nature there are many species that are completely autogamous or allogamous, as expected, but also a multitude of species with intermediate levels of autogamy that may be in the evolution phase to complete autogamy or are maintained with stable autogamy rates by the action of various mechanisms (Goodwillie et al., 2005).

8.4 Crossbreeding and Heterosis

We saw in Section 8.2.1 that when a set of lines of reduced census size is established from a large base population, the variance of an additive and neutral trait is redistributed within and between lines producing a dispersion of the mean of the lines with a variance that is predicted by equation (8.10). If there is no inbreeding depression, the average of the trait in the set of lines will remain constant and equal to that of the base population. With inbreeding depression, however, the mean will decay with respect to the initial one. If, at any given time, the lines meet to form a panmictic population, the mean will revert to its initial value. This unique population is called synthetic and the recovery of the mean that occurs when inbreeding is destroyed is known as heterosis. Heterosis can be considered as the reverse process to inbreeding depression. If lines carrying deleterious recessive alleles fixed in homozygosis for different loci are crossed, the resulting offspring will be heterozygous for those alleles, showing greater fitness than the average of the parental lines and, sometimes, exceeding that of any of them. For example, if in an inbred line the genotypic constitution of the individuals for two loci is $aaBB$ and in another is $AAbb$, where the alleles designated with lowercase reduce the fitness of the individual in homozygosis while those identified by capital letters do not reduce it and are dominant, the cross of the lines will produce another one with genotype $AaBb$ and superior fitness to the parental lines. However, in general, heterosis usually only implies an average fitness of the hybrid higher than the average of the two parents.

Let us assume the model of a locus with effects a, h and d presented in Table 8.1 with $F = 0$, given that panmixia is assumed in each line, and consider two lines (1 and 2) with allele frequencies p_1 and q_1 and p_2 and q_2, respectively. The contribution of the locus to the mean of each of them is $M_1 = aq_1 + 2dp_1q_1$ and $M_2 = aq_2 + 2dp_2q_2$, with an average of $\overline{M} = (M_1 + M_2)/2 = (a\overline{q} + 2d\overline{pq}) - (dy^2/2)$, where $\overline{q} = (q_1 + q_2)/2$ and $y = q_1 - q_2$ is the difference of the allele frequencies between the two lines. From the cross between individuals belonging to different lines descendants will be obtained with genotypes AA, Aa and aa and frequencies p_1p_2, $p_1q_2 + p_2q_1$ and q_1q_2, respectively, so the average of the generation F_1 will be $M_{F_1} = (a\overline{q} + 2d\overline{pq}) + (dy^2/2)$. Therefore, the heterosis in the F_1 generation will be

$$Het(F_1) = M_{F_1} - \overline{M} = dy^2, \tag{8.16}$$

that is, it will depend on the existence of dominance ($d \neq 0$) and the squared difference in allele frequency between the lines ($y \neq 0$). If a subsequent generation of panmixia occurs, the average contribution of the locus in generation F_2 will be $M_{F_2} = a\overline{q} + 2d\overline{pq}$, so that heterosis in the second generation will be

$$Het(F_2) = M_{F_2} - \overline{M} = dy^2/2, \tag{8.17}$$

that is, half of that observed in the F_1 generation. The reason for the reduction in heterosis between the first and second generations is the inbreeding depression that occurred as a consequence of the fact that a proportion of the recessive alleles present as heterozygotes in F_1 are manifested in homozygosis in F_2. In successive panmictic generations, the mean is not expected to change with respect to that of F_2, except for genetic drift or selection, since the locus in question has reached Hardy–Weinberg equilibrium. When more than one locus is considered in the absence of epistasis, the total heterosis will be the sum of those corresponding to all loci (i), that is, $\Sigma d_i y_i^2$. As with inbreeding depression, if there is no directional dominance (values of d mostly of the same sign for the different loci that affect the trait), the contributions of different loci could be compensated and heterosis would not be manifested. Again, most of the traits related to fitness, which show inbreeding depression, also present heterosis, while many morphological traits do not present either of them.

As we saw in Chapter 3, some traits have maternal effects, that is, the genotype of the mother has an effect on the phenotype of the offspring. In this case, the effects of heterosis (and inbreeding depression) will be manifested with one generation of delay, that is, the maternal component of the heterosis in F_1 will be manifested in F_2, so it can sometimes be observed that the mean of F_2 is higher than that of F_1, contrary to that indicated by equations (8.16) and (8.17). The genotype–environment interaction generated in the crossing of lines may also be important. As already mentioned earlier in this chapter, it is possible that if two lines have been adapted to specific environmental conditions, the cross between the two may produce allelic combinations that are not adapted to any of the environments in which the parental lines live. This may result in the absence of heterosis or a decline in the mean of fitness components (outbreeding depression).

The epistatic interactions between loci can imply changes in the mean in any direction with respect to the previous predictions. Although dominance is a common factor in both inbreeding depression and heterosis, it is not necessary for heterosis, as this may occur from additive × additive epistasis without dominance. Inbreeding depression is also a function of additive × dominance and dominance × dominance epistasis, but not heterosis (Walsh and Lynch, 2018, p. 258).

Crossing and heterosis constitute, together with artificial selection, the basis of plant and animal breeding. Artificial selection, as we will see in the next chapter, makes use of the additive variance of the selected trait, while crossing and heterosis make use of the genotypic variance. The idea is to produce multiple inbred lines in allogamous plant species, select them for certain traits of interest and cross them to obtain heterotic offspring by dominance and, if possible, by epistasis. The procedure is not usually applied in animals, since in these it is not usually feasible to obtain lines that are highly consanguineous due to inbreeding depression.

8.4.1 General and Specific Combining Abilities

The result of crossing between two lines can be predicted by the general and specific combining abilities that are explained below. Suppose that, starting from a base population, a high number (n) of inbred lines is established, which are maintained independently until they reach an average coefficient of inbreeding F and are then crossed each to one another in all possible combinations, which is usually feasible when working with plants (Figure 8.11).

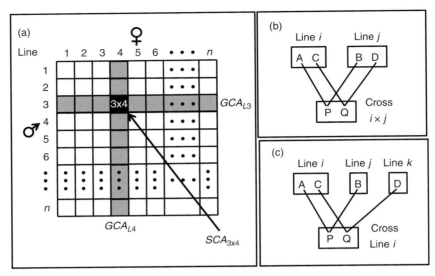

Figure 8.11 Scheme of the procedure for obtaining inbred lines and their crosses in plants. (a) From a base population, n inbred lines are obtained and crossed to one another. The general combining ability (GCA) of each line is the average of the trait in all the crosses in which it is involved (in the figure, only the crosses in which the line acts as a male or female parent are shown in grey). The specific combining ability (SCA) is the deviation of a specific cross from its expected value. (b, c) Schemes used to establish which components of the genotypic variance contribute to general combining abilities and to specific combining abilities of the different crosses (see text).

Suppose that the mean of the trait is evaluated for each cross and the average of all of them is obtained. The general combining ability of a given line (GCA) is defined as the average of the trait in all crosses in which that line is involved, deviated from the general mean. For example, Figure 8.11 shows the row whose average provides the GCA of line 3 (GCA_{L3}) and the column that gives that of line 4 (GCA_{L4}). The mean observed in the cross ($M_{3\times4}$) may deviate from its expected value ($GCA_{L3} + GCA_{L4}$) and the deviation is called the specific combining ability (SCA) of that cross. Therefore, the average of the cross is $M_{3\times4} = GCA_{L3} + GCA_{L4} + SCA_{3\times4}$. (Only the 3 × 4 cross is considered, in which line 3 acts as a male parent and 4 as female to simplify, but the rows and columns corresponding to the reciprocal crosses should also be averaged.)

Since the lines are crossed in all possible combinations, the genotypic variance, with its additive, dominance and epistatic components, $V_G = V_A + V_D + V_{AA} + V_{AD} + V_{DD} + \ldots$, is equal to the sum of the variances within and between crosses. However, the interest is focused on finding crosses that stand out from the rest because of their mean, which is why the between-crosses variance component (V_{BC}) is of interest. The contribution of the different components of the genetic variance to V_{BC} can be obtained by recalling from Chapter 6 that the covariance between relatives is given by $\text{cov} = rV_A + uV_D + r^2V_{AA} + ruV_{AD} + u^2V_{DD} + r^3V_{AAA} + r^2uV_{AAD} + ru^2V_{ADD} + u^3V_{DDD} + \ldots$ (equation (6.9)), where r is the coefficient of additive relationships and u is that of dominance relationships among relatives. To determine the components of the genetic variance that contribute to the variance of the means of the crosses, consider panel (b) of

Figure 8.11. Individuals A and C of line i are crossed with individuals B and D, respectively, of line j to obtain the individuals P and Q from the cross of both lines. Recalling from Chapter 6 how to calculate the coefficients r and u from the coancestry coefficient between individuals (f), we obtain $r = 2f_{PQ} = 2\,\frac{1}{4}\,[f_{AC} + f_{AD} + f_{BC} + f_{BD}] = 2\,\frac{1}{4}\,[F + 0 + 0 + F] = F$, since the expected coancestry between two individuals of a line is the average inbreeding coefficient of the line, which is F, and it is assumed that there is no coancestry between individuals of different lines. On the other hand, $u = (f_{AC} \times f_{BD}) + (f_{AD} \times f_{BC}) = (F \times F) + (0 \times 0) = F^2$. Therefore, the variance of the means of the crosses is

$$V_{BC} = FV_A + F^2 V_D + F^2 V_{AA} + F^3 V_{AD} + F^4 V_{DD} + \ldots, \tag{8.18}$$

that is, its terms depend on the inbreeding coefficient of the lines. Therefore, if the objective is to maximize the variance between lines, one should try to achieve the maximum inbreeding in these. Note also that the dominance and epistatic components of the variance are weighted by powers of F, so their contribution will be small unless F is high. The corresponding variance within crosses (V_{WC}) would be obtained by subtraction as $V_{WC} = V_G - V_{BC}$, that is,

$$V_{WC} = (1 - F)V_A + (1 - F^2)V_D + (1 - F^2)V_{AA} + (1 - F^3)V_{AD} + (1 - F^4)V_{DD} + \ldots \tag{8.19}$$

Note that if $F = 1$ all the genotypic variance is ascribed to variance between crosses.

The variance between crosses, in turn, can be decomposed into the variances of general and specific combining abilities. To calculate the components of the variance of the general combining ability of the lines, consider panel (c) of Figure 8.11. The males A and C of any line i are crossed with females of two other lines (j and k) producing descendants P and Q. Calculating the coefficients r and u as before, we have $r = 2f_{PQ} = 2\,\frac{1}{4}\,[f_{AC} + f_{AD} + f_{BC} + f_{BD}] = 2\,\frac{1}{4}\,[F + 0 + 0 + 0] = \frac{1}{2}F$ and $u = (f_{AC} \times f_{BD}) + (f_{AD} \times f_{BC}) = (F \times 0) + (0 \times 0) = 0$. Therefore, the variance of the general combining ability of the lines acting as male parent is $V_{GCA,male} = (1/2)FV_A + (1/4)F^2 V_{AA} + (1/8)F^3 V_{AAA} + \ldots$, and if it were acting as a female parent it would occur similarly, so, considering both sexes,

$$V_{GCA} = FV_A + \frac{1}{2}F^2 V_{AA} + \frac{1}{4}F^3 V_{AAA} + \ldots, \tag{8.20}$$

that is, the variance of the general combining ability depends only on the additive variance and the additive $\times$ additive epistatic components. The variance of the specific combining ability can be obtained by subtracting V_{GCA} from V_{BC}, that is, $V_{SCA} = V_{BC} - V_{GCA}$, so that

$$V_{SCA} = F^2 V_D + \frac{1}{2}F^2 V_{AA} + F^3 V_{AD} + F^4 V_{DD} + \ldots \tag{8.21}$$

Therefore, for a cross $M_{i \times j} = GCA_i + GCA_j + SCA_{i \times j}$, the first two terms depend on the additive variance and the additive $\times$ additive epistatic components, while the latter depends on the dominance variance and the rest of the epistatic components. If there were within- and between-locus additivity this term would be null.

The idea that underlies this method is to obtain crosses with greater heterosis, which are those that are commercialized with remarkable success in cross-pollinated species such as corn or rye. The lines can also be previously selected for different attributes of economic or reproductive interest (resistance to pathogens or cold, germination capacity, etc.) that could be combined in the crosses. Obviously, in the case of self-pollinated species, inbred lines would be commercialized without carrying out the crosses.

The method to obtain inbred lines also has several additional advantages. First, the optimal genotype (the best line or the best cross) can be regenerated indefinitely as long as the corresponding parental lines are available, although in the case of the hybrid it can only be used to the maximum advantage in one generation, since in the next generation we would lose part of its attributes by segregation and inbreeding depression. Second, the fact that the individuals of the lines or of the cross are very uniform phenotypically (what is achieved if the inbreeding reached in the lines is high) is a practical advantage since, for example, mechanical harvesting is facilitated if all plants have the same height and the ripening of the fruits occurs approximately simultaneously in all of them. On the other hand, the uniformity between plants allows the evaluation of the phenotypic average of the line or crosses with little error, estimating its mean genotypic value instead of its additive value. A disadvantage of the method is that it requires a considerable amount of time until the desired high inbreeding level in the lines is reached and inbreeding depression can be a problem in the maintenance of the lines or their viability in the case of cross-pollinated species.

The process of improvement by crossing, however, is carried out only in some cultivated plants, mainly in corn. When hybrid maize started in the 1920s, it was considered that the recessive gene action was almost universal and it was decided to obtain hybrids of pure lines, following the experimental series started by Mendel. By 1950 it was experimentally proved that the heritability of production was high and that artificial selection was effective. In this context, a cycle of inbreeding and subsequent crossing of lines that would span several generations would result in lower production than the practice of selection during the same period, unless the genetic variance of the production was fundamentally non-additive. The selected lines had advantages and disadvantages. On the one hand, their profits were permanent or almost so, which would allow the farmer to use part of the seed to replant, at least for a few years, although after a few generations it would be necessary to buy more selected seeds. On the other hand, the conformation of the plants obtained by selection would not be uniform, which would make mechanized harvesting less efficient. The hybrids, however, are uniform in conformation, but re-sowing would entail the loss of at least half of the original heterosis. Probably, if selection had been practised from the beginning, crossing pure lines would not have been the method of choice, but the genetic knowledge at that time did not lead to this option. In fact, the existence of additive genetic action was something scarcely considered by the genetists of the moment, knowledge that had to wait until the 1960s. The existence of a practical monopoly in seed suppliers, circumscribed to a small number of companies, imposed the use of hybrids.

Crossbreeding in animals aimed at improving their production by heterosis is also very common, although it is done in a different way. Normally, a few lines are kept artificially selected for different characteristics without reaching too high inbreeding coefficients, and then they are crossed in order to combine the selected characteristics (complementarity) in both lines and take advantage of the heterosis of the cross. For example, a selected line can be crossed for prolificity in pigs with another selected for meat quality with the intention that the cross between both presents an acceptable average for both traits. In order to take advantage of the heterosis of maternal characters, three- or four-way crosses are sometimes made, that is, with three or four lines. For example, lines A and B can be crossed and hybrid females, which can show maternal heterosis, can be crossed with males from a third line C to obtain the commercial product. Another alternative, when only two lines are available, is to carry out the so-called reciprocal recurrent selection (Comstock et al., 1949), in which the general and specific combining abilities are selected simultaneously, as well as the so-called rotational crosses, where lines A and B are crossed recurrently. For example, in the latter, the two lines are crossed first, and then AB hybrid females are crossed with males of line A obtaining individuals [(AB)A], with 3/4 of the genome of A and 1/4 of that of B, which are now crossed with individuals of line B producing hybrids [((AB)A)B], with 3/8 of the genome of A and 5/8 of that of B, and so on.

An application of the crosses, very frequently used in the past, is the introgression of genes of interest from one variety into another, which has been applied mainly in plants. The method consists of crossing the commercial variety with another carrier of a desired genetic attribute controlled generally by a single locus. Individuals of the offspring are selected for the attribute and backcrossed with the commercial variety repeatedly over multiple generations (always selecting for the attribute of interest in the offspring) in order to achieve a final product with the greater part of the genome of the commercial variety and the gene of interest fixed. The current techniques of molecular genetic engineering allow introgression to take place almost immediately, replacing this crossing technique.

Sometimes the need is the opposite, that is, to eliminate certain unwanted allelic variants or exogenous genomic regions that have been introgressed by crossing in a given population. This can be achieved if the complete genealogical information is available or that of molecular markers that allow the foreign regions of the genome to be found (Amador et al., 2013).

8.5 Applications in Conservation

The extinction of species has been accelerated substantially in the last century due to environmental deterioration, overexploitation, the invasion of foreign species and climate change, among other factors, and human intervention is increasingly needed to guarantee their persistence (Frankham et al., 2010; Allendorf et al., 2013). The populations of species in danger of extinction or kept in captivity are characterized by invariably having reduced population census sizes, either temporarily or permanently, as well as geographic fragmentation and high levels of isolation. As a result, in these populations there is a loss of diversity due to genetic drift and, consequently, a reduction of their adaptive potential in the face of new environmental challenges. In addition, there is an increase in the rates of inbreeding, with consequent inbreeding depression and fixation of deleterious recessive mutations, accelerating the deterioration of their reproductive capacity. There are several aspects of conservation genetics to which quantitative genetics can make important contributions, which are summarized in the following sections.

8.5.1 The Minimum Effective Size of a Viable Population

A matter of great interest in conservation genetics is the minimum effective size that a population must have to persist for a certain period of time under the threat of inbreeding depression and the loss of variation due to genetic drift. The prediction distinguishes between short-term (a few generations) and long-term processes (for example up to 100 generations). For the first scenario, the consensus response comes from animal breeding programmes. In these programmes it is usually considered that the maximum increase in inbreeding allowed is 1% per generation, which implies (Chapter 5) a minimum effective population size of $N_e = 1/(2\Delta F) = 1/0.02 = 50$ individuals (Franklin, 1980; Soulé, 1980). However, other more specific proposals have been suggested, such as the establishment of the minimum size of a viable population in the short term that guarantees that the population suffers a maximum decline of 10% in the first five generations (Frankham et al., 2014). For example, assuming a total deleterious load in vertebrates of about six recessive lethal equivalents (the average obtained for mammals and wild birds for fecundity and viability traits, Figure 8.3), the minimum effective size can be obtained by clearing F from equation (8.5) with $B = 6$ and $W_F = 0.9W_0$, and substituting the result in $F_5 = 1-(1-1/2N_e)^5$. Solving for N_e, we obtain $N_e = 142$, which must be an overestimation, by not including the effects of purging (equation (8.15)).

The second prediction corresponds to the minimum effective size of a viable population sufficient to retain the long-term evolutionary potential, defined as the effective size that would be needed to maintain the heritability of a neutral additive quantitative trait. Recall that, with a genetic variation model at mutation–drift equilibrium for a quantitative character, the equilibrium additive variance is $\hat{V}_A = 2N_eV_M$ (equation (7.7), Chapter 7). Since the mutational variance takes values around $10^{-3}V_E$, this means that for a neutral quantitative trait with heritability $h^2 = 0.5$ (that is, with $V_A = V_E$), the critical effective size should be around 500 individuals (Franklin, 1980). When carrying out simulations that consider other additional factors such as mutational meltdown, it follows that the minimum effective size to guarantee the survival of the population over a period of about 100 years should be at least 1000 individuals (Lynch et al., 1995; Lynch and Lande, 1998).

In captive breeding programmes it has been proposed that the minimum feasible effective size of a population is the one that would be needed to maintain at least 90% of the genetic variation over a period of 100 years (Frankham et al., 2014). The analysis of 188 studies of *ex situ* conservation programmes where estimates of genetic variation were available indicated that to meet the above criterion a minimum number of 15 founders and a population size of about 100 individuals would be required (Witzenberger and Hochkirch, 2011).

In summary, the classic recommendation on the minimum effective census size of a viable population is the so-called 50/500 rule for the short/long term, although some studies suggest that these figures should be increased to 100/1000 (Frankham et al., 2014). In any case, the rule refers to the effective size of the population (N_e), not to the actual census size of breeding individuals (N). Since, as we saw in Chapter 5, it has been estimated that the N_e/N ratio may be of the order of 0.1, the minimum census sizes should be of the order of 1000/10,000 individuals. Nevertheless, this recommendation should be taken with caution, as it is limited by the assumptions of the models and by the fact that the risk of extinction depends on various factors that can interact with each other (Flather et al., 2011). It is, therefore, only an orientation rule. Population viability analyses (Traill et al., 2007) consider genetic and demographic factors of the population and can provide more specific recommendations to calculate the probability of persistence of a population during a specific period of time.

8.5.2 Consequences of Conservation Methods

From the genetic point of view, the priority objective of conservation programmes is to maintain the maximum possible levels of genetic variation to guarantee the evolutionary potential, as well as to avoid inbreeding depression and the adaptation of the population to environmental conditions of captivity. As we saw in Chapter 5, there are several procedures recommended to maintain the maximum possible effective size in the population, among which the equalization of parental contributions to offspring stands out, which reduces the rates of genetic drift and inbreeding to half that expected without this type of management. The optimization of this method, which can be carried out using genealogical information or that derived from molecular markers (de Cara et al., 2013), consists in finding the optimal contributions of the parents that minimize the average coancestry in the offspring (Chapter 5). When the individuals are not related or the coancestries in the population are uniform (as occurs after a few generations in an isolated population), the method is reduced to equalizing the parental contributions.

The procedures of equalization of contributions or contributions of minimum coancestry maintain the maximum genetic diversity, but have a series of collateral consequences on adaptive variation. If couples are forced to contribute two offspring, natural selection for fecundity is avoided, except for mating failures, so that selection only occurs for viability within families. This implies a relaxation of natural selection with both positive and negative consequences. On the one hand, by reducing the intensity of selection, the possibility of adaptation of the population to the conditions of captivity is reduced, which is a desirable aspect if the final objective is the reintroduction of the captive population in its natural environment (Williams and Hoffman, 2009). On the other, the relaxation of selection induced by the method reduces the genetic purging and increases the fixation of deleterious mutations with regard to fecundity, reducing the mean fitness of the population in the long term. Despite this disadvantage, analytical and simulation studies indicate that this method provides greater fitness than an unmanaged procedure in small populations (less than 50 individuals) and in the short term up to about 10–20 generations (see, for example, García-Dorado, 2012), and the experimental results confirm these expectations (see, for example, Rodríguez-Ramilo et al., 2006). Figure 8.12 illustrates these results. Equalization of the parental contributions allows the restriction of the average inbreeding (F) of the population maintaining greater heterozygosity (H) in comparison with the case in which no management is done in the population. However, the inbreeding load (B) is purged less efficiently. In spite of this, the average fitness of the population (W) is only lower than that of the scenario without management from about 40 generations with the population size considered.

Most populations of endangered species, both in natural environments and in captivity, are characterized by being structured in reproductive groups with a certain degree

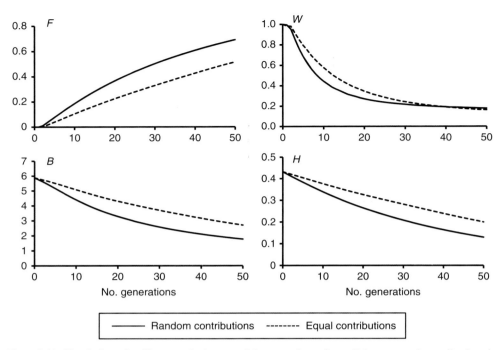

Figure 8.12 Simulations that illustrate the impact of the procedure of equalizing parental contributions in conservation programmes (broken lines) compared to the unmanaged scenario (random parental contributions). F, Genealogical inbreeding coefficient. W, Average fitness of the population. B, Inbreeding load. H, Average expected heterozygosity of neutral alleles. Simulations based on 1000 replicates of a population with size $N = 20$ obtained from a population base of high census size at deleterious mutation–selection equilibrium considering a model of deleterious mutations with rate $U = 0.5$, effect of the mutations with exponential distribution and mean $s = 0.05$ in homozygosis and mean dominance coefficient $h = 0.2$ following the model of Figure 7.5. The inbreeding load is $B = 6$, affecting 1/3 of it to fecundity and 2/3 to viability, according to the data of O'Grady et al. (2006).

of isolation, which generates local inbreeding and differentiation between groups. Conservation methods should take into account this subdivision of the population. One of the basic rules, based on Wright's island model (1951), is to allow a migrant per generation and sub-population (Mills and Allendorf, 1996), which restricts the increase in local inbreeding, maintaining a certain differentiation between sub-populations, which is often desirable if the sub-populations are adapted to their local environments. Nevertheless, there are more complex methods to establish the most appropriate migration rates that make use of demographic (Wang, 2004), genealogical or molecular information (Fernández et al., 2008), and that can provide much more effective control of inbreeding and genetic diversity.

8.5.3 Molecular Variation as a Complement to Quantitative Genetic Variation

The primary objective of conservation programmes from the genetic point of view should be the maintenance of variation for quantitative characteristics, especially those related with fitness. However, its evaluation is often an arduous task and, alternatively or complementarily, molecular information, particularly from neutral molecular markers, can be used as a tool for monitoring and managing genomic diversity (de Cara et al., 2013) and the evaluation of inbreeding depression (Slate and Pemberton, 2002). The observed empirical relationship between molecular variability and that of quantitative traits, however, is only moderate. For example, the mean correlation between molecular marker heterozygosity and genetic variation estimated in two meta-analyses that included 71 and 34 studies for a number of quantitative traits was 0.22 and 0.43, respectively (Reed and Frankham, 2001, 2003). However, a later analysis with 2120 estimates of genetic variation of quantitative traits and 5046 estimates of molecular variation provided a high confidence limit of the correlation of 0.5 (Mittell et al., 2015). On the other hand, clear experimental associations have been found between neutral variation of markers and the adaptive changes in quantitative characters. For example, Santos et al. (2012) detected a significant association between the variation of microsatellite markers and the changes observed in fitness components as a result of adaptation to laboratory conditions. On the other hand, several simulation and experimental studies have shown that neutral variation is a good predictor of the adaptive potential of populations and, in particular, measures of variation of allele frequencies of molecular neutral markers are good predictors of the short-term response for quantitative traits, while those of allelic diversity are good predictors of the long-term response (Caballero and García-Dorado, 2013; Vilas et al., 2015).

Problems

8.1 The mean fitness of a panmictic population with a large census size is 0.8. A small size population is founded and when an average inbreeding coefficient of 0.12 is reached, the population's fitness is 0.5. (a) What are the estimates of the expressed load and the inbreeding load in the starting population? (b) What will be the expected mean fitness of the population of small size when the inbreeding coefficient reaches the value of 0.2?

8.2 The rate of deleterious genomic mutation estimated in a species is $U = 0.2$ and the harmonic mean of the dominance coefficient of the mutations is 0.1. In a very large census size population, the estimated mean fitness is 0.62. (a) What is the expected value of the inbreeding depression rate in that population? (b) Is the estimated value of the harmonic mean of the dominance coefficient compatible with the estimate obtained from the inbreeding depression rate?

8.3 A set of lines has been created that are maintained each generation with 6 males and 24 females. What is the expected value of the within- and between-line genetic variances for a neutral additive trait with initial genetic variance 3.5 after 10 generations?

8.4 The shell size of 10 embryos of each of 50 *Littorina* females from three localities has been analysed. It is assumed that the embryos come from a single fertilization, the reason why it is considered that the families are groups of full sibs. The results of an analysis of variance show the following values of variance components: 0.45 between localities, 0.11 between families within a locality, and 0.43 between sibs within a family. The corresponding analysis of a set of neutral molecular markers has also allowed an estimate of the allele frequency differentiation between localities of $F_{ST} = 0.023$ to be obtained. (a) Calculate the heritability of the character and the index of quantitative trait differentiation, Q_{ST}. (b) What does the comparison of this index with the available value of F_{ST} suggest?

8.5 A line with a census size of 20 individuals is founded from a population of large census size for which the mean fitness is 0.86 and an inbreeding load of 2.4 has been estimated. Assuming that the deleterious mutations that segregate in the starting population have an average purged inbreeding coefficient of 0.2, (a) what would be the expected value of fitness and the remaining inbreeding load in the line after 20 generations? (b) If there had not been purging, what would those values have been?

Self-Assessment Questions

1 Inbreeding depression always implies a reduction in the mean of the trait with the increase in inbreeding.

2 There may be absence of inbreeding depression for a trait even though the loci that affect it are not additive.

3 The additive variance is always increased with inbreeding.

4 The inbreeding load in humans is approximately $B = 1$ lethal equivalents, therefore, in the offspring of full sibs an increase in mortality of 22% is expected.

5 If in an infinite population with high inbreeding due to autogamy there is a generation of panmixia, the inbreeding reached by the population disappears.

6 Dominance and epistasis imply that $Q_{ST} > F_{ST}$ so they can hide the footprint of diversifying selection.

7 The evolution of allogamy to autogamy in a population requires that the number of lethal equivalents be less than or equal to 2.

8 Heterosis is the expected frequency of heterozygotes.

9 Genetic improvement through crosses is more effective if the gene action of the character to be improved is not additive.

10 The method of equalizing parental contributions in conservation programmes may produce a lower long-term fitness than in the case where parental contributions are random, by reducing the intensity of natural selection.

9 Artificial Selection

Concepts to Study

- Selection differential and selection intensity
- Direct and correlated responses to selection
- Realized heritability
- Bulmer effect
- Accuracy of selection
- Selection within families, between families, index and BLUP selection
- Marker-assisted selection

Objectives for Learning

- To understand the concepts of selection differential and selection intensity and the relation between these and the selection response
- To know how a correlated response occurs and to understand the concept of coheritability
- To learn how to estimate the realized heritability and understand its meaning
- To recognize the Bulmer effect and its implications in the response to selection
- To understand the concept of selection accuracy
- To learn how to predict the response to selection within and between families
- To understand the concept of marker-assisted selection

9.1 Principles of Artificial Selection and Its Applications

Artificial selection is perhaps the most important application of quantitative genetics, being the main agent for the genetic improvement of plants and animals. Carried out, sometimes inadvertently, from the start of domestication, artificial selection consists of using only the most suitable individuals as reproducers in relation to a specific objective, generally of interest for human consumption or welfare. Artificial selection is the tool with which most of the increase in the production of domestic animals has been achieved, such as the duplication of milk production in cattle and the four-fold increase of the weight in chickens in the last 50 years (Hill, 2014). But artificial selection is also a basic experimental tool for genetic research (Hill and Caballero, 1992). For example, with artificial selection it has been shown that the phenotypic values of almost any quantitative trait can be altered permanently, far exceeding the initial ranges of variation, and that most of the response to selection is due to the change in the allele frequencies of the loci that affect the trait in the population. Likewise, artificial selection experiments have allowed a large amount of information to be obtained about the nature of those loci (Hill and Bunger, 2004). The initial response to artificial selection can be used to estimate the genetic variance of the selected trait in the population. On the other hand, it usually produces genetic changes in other correlated traits, allowing the estimation of genetic covariances between different characters or the same character in different environments. From the long-term response, the number of genes involved and their initial frequencies can be inferred, as well as the possible interactions between artificial and natural selection. Likewise, the cross of lines selected in divergent directions is an important source of information to study the physiological and biochemical characteristics of the traits under study and can be used to identify and map the genes involved. Finally, the selection exercised in a population initially lacking in genetic variation allows us to estimate the mutation rate with which new alleles appear affecting the trait, its effects and dominance relationships, as well as the pleiotropic effects on other traits.

The simplest form of artificial selection is to apply it directly on the phenotypic values of individuals, whose general principles will be studied in the next section. Later we will see that the response to selection can be improved if information from relatives and molecular genetic markers is added to the individual information, and that several traits can also be selected simultaneously. Also, given that the number of selected parents is generally small, artificial selection usually leads to an increase in the inbreeding of the population, with the consequences described in Chapter 8. The study of the bases of artificial selection facilitates the understanding of natural selection, which we will see in the next chapter.

9.1.1 Response to Selection and Its Prediction

Let us first consider the selection of a quantitative trait based on the phenotypic values of individuals ignoring, for the moment, the side effects of genetic drift associated with reduced census sizes, as well as the possible opposition of natural selection. Since both the number and the effects and frequencies of the loci that affect the character under selection are unknown, the infinitesimal model described in Chapter 1 is adopted. For this, a large number of loci of small and additive effect control the character, so that the distribution of phenotypic values is normal with phenotypic mean $\overline{X}$ and variance σ_P^2 (we will use this notation for the variances, as well as the usual V notation, as appropriate for the development of equations). The response to selection will then be measured as the change in the population mean after one generation, whose basic expression is explained below.

Figure 9.1 presents the phenotypic values of a trait for individuals of a population in two consecutive generations, representing on the abscissa axis the average values of the character for pairs of parents and on the ordinate those of their corresponding offspring. This figure is analogous to Figures 1.1 (Chapter 1) and 6.1 (Chapter 6) referring to human height. The origin of coordinates indicates the average of the parental population $(\overline{X})$, which we assume is equal to the average of the progeny, and we also assume that the parents have been paired at random. As we saw in Chapter 6, the slope of the regression line of the average value of the offspring on the average parental value is an estimate of heritability, that is, $b_{O\overline{P}} = h^2$, if it is assumed that there are no causes of environmental similarity between parents and offspring for the character. Suppose now that the group of parents with the highest phenotypes for the character is selected (black circles) by truncating the distribution as indicated in the figure. Truncation selection is the simplest, but not the only, way to carry out artificial selection. The mean of the character of the selected parents deviated from the population mean is called the selection differential (S), and the mean of the offspring of the selected parents, also deviated from the population mean, is the response to selection (R)

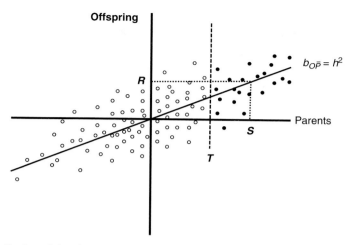

Figure 9.1 Distribution of the phenotypic values of a character in a population in two consecutive generations. The abscissa axis indicates the mean of the character for each pair of parents and the ordinate the mean of their offspring. The average of the parental population, which is assumed to be equal to that of the offspring, is the origin of coordinates. The regression of the mean of the values of the offspring on that of their parents $(b_{O\bar{P}})$ estimates the heritability (h^2) of the trait. If a truncation is carried out in the distribution of parents at point T, the means of parents and offspring, deviated from the population mean, are S and R, respectively, from which it follows that $R = h^2S$.

resulting from the pressure imposed. Since the regression $b_{O\bar{P}} = h^2$ applies equally to this portion of the data, it follows that $R = b_{O\bar{P}}S$, and therefore,

$$R = h^2S, \tag{9.1}$$

an expression originally proposed by Lush (1945), which is sometimes referred to as the breeder's equation.

We saw in Chapter 3 that the regression of the additive value (A) of an individual on its phenotypic value (P) is the heritability and, therefore, the expectation of the additive value is $E[A] = h^2P$ (equation (3.17)). This expression is equivalent to (9.1), since the mean of the offspring, deviated from the population mean is, by definition (Chapter 3), the additive value of the parents. On the other hand, the phenotypic value of the parents, deviated from the population mean, is equal to S.

Figure 9.1, however, does not illustrate the process of artificial selection, which involves the mating of the selected individuals to obtain the offspring generation. This illustration is shown in Figure 9.2, where the upper distribution is the parental generation and the lower is the filial one, obtained by mating the parents selected by truncation, and whose variance is assumed to be equal to that of the parental generation. The mean of the selected individuals $(\bar{X}_S)$ deviated from the mean of the parental generation $(\bar{X})$ is the selection differential $S = \bar{X}_S - \bar{X}$, and the difference between the means of the offspring and parental generations is the response to selection, $R = \bar{X}_O - \bar{X}$. Note that the average of the offspring is lower than that of the selected parents. This is the effect referred to by Galton (1889) as a regression to mediocrity mentioned in Chapter 1, which occurs as a consequence of the fact that heritability is less than unity, and part of the superiority of the selected parents is due to the environment.

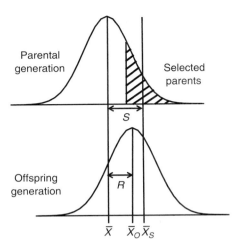

Figure 9.2 Illustration of the artificial selection process. The parental distribution, with mean $\overline{X}$, is truncated, having the selected individuals a mean $\overline{X}_S$, whose deviation with respect to $\overline{X}$ is the selection differential, S. The selected parents are crossed to obtain the filial generation whose mean is $\overline{X}_O$, and the difference between this and the parental generation is the selection response, R.

Equation (9.1) indicates that the response to selection of a trait depends on its heritability and the selective pressure applied. The smaller the proportion (ϕ) of individuals selected as parents, the further apart their mean will be from the population mean and, therefore, the greater the selection differential S. This also depends on the variance of the phenotypic distribution since, for the same selected proportion, the value of S will be higher the higher the variance. The response to selection can be predicted without the need to actually carry out selection if estimates of heritability and phenotypic variance are available and normality of the phenotypic distribution is assumed. By standardizing the expression (9.1) with the phenotypic standard deviation, σ_P, we have, $R/\sigma_P = h^2(S/\sigma_P) = h^2 i$, where i is the standardized selection differential or selection intensity, from which it is obtained

$$R = ih^2\sigma_P = i\sigma_A^2/\sigma_P = ih\sigma_A,\tag{9.2}$$

where σ_A^2 is the additive variance of the trait.

The value of the selection intensity can be obtained from the tables of the normal distribution. Consider the standardized normal $N(0, 1)$ of Figure 9.3 where T is the truncation point for a selected proportion ϕ and z the value of its ordinate. The average value of the selected individuals is i that, with the normalized density function of the normal distribution, is obtained as

$$i = \frac{1}{\phi}\frac{1}{\sqrt{2\pi}}\int_T^{\infty} x e^{-x^2/2}dx = \frac{1}{\phi}\frac{1}{\sqrt{2\pi}}[-e^{-x^2/2}]_T^{\infty} = \frac{1}{\phi}\frac{1}{\sqrt{2\pi}}e^{-T^2/2} = \frac{z}{\phi}.\tag{9.3}$$

Therefore, from the ϕ value, z can be obtained from the tables of the normal distribution and used to predict i in a population of infinite census size. When the census size is finite, the selection intensity is somewhat lower than that obtained with equation (9.3).

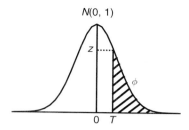

Figure 9.3 Standardized normal distribution in which a proportion ϕ of the distribution is selected by truncation at point T, whose ordinate is z.

For example, if we select an individual of 10, two of 20, four of 40 or ten of 100, the intensities are $i = 1.54, 1.64, 1.69$ and 1.73, respectively, while the value for an infinite population is 1.75. As we will see later, the inbreeding generated by a small number of parents implies a reduction in the long-term response.

    If the intensities of selection are different in each sex, they are averaged without the need to weight the number of individuals of each sex, that is, $i = (i_m + i_f)/2$, since half of the genes of the offspring come from male parents and the other half from female parents regardless of how many they are. Thus, if the trait is only selected in a sex, for example females in the case of characters that are only expressed in these, the applied selection intensity would be $i = i_f/2$.

    Equation (9.2) indicates the keys to increase the response to selection. The first is to increase the selection intensity (i) by choosing a low proportion (ϕ) of individuals as parents. This is usual in the genetic improvement of domestic animals, where artificial insemination techniques and other technologies allow very few sires and many dams to be used. In any case, the possible values of i range in a reduced interval, between 0.7 ($\phi = 0.5$) and approximately 2 (ϕ somewhat less than 10%). Another way to increase the selection response is to reduce the phenotypic variance of the population, which can be achieved by reducing the environmental variation or the experimental error in the evaluation of phenotypes, if possible with repeatable measures (Chapter 3).

    In situations in which generations overlap, the response can be presented per year, as an annual response, $R_a = R/I_g$, where I_g is the generation interval, that is, the age of the parents when the offspring are born. Note that the response per generation is maximized with a maximal selection intensity (i), whereas the annual response is maximized when i/I_g is maximal.

    Extensions for the prediction of selection response for inbred populations are given by Walsh and Lynch (2018, chapter 23).

9.1.2 Correlated Response

When the selected character presents a genetic correlation (r_A) with others, a correlated response can occur in these. Let us suppose that trait X is selected, producing a response $R_X = i_X h_X \sigma_{AX}$ (equation (9.2)), and that this character presents phenotypic and genetic covariances cov_P and

cov_A, respectively, with trait Y. Selection on X will produce a correlated selection differential on Y that would be $S_{CY} = cov_A$. Since, as we have seen in the previous section, the response to selection is equivalent to the additive value of the parents, the correlated response on Y will depend on the regression of the additive values of Y on X ($b_{A,Y.X}$), that is, $R_{CY} = b_{A,Y.X}R_X = (cov_A/\sigma^2_{AX})i_X h_X \sigma_{AX}$. Since $r_A = cov_A/(\sigma_{AX}\sigma_{AY})$, we get that

$$R_{CY} = i_X r_A h_X h_Y \sigma_{PY}, \qquad (9.4)$$

where the product term $r_A h_X h_Y$ is called coheritability.

9.1.3 Measure of the Response

The response to selection predicted by equation (9.2) is strictly valid for one generation, since the change in allele frequencies caused by selection will alter the population parameters (h^2 and σ^2_P) in successive generations. However, in practice, the prediction is usually applicable to a few generations and the change in the mean by artificial selection presents an approximately linear trajectory in this period. This is illustrated in Figure 9.4, which shows the results of a selection experiment for increasing and decreasing the number of sternopleural bristles in *Drosophila melanogaster* (Vilas et al., 2015). The means of the selected lines can show jumps from one generation to another depending on the genetic composition of the selected population, on chance (genetic drift), measurement errors and environmental factors that affect the mean of the trait. If these jumps are erratic they will simply produce noise in the response but no bias. However, sometimes there are sustained environmental changes that occur over time creating trends that distort the response, or there even may be unwanted genetic changes such as adaptation by natural selection of the selected population to the culture medium or the environmental conditions in which individuals grow. In order to reduce these latter effects as much as possible, it is common in experimentation to maintain, simultaneously with the selected line or lines, one or more control lines with the same number of parents and offspring but without selection. Alternatively, when appropriate, selection can be carried out in both directions (as in the case of Figure 9.4) and use the divergence between both (panel (d)) as a measure of the response corrected for environmental biases or genetic changes that affect equally both types of lines.

If the response is approximately linear during the period in which selection is carried out, as in the example of Figure 9.4 (although there are signs of non-linearity from the third generation), the response can be quantified as the regression of the mean of each generation on the corresponding ordinal. This provides a descriptive summary of the response produced but does not take into account the magnitude of the selection that has been applied. To include this last factor, the total response is usually calculated as the regression coefficient of the cumulative response on the selection differential accumulated over generations (if the change is approximately linear). With this is obtained (equation (9.1)) an estimate of the heritability in the starting population, which is called realized heritability (h^2_r) (Falconer, 1952). This representation is illustrated in panel (c) of Figure 9.4 for each direction of selection (and in panel (d) for the divergent response). Note that the realized heritabilities are somewhat different in each direction of selection and that both are, in turn, somewhat lower than those

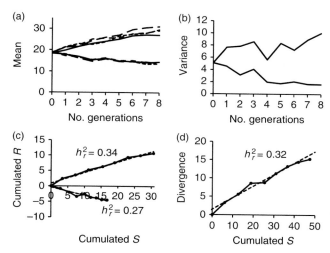

Figure 9.4 Results of an artificial selection experiment in four lines for increase and four for decrease of the number of sternopleural bristles (sum of the two sternopleural plates) in *Drosophila melanogaster* (data from Vilas et al., 2015). In each generation, 50 males and 50 females were evaluated per line, selecting the 10 males and 10 females with more extreme values of the trait (increase or decrease depending on the case). The selected individuals were mated in bottles (what is called mass selection). (a) Change in the mean number of bristles in each generation and selected line. (b) Phenotypic variance averaged over lines. (c) Accumulated response (R) against the accumulated selection differential (S) (each point indicates a generation). The slope of the adjusted regression line (dotted line) is an estimate of the realized heritability (h_r^2). (d) Same as panel (c), but using the divergence between the means selected in both directions.

obtained from the base population through a design of sib families ($h^2 = 0.38$; Vilas et al., 2015). The figure also shows that the selection differential applied in the selection for decreasing the number of bristles was lower in the last generations than in the first ones, whose causes will be discussed in the next section.

As already indicated, the response to selection may vary between lines with the same number of individuals scored and selected from the same base population for various reasons, including genetic drift, measurement errors and differences in the genetic constitution of the lines. If several selected lines are available, as in the case of Figure 9.4, an empirical variance of the response can be obtained. For example, in the last generation of selection the variances of the response observed in the selection lines are $V(R) = 1.76$ and 0.04 for the lines of increase and decrease, respectively. If there are no replicates it is possible to obtain a theoretical approximation of this variance when it is due to genetic drift and sampling error. Recall from Chapter 8 (equation (8.10)) that the variance of line means for an additive trait under a neutral model is $2F_t V_A = 2[1-(1-1/2N_e)^t]V_A$, which can be approximated by tV_A/N_e when $t \ll 2N_e$. On the other hand, the variance of the measurement error is equal to the phenotypic variance (V_P) divided by the number of individuals evaluated (n). The variance of the response can then be approximated by

$$E[V(R)] \approx \frac{tV_A}{N_e} + \frac{V_P}{n} = V_P\left(\frac{th^2}{N_e} + \frac{1}{n}\right) \tag{9.5}$$

(Hill, 1980), which can also be used to obtain the sampling variance of the realized heritability. In the case of the experiment in Figure 9.4, the number of individuals evaluated per line was $n = 100$. If we approximate the effective size of the lines by $N_e \approx 15$ and consider the estimates of h^2 in each set of lines (0.34 and 0.27 for increase and decrease, respectively) and the approximate phenotypic variances in the last generations of selection (8 and 2 for increase and decrease, respectively), we have that $E[V(R)] = 1.5$ and 0.31 for increase and decrease, respectively, that can be compared with the observed values (1.76 and 0.04, respectively).

The standard error of the estimated realized heritability can also be obtained from the standard error of the regression, but this is generally an underestimate because it ignores that the means of different generations are correlated (Hill, 1971). However, estimations of the realized heritability can be obtained through REML with the animal model (Chapter 6), which considers this correlation between generations. In addition, REML takes account of fixed effects and the genetic changes that occur throughout the process, and allows for dominance effects (Smith and Maki-Tanila, 1990).

When there is epistatic variation in the quantitative trait, the realized heritability and, therefore, the selection response are a function of the additive $\times$ additive components of variance, such that $h_r^2 = (V_A + \frac{1}{2}V_{AA})/V_P$ if other higher order interactions are ignored (Mackay, 2014).

9.1.4 Asymmetry of the Response

Figure 9.4 also illustrates that the response to selection in both directions can be asymmetric. There are several causes for this to occur, the most important of which are the following:

1 A possible cause is genetic drift, particularly if the effective size of the selected lines is low and there are no replicates in each direction of selection.

2 The genetic basis of the character may also be responsible for the asymmetry. If the variation is due to few loci and/or there are some of great effect in the base population that contribute to a large part of the variation, when their allele frequencies are not intermediate, the response may be greater in one direction than in the other. Likewise, dominance may contribute to asymmetry, since the change in gene frequency of recessive alleles at low frequency or of dominant alleles at high frequency is much slower than in the opposite scenario (see Figure 2.8 of Chapter 2).

3 If the trait shows inbreeding depression, the change produced will be added to that of the response to artificial selection in the direction of decrease but will be subtracted in that of increase. The effect of inbreeding depression can be partially eliminated if an unselected line with the same census size as the selected one is used as control, but it will not be completely avoided because the parents of the selected lines will tend to be more consanguineous than those of the control line (as we will see in Section 9.2.2 below).

4 There is also the possibility that the character presents maternal effects that affect the response positively or negatively in a given direction. For example, the weight at weaning in mammals depends on the production of breast milk, which, in turn, depends on the mother's weight, so that the selection for weight gain or decrease may affect the maternal component in a different way in the offspring.

5 If the phenotypic variance changes according to the direction of selection, for example, by scaling effects, the selection differential will also be affected although the realized heritability estimate will correct, at least partially, this effect.

6 Finally, a usual cause of asymmetric responses is the opposition of natural selection to the change in the mean in one direction of selection but not in the other. For example, if there are loci that affect the selected character that also reduce the value of some fitness component, its increase in frequency will be restricted by selection.

If selection implies a lower fecundity of the more extreme parents or a greater mortality of their offspring, this will produce selection differentials lower than expected. One way to detect this effect is to calculate the selection differential weighting the value of the parents by their contribution to offspring. If selection acts in the indicated manner, the selection differential calculated in this way will be greater than that obtained without weighting.

When the selected trait is a component of fitness, asymmetry is the most general result (Frankham, 1990) since the response to decrease the mean is accompanied by inbreeding depression and, in addition, deleterious alleles are expected to be at low frequencies in the starting population so they will show a larger response to selection. An example is presented in Figure 9.5, which shows the response to increase and decrease in productivity (number of pupae per pair) in *Drosophila melanogaster* deviated from the average productivity of unselected controls (Ávila et al., 2013). The realized heritability corresponding to the selection for reduction in productivity was 0.64 ± 0.08 whereas the corresponding one for increase was 0.04 ± 0.03.

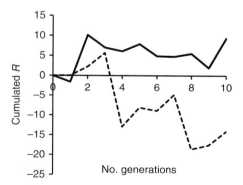

Figure 9.5 Accumulated response to selection for increased and decreased productivity (number of pupae per couple) in *Drosophila melanogaster* throughout generations (data from Ávila et al., 2013). The results are the average of four replicates. In each generation, 25 pairs were established ($N = 50$), and the 5 with greater or lower productivity were selected. Five males and five females of the offspring of each of the selected pairs were mated randomly to form the 25 parental pairs of the next generation. The means of the selected lines were deviated from those of four control lines, maintained in the same way in the absence of selection, in order to correct the response for environmental changes that could affect the set of lines.

Let us now return to the example in Figure 9.4. In this, a remarkable asymmetry of the response is observed. The number of sternopleural bristles is a character that is considered close to neutrality, without inbreeding depression and, although some genes that affect the trait also have deleterious pleiotropic effects (for example, often lethal alleles are observed with effect on the trait), it is unlikely that the observed short-term asymmetry is due to the effect of natural selection. The weighted selection differential mentioned above could not be calculated in this case because the individuals were mated in groups (what is called mass selection) instead of pairs (individual selection), not being able to find out the contribution of each pair to offspring. In this case, the most likely explanation is that the asymmetry is due to changes in the phenotypic variance. As shown in panel (b) of Figure 9.4, the variance increased with generations in the lines selected for high bristle number and decreased in those selected for low bristle number, which would imply an increase and a reduction of the selection differential, respectively, as observed in panel (c) in the case of the second ones.

9.1.5 Change of Allele Frequency Due to Selection

The change in the population mean produced by artificial selection is due to the change in the allele frequencies of the loci that affect the character. Since it is not usually possible to identify such loci, changes in frequency cannot be followed. However, it is possible to relate the effect of an allele on a quantitative trait with the expected change in its allele frequency by selection. In Chapter 2 we studied the allele frequency changes by natural selection of an allele with selective advantage s (Tables 2.6 and 2.7) if the average fitness approaches unity. In the case of an additive locus with frequencies p and q, this change is $\Delta q = spq/2$. To establish the aforementioned relationship, consider the model in Table 9.1, where the A_2 allele has an effect a on a trait subjected to artificial selection and we will see the equivalence with the fitness model with selective advantage s in the homozygote.

Table 9.1 shows the phenotypic values (P), which are assumed to be equal to the genotypic ones (G) and the frequency of the gene dose of each genotype (Q). To predict the change in allele frequency given a phenotypic change by selection, consider the regression coefficient of Q on P, $b_{Q,P} = \text{cov}_{Q,P}/\sigma_P^2$, where $\text{cov}_{Q,P} = pqah + q^2a - q(aq + 2dpq) = pq\alpha$, and α is given by equation (3.6). If a selection differential $S = i\sigma_P$ is applied, the expected change in allele frequency is $\Delta Q = b_{Q,P}S = (pq\alpha/\sigma_P^2)i\sigma_P = ipq\alpha/\sigma_P$. Equating ΔQ to the change in allele frequency in the natural selection model, $spq/2$, and clearing s we have that the

Table 9.1 *Model for an additive locus with effect on a quantitative trait subjected to artificial selection and its equivalence with a model of natural selection on fitness*

Genotype	A_1A_1	A_1A_2	A_2A_2	Mean
Frequency	p^2	$2pq$	q^2	
Fitness (w)	$1-s$	$1-sh$	1	
Value of the trait ($P=G$)	0	ah	a	$aq+2dpq$
Frequency of allele A_2 (Q)	0	$1/2$	1	q

selection coefficient against the genotype A_2A_2 is $s = 2\alpha i/\sigma_P$. If there is additive gene action at the locus ($h = 0.5$), $\alpha = a/2$ and

$$s = \frac{ai}{\sigma_P} \tag{9.6}$$

(Haldane, 1931; Robertson, 1960). Expression (9.6) is also valid with dominance. For example, if $h = 0$ (allele A_2 is recessive) the change in expected allele frequency is spq^2 (Table 2.7, Chapter 2) and $\alpha = aq$. Equalling the expressions, $spq^2 = ipqa/\sigma_P$ and, clearing, we obtain again equation (9.6).

Expression (9.6) is sufficiently accurate for values of $s \leq 0.5$ (Latter, 1965) and can be obtained in another way that is perhaps more intuitive (Falconer and Mackay, 1996, p. 202). Let us assume that the phenotypic distribution of the individuals with genotype A_2A_2 is that illustrated in Figure 9.6, the variation σ_P^2 around the mean being due to the effect of other loci and the environment. Suppose that the proportion of the chosen distribution when selecting by truncation at point T is ϕ (striped region) and z is the ordinate corresponding to T. Consider now the distribution of A_1A_1 individuals whose mean is displaced to the left in relation to that of A_2A_2 by a magnitude a. If the truncation point on A_1A_1 is moved to the left in a units, the proportion of A_1A_1 cut will also be ϕ, so that if a is small, we can obtain the selected proportion of the distribution of A_1A_1 with the truncation point T, which will be approximately $\phi - az/\sigma_P = \phi - ai\phi/\sigma_P$, since $z = i\phi$ (equation (9.3)). From Table 9.1, the fitness of A_1A_1 is $1 - s = w_{A_1A_1}/w_{A_2A_2} = \phi[1 - (ai/\sigma_P)]/\phi$, from which expression (9.6) is obtained again.

The approximation (9.6) allows us to calculate the probability of fixation of an allele with artificial selection using the Kimura equation (equation (7.1), Chapter 7) or its approximations (Table 7.1). For example, in a selected population with $N = 40$ selected parents from 200 evaluated each generation ($i = 1.4$) and effective size $N_e = 30$, a mutant with effect $a/\sigma_P = 0.2$ at an additive locus ($h = 0.5$) would have selective advantage $s = ai/\sigma_P = 0.2 \times 1.4 = 0.28$ and its probability of fixation (Table 7.1) would be $P_f \approx 2N_e sh/N = N_e s/N = 30 \times 0.28/40 = 0.21$ if it appears in one of the selected parents, or $P_f \approx 30 \times 0.28/200 = 0.042$ if it appears in one of the evaluated individuals.

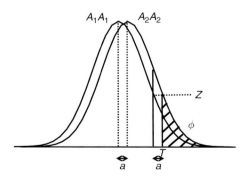

Figure 9.6 Phenotypic distribution for individuals with genotypes A_1A_1 and A_2A_2 for the model in Table 9.1. The effect of the A_1A_1 genotype relative to that of A_2A_2 is a. The point of truncation in the distribution is T with value z on the ordinate axis, and ϕ is the proportion of the distribution that is selected (striped region).

Consider now the contribution to the mean of a locus expressed in equation (3.3) (Chapter 3) which, for an additive locus and the values in Table 9.1, is $M = aq$. After one generation of selection the allele frequency will change from q to q', so that the change in mean will be $\Delta M = aq'-aq = a\Delta q$. As $\Delta q = spq/2$, substituting for equation (9.6), $\Delta q = aipq/(2\sigma_P)$ and $\Delta M = a^2 ipq/(2\sigma_P)$. Since the additive variance is $V_A = \sigma_A^2 = a^2 pq/2$ (equation (3.9), Chapter 3), for an additive locus we have that $\Delta M = i\sigma_A^2/\sigma_P$, that is, equation (9.2) is obtained as expected.

The previous deductions were developed assuming a panmictic population, but can also be obtained for a non-panmictic one. Recall that, in this case, it is necessary to distinguish between the average allelic excess (e) and the average effect of allelic substitution (α), which coincide in a panmictic population (Table 8.1). The change in allele frequencies promoted by selection is a function of the average excess e, while the change in the average phenotypic value of the population is a function of the average effect α. Thus, the change in allele frequency (ΔQ) predicted from the phenotypic change would be obtained as before, $\Delta Q = b_{Q.P}S$, so that $\Delta Q = (pqe/\sigma_P^2)S = ipqe/\sigma_P$. On the other hand, the phenotypic change (ΔP) can be obtained as $\Delta P = b_{P.Q}\Delta Q$, where $b_{P.Q} = 2\alpha$ by definition, that is, $\Delta P = 2\alpha\Delta Q$. Substituting ΔQ in the last expression we have that $\Delta P = R = 2pqe\alpha i/\sigma_P$. We saw in Chapter 8 that in the absence of panmixia, $\sigma_A^2 = 2e\alpha pq$ (equation (8.2)), therefore, substituting we obtain $R = i\sigma_A^2/\sigma_P$, that is, again equation (9.2).

9.2 Effect of Selection on Genetic Variance

9.2.1 The Bulmer Effect

During the selective process the genetic variance is progressively reduced as the alleles tend to extreme frequencies by selection and drift. There is also a reduction in genetic variance, called the Bulmer effect, which occurs only during the initial generations of selection, particularly in the first, even in infinite populations. The reason for this reduction is that the selected individuals correspond to the extremes of the phenotypic distribution, so that their variance must be lower than that of the population as a whole. Bulmer (1971) showed that the phenotypic variance of the selected parents is $V_P^* = V_P(1 - k)$, with $k = i(i - x)$, where i is the selection intensity and x the deviation of the truncation point with respect to the population mean in phenotypic standard deviations. Since the additive variance is $V_A = h^2 V_P$, its reduction is

$$V_A^* = V_A(1 - kh^2). \tag{9.7}$$

This decrease in the additive variance can be interpreted in terms of negative gametic disequilibrium. As we have said, the selected individuals, coming from the extreme of the distribution, will be more similar to each other than those taken at random from the population. The fact of being similar implies that if one carries alleles that increase the trait, it must also carry alleles that reduce it, and the same will occur to others. Therefore, there is a negative correlation of allelic effects within individuals that generates a negative disequilibrium whose magnitude is $-kh^2$. This disequilibrium is reduced by half in the subsequent generation for the following reason. If the selected parents produce full-sib families (whose genetic correlation is $r = 1/2$), half of the genetic variance in the offspring will be within families ($V_{AW} = \frac{1}{2}V_A$, which is not affected by the parents' phenotypic value in the absence of linkage) and the other half between families, which will be affected by the aforementioned reduction ($V_{AB} = \frac{1}{2}V_A^*$ $= \frac{1}{2}V_A[1 - kh^2]$). Therefore, the total variance of the offspring will be $V_{AW} + V_{AB} = V_A(1 - \frac{1}{2}kh^2)$, that is, the negative disequilibrium will have been reduced by a half ($-\frac{1}{2}kh^2$). This will occur again in successive generations, disappearing the disequilibrium in practical terms after a few. In general, the additive variance in generation t can be obtained as

$$V_{A,t} = \frac{1}{2}\left(1 - kh_{t-1}^2\right)V_{A,t-1} + \frac{1}{2}V_{A,0}. \tag{9.8}$$

For example, if $V_{A,0} = V_E = h^2 = 0.5$ and 20% of the population is selected ($\phi = 0.2$), we have that $i = 1.4$, $x = 0.842$ and $k = 0.781$. The additive variance expected in generations 0 to 4 with equation (9.8) will be 0.5, 0.402, 0.381, 0.376 and 0.375, respectively. That is, the reduction is virtually non-existent from the third generation.

9.2.2 The Effective Population Size with Artificial Selection

In Chapter 5 we saw that the effective population size can be predicted in the case of a selected population with equation (5.13) which, in the simplest case of random mating between individuals ($\alpha = 0$) and Poisson variance of the family contributions due to non-heritable causes ($S_k^2 = 2$), is reduced to

$$N_e \approx \frac{N}{1 + Q^2 C^2} \tag{9.9}$$

(Robertson, 1961), where C^2 is the variance of the selective advantages (standardized so that the average selective advantage is equal to one) and Q is the term that determines the drag of the selective advantages along generations whose accumulated magnitude is $Q \approx 2/(2 - G)$ if mating is, as assumed, random ($r = 0$ in equation (5.14)). In this expression, G is defined as the proportion of the variance remaining after its decay by genetic drift and selection.

In the context of artificial selection by truncation, if we ignore genetic drift we have, from equation (9.7), that $G = 1 - kh^2$. On the other hand, Robertson (1960) showed that for truncation selection $C^2 = i^2 t$, where t is the intraclass correlation of family members. This can be demonstrated as follows. The standardized variance is the variance of the selective effects $C^2 = V(s)$ and, by equation (9.6), $s = ai/\sigma_P$. Therefore, $V(s) = V(ai/\sigma_P) = (i^2/\sigma_P^2)V(a)$ and, since the variance of the effects would be equivalent to the variance between families for the selected trait (σ_B^2), we have that $C^2 = i^2(\sigma_B^2/\sigma_P^2) = i^2 t$.

Applying these expressions in the previous example, which considered a population with $h^2 = 0.5$, $i = 1.4$ and $k = 0.781$, we have that $G = 0.6096$ and $Q = 2/(2 - G) = 1.438$. If we consider full-sib families and ignore effects of dominance and common environment, $C^2 = i^2 t = i^2 h^2/2 = 0.49$. Substituting in equation (9.9), we obtain $N_e \approx N/2$, that is, in this example, the asymptotic effective size with selection is half the number of breeders.

If mating of individuals is not random, but there is a β proportion of self-fertilized matings or mating between sibs, the relevant equations to predict the effective size are (5.13) and (5.14) with $r \approx \beta$ and intraclass correlation $t = h^2 G(1 + \beta)^2/4$ and $t = h^2 G(2 - \beta G)$, respectively (Santiago and Caballero, 1995). In the second case, in addition, it is necessary to substitute $(1 + \alpha)$ for $(1 + 3\alpha)$ in equation (5.13).

9.2.3 Combined Effect of Genetic Drift and Selection on Genetic Variance

The previous sections ignore the fact that genetic variance and, hence, the selection response, are reduced by genetic drift. This is particularly important in selected populations, since their number of breeding individuals is usually small and selection reduces the effective size, as explained in the previous section.

Consider first the impact of genetic drift ignoring that of selection. We have indicated that in a design with full-sib families, half of the additive variance is assigned to the component of within-family variance (V_{AW}) and the other half to the between-family variance (V_{AB}). If the number of males is N_m and that of females N_f, these variances in generation t are obtained by $V_{AW,t} = \frac{1}{2} V_{A,t-1}$, $V_{ABm,t} = \frac{1}{4} V_{A,t-1}[1 - 1/(2N_m)]$ and $V_{ABf,t} = \frac{1}{4} V_{A,t-1}$ $[1 - (1/2N_f)]$, where the last two equations express the components assigned to males and females, respectively. Adding the latter, we obtain the global between-family component, $V_{AB,t} = \frac{1}{2} V_{A,t-1}(1 - 1/N_e)$, where $1/N_e = 1/(4N_m) + 1/(4N_f)$ (equation (5.3)). Therefore, the total genetic variance in generation t is

$$V_{A,t} = V_{A,t-1} \left(1 - \frac{1}{2N_e} \right). \tag{9.10}$$

If we consider now the reduction in variance by selection (Bulmer effect), the within-family component remains unchanged, $V^*_{AW,t} = \frac{1}{2} V_{A,t-1}$, and would be obtained by means of equation (9.10), but the between-family components would be $V^*_{ABm,t} = \frac{1}{4} V^*_{A,t-1}[1 - 1/(2N_m)](1 - k_m h^2_{t-1})$ and $V^*_{ABf,t} = \frac{1}{4} V^*_{A,t-1}[1 - 1/(2N_f)](1 - k_f h^2_{t-1})$ for males and females, respectively, or $V^*_{AB,t} = \frac{1}{2} V^*_{A,t-1}(1 - 1/N_e)(1 - k h^2_{t-1})$ for the global component, where $k = (k_m + k_f)/2$. The additive variance with selection per generation can be obtained, therefore, as

$$V^*_{A,t} = \tfrac{1}{2} \, V^*_{A,t-1}(1 - 1/N_e)\left(1 - k h^2_{t-1}\right) + \tfrac{1}{2} \, V_{A,t-1}, \tag{9.11}$$

where the last term is calculated, as we have said, by means of equation (9.10), and indicates the reduction in within-family variance due to the inbreeding of parents. Note that equation (9.10) expressed as a function of the initial variance is $V_{A,t} = V_{A,0}(1 - F_t)$ where F_t is the coefficient of inbreeding. In the case of a particular family, the expected value of the additive variance of the offspring would be obtained as $V_{AW} = V_{A,0}[1 - \frac{1}{2}(F_m + F_f)]$, where F_m and F_f are the coefficients of inbreeding of the male and female parents, respectively. The bracketed term refers to the Mendelian segregation of the progeny, as indicated in Chapter 4 (see equation (4.5)). In the context of selection, equation (9.10), which excludes the effect of selection, is often called additive 'genic' variance, to distinguish it from the additive genetic variance expressed by equation (9.11) (see also Section 3.3.1 of Chapter 3). Finally, it should be noted that the effective size strictly applicable to equations (9.10) and (9.11) must be that corresponding to each generation since, as explained in Chapter 5, the effective size with selection is progressively reduced over generations until reaching an asymptotic value determined by equation (9.9).

Figure 9.7 illustrates the decay of the heritability with generations considering genetic drift and the Bulmer effect (solid line) or only genetic drift (dashed line), making it clear that the decline due to the Bulmer effect occurs, as previously indicated, in the initial generations.

9.3 Long-Term Response

The short-term response depends only on the intensity of selection, the heritability and the phenotypic variance of the character, as can be deduced from expression (9.2). However, the long-term response depends on more factors, such as the initial frequencies and effects of genes, the effective population size and mutation. For example, the long-term response is expected to be greater the larger the number of loci that affect the trait segregating in the base population. In fact, under the simplest assumptions it would be possible to make a rough estimate of this number from the divergence between two lines from the same base population selected in opposite directions, assuming the base population was obtained by the cross between two inbred lines (Figure 9.8). If we assume that those inbred lines are very different for the trait in question, it is likely that there will be different alleles fixed in each of them, so the cross will be heterozygous for these loci. Now suppose that the cross is selected for increase and decrease of the trait and that, for the n heterozygous loci that it carries, the alleles that increase the trait are fixed in the line selected upwards and the same occurs with those that reduce the trait in the line selected downwards. In other words, we assume that the effective size in each selected line is very high and there is no loss of alleles by drift. If we assume that for all loci the difference between the two homozygotes is the same, a, the total divergence reached in the selection limit will be $R_T = na$. On the other hand, the initial additive variance of the cross will be $\sigma_A^2 = \sum_n (a^2/2)\frac{1}{2}\frac{1}{2} = na^2/8$, since the initial allele frequencies are 1/2 for each of the n loci. Therefore, we can notice then that $R_T^2/(8\sigma_A^2) = n$. That is, knowing the magnitude of the response and the initial additive variance the number of loci could be estimated assuming equal effects, which is called the effective number of loci or factors. In general, this number will be an underestimation of the number of loci in the base population if the effects are not equal and/or if all the alleles are not fixed, or an overestimation if the initial frequencies are not intermediate (they would always be if the base population is the F_1 of a cross, as in Figure 9.8). This procedure is called the Castle–Wright method (Castle, 1921; Wright, 1952), and is unreliable as an estimator of the number of loci due to the multiple assumptions on which it is based, in addition to ignoring linkage and epistasis (Zeng et al., 1990). However, it has some utility on some occasions. For example, prior to starting a mapping analysis of quantitative trait loci from the cross of two inbred lines that differ in the mean of the trait (see Chapter 11), the method can be used to deduce the presence or absence of loci with large effects in the cross, although the estimated effective number of factors is not too reliable (Lander and Botstein, 1989).

9.3.1 Prediction of the Long-Term Selection Response

Under the assumption of the infinitesimal model it is possible to obtain approximate predictions of the long-term response (Robertson, 1960). Since this must be greater the larger the

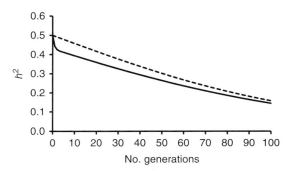

Figure 9.7 Reduction in heritability (h^2) due to genetic drift and selection (Bulmer effect) (continuous line) or only to genetic drift (dashed line). The example corresponds to a population with initial heritability $h^2 = 0.5$ ($V_{A,0} = V_E = 0.5$), selection intensity $i = 1.4$, $k = 0.781$ and effective size $N_e = 30$. Values obtained with equations (9.10) and (9.11).

number of loci, the infinitesimal model, which assumes an indefinitely large number, provides the theoretical maximum response. Let us first consider the case in which mutation is ignored and, for simplicity, we also ignore the Bulmer effect, since this only has a short-term impact. In this scenario, the reduction in additive variance is exclusively due to genetic drift and is given by equation (9.10), there being no source of increase in it. The phenotypic variance remains constant throughout generations as a consequence of the adoption of the infinitesimal model. Thus, the cumulative response to generation t would be $R_t = R_1 \sum_{g=0}^{t} [1 - 1/2N_e]^g$, where $R_1 = i\sigma_{A,0}^2/\sigma_P$ is the response in the first generation (equation (9.2)). Since the sum is for a geometric series of ratio $1 - [1/(2N_e)]$, we have that

$$R_t = R_1 \frac{(1 - 1/2N_e)^t (1 - 1/2N_e) - 1}{(1 - 1/2N_e) - 1} \approx 2N_e R_1 \{ 1 - (1 - 1/2N_e)^t \},$$

that is to say,

$$R_t = 2N_e R_1 F_t, \tag{9.12}$$

where F_t is the expected inbreeding coefficient in generation t. Over generations, the cumulative response in the limit will be

$$R_\infty = 2N_e R_1, \tag{9.13}$$

that is, $2N_e$ times the initial response. The above equation implies additivity. If the selected genes were recessive, the limit could be much higher. The equation also illustrates the antagonism between short- and long-term response. The response in the first generation depends fundamentally on the selection intensity. This is usually increased by reducing the number of parents and, therefore, reducing the effective size, which restricts the long-term response. On the other hand, given a certain number of individuals evaluated (M), the maximum final response should occur when a ratio $\phi = 0.5$ is selected. If we assume that $N_e \approx M\phi$, we have $N_e i = M\phi i = M\phi(z/\phi) = Mz$. Since z is the ordinate of the truncation point, this is maximum when the distribution is truncated by half, that is, when 50% of the individuals are selected as parents. It is possible, however, to obtain the optimal selected

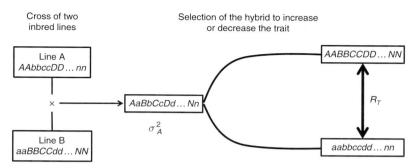

Figure 9.8 Experimental design to estimate the effective number of factors that segregate in a hybrid population obtained by crossing two highly inbred lines with the Castle–Wright method (Castle, 1921; Wright, 1952). Alleles in uppercase and lowercase increase and reduce the character, respectively. The ratio between the square of the response to divergent selection of the hybrid (R^2_T) and eight times the value of its additive variance σ^2_A is an estimate of the number of loci if these were to have the same effect.

proportion that produces the maximum response after a certain number of generations (Jódar and López-Fanjul, 1977).

An illustrative example of the long-term response expressed by equation (9.12) and its limit is presented in Figure 9.9 (broken line). The half-life of the response, that is, the number of generations elapsed for the cumulative response to be half the response in the limit, can be obtained by approximating (9.12) by $R_t \approx 2N_eR_1[1 - e^{-t/2N_e}]$, valid when $t < 2N_e$. Equalling the response to half the limit, N_eR_1, and clearing t, we obtain that $t_{half-life} = (2\ln2)N_e \approx 1.4N_e$ generations, which is also the half-life for the decay of neutral variation (Fisher, 1930). In the case of the broken line in Figure 9.9, the half-life would be $1.4 \times 30 = 42$ generations. For recessive genes, the half-life can be much longer, up to $2N_e$ generations (Robertson, 1960).

Now consider the impact of mutation on the response to selection (Hill, 1982). First, we will deduce the response to selection due exclusively to one generation of mutation. We saw in Chapter 7 that the increment in additive variance per generation by mutation is the mutational variance (V_M or σ^2_M). The response due to one generation of mutation is, then, using equation (9.2), $R_{M1} = i\sigma^2_M/\sigma_P$, and the cumulative response over t generations would be, as before, $R_{M1,t} = R_{M1}\sum_{g=0}^{t}(1 - 1/2N_e)^g = 2N_eR_{M1}[1 - (1 - 1/2N_e)^t] = 2N_eR_{M1}F_t$. When $t = \infty$,

$$R_{M1,\infty} = 2N_eR_{M1} = 2N_ei\sigma^2_M/\sigma_P = i\sigma^2_{A,\infty}/\sigma_P, \tag{9.14}$$

where $\sigma^2_{A,\infty} = 2N_e\sigma^2_M$ is the additive variance in the mutation–drift equilibrium (equation (7.7), Chapter 7). Equation (9.14) shows that, for a given trait and population (fixed σ^2_M and σ^2_P) and the same selection intensity (i), the response to selection is proportional to N_e, as shown experimentally by López and López-Fanjul (1993a).

Expression (9.14) is also the asymptotic response due to new mutation, that is, the response that would occur per generation by the constant input of variation by mutation. This can also be demonstrated in another way. The probability of fixation of an additive mutation is approximately $P_f \approx N_es/N$ (Table 7.1, Chapter 7). Substituting expression (9.6) we have

$P_f \approx N_e a i / N \sigma_P$. When the mutation is fixed, the expected response is $P_f \times a = N_e a^2 i / N \sigma_P$. If U is the haploid mutation rate, the total number of mutations that appear each generation is $2NU$, and the expected total response will be $R = 2NU \times N_e E[a^2] i / N \sigma_P$. Recall from Chapter 7 that, for additive mutations, the mutational variance is $\sigma_M^2 = U \overline{a^2}/2$ and that, if we assume that in the distribution of mutational effects on the trait half are positive and the other half negative, the value of $E[a^2]$ relevant to artificial selection in a direction of selection is $\overline{a^2}/2$. Therefore, substituting, we obtain equation (9.14).

Let us now consider the cumulative response after t generations of mutation and selection, which will be

$$R_{M,t} = \sum_{g=1}^{t} R_{M1,\infty}[1 - (1 - 1/2N_e)^g] = R_{M1,\infty}[t - \sum_{g=1}^{t}(1 - 1/2N_e)^g].$$

Developing the sum of the terms of the geometric series, it is obtained that $R_{M,t} = R_{M1,\infty}\{t - 2N_e[1 - (1 - 1/2N_e)^t]\}$, that is,

$$R_{M,t} = 2N_e(i\sigma_M^2/\sigma_P)(t - 2N_e F_t). \tag{9.15}$$

During the first generations ($t < N_e$), equation (9.15) can be approximated by $R_{M,t} \approx i\sigma_M^2 t^2 / 2\sigma_P$, which indicates that the cumulative response from new mutation is quadratic under the infinitesimal model.

Finally, we can combine the response due to initial variation (equation (9.12)) and that due to new mutation (9.15), that is,

$$R_T = R_t + R_{M,t} = (2N_e i/\sigma_P)[\sigma_{A,0}^2 F_t + \sigma_M^2(t - 2N_e F_t)]. \tag{9.16}$$

Figure 9.9 shows the prediction of the total cumulative response expressed by equation (9.16) (solid line). When the initial additive variance is exhausted, that is, a *plateau* of response is reached, the further response is only due to new mutation and becomes linear, with a rate quantified by equation (9.14). If there is no initial variation or, in general, when $\sigma_{A,0}^2/\sigma_{A,\infty}^2 < 1$, the selection response is initially convex instead of concave and, later, linear. Wei et al. (1996) show more general predictions of cumulative response, incorporating the Bulmer effect and inbreeding depression.

9.3.2 Comparison with Empirical Data

The infinitesimal model evidently does not fit to reality, since the number of loci is not indefinitely large, nor are all of them additive and of infinitesimally small effect. In general, the observed limits and the half-life of the response to the limit are usually smaller than those predicted by the infinitesimal model, which occurs, for example, if there are few loci with large effects. In general, the greater the leptokurtosis of the distribution of effects of the loci that affect the trait (there are many genes of small effect and some of very large effect) the response in the limit will be lower than that predicted by the infinitesimal model (Hill and Rasbash, 1986). In this case, the response will also be irregular, with drastic jumps in some generations, and there will be a notable variation between the replicates of the selected line.

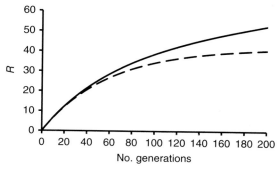

Figure 9.9 Cumulative response throughout generations with an infinitesimal model that excludes (broken line) or includes (continuous line) mutation. The predictions assume the same parameters as those in Figure 9.7. The non-mutation response is obtained with equation (9.12) and the response with mutation with equation (9.16), assuming a mutational variance $\sigma_M^2 = 0.001$.

However, the predictions of the infinitesimal model are very useful and correspond, at least qualitatively, with the observations. For example, theoretical predictions about the impact of an initial bottleneck (Robertson, 1960; James, 1970) fit well with observations (Frankham, 1980). Thus, a bottleneck of a single pair reduces the response by approximately 30%, which would be the expectation with the infinitesimal model. On the other hand, the limit to the selection response established by the infinitesimal model must be a function of the effective population size (equation (9.13)). In fact, the ratio between the responses in generation 50 and the first generation of selection of various experiments shows a clear relationship with $2N_e$ (Weber and Diggins, 1990; Weber, 2004) as illustrated in Figure 9.10. However, it is difficult to discriminate between the predictions of the infinitesimal model and other models, for example one in which there is a finite number of loci with allele frequencies with a U-shaped distribution, that is, with extreme frequencies more frequent than the intermediate ones (Zhang and Hill, 2005).

The idea of the limits to selection with the infinitesimal model has also been extended to linked loci (Hill and Robertson, 1966), in which case there is a reduction in the selection response due to the interference between loci discussed in Chapter 7. However, this reduction is only pronounced if the genome size is very small and, therefore, linkage is considerable. For example, in a selection experiment in which recombination was suppressed in chromosomes II and III of *Drosophila melanogaster* (in males there is no recombination), the response in the limit was only reduced by 25% in relation to that observed with recombination (McPhee and Robertson, 1970).

An example of a long-term response is illustrated in Figure 9.11, which presents the results of a selection experiment in mice for 60-day weight started in 1930 and continued for 84 generations (Goodale, 1938; Wilson et al., 1971). The initial population consisted of 8 males and 28 females. Each generation an average of 649 mice was evaluated but with great variability (between 75 and more than 1000), an average of 37 males and 100 females being selected as reproducers, also variable. The figure shows that the response reached a limit approximately towards generation 50, with large fluctuations in the mean from that moment.

Another observation of this experiment was an increase in the phenotypic variance over generations, so that the phenotypic variation in the first 10 increased to almost quadruple in the last 20 (Wilson et al., 1971). This is also a very common observation in selection experiments, even though the additive variance is expected to be reduced by selection and drift. The increase is usually due, at least in part, to a scale effect, and is corrected by calculating the coefficient of phenotypic variation (the phenotypic standard deviation divided by the mean). In Goodale's experiment, part of the increase was corrected in this way but even so, the coefficient of variation increased by 50% between the initial and final generations. The reason for the increase not explained by a scale effect is usually attributed to the higher environmental sensitivity of the homozygous genotypes (frequent in populations selected over many generations) mentioned in Section 3.4.1 (Chapter 3). It is also possible that it is due to the existence of genotype–environment interaction (Section 3.5.2, Chapter 3). There may be individuals whose genotype has little environmental sensitivity, that is, their phenotype changes little in different environments, and others with great sensitivity, that is, their genotype changes substantially with the environment. For example, if the phenotype of individuals with greater sensitivity increases in the best environments, artificial selection will tend to select them to a greater extent, since the best phenotypes correspond to the best genotypes and a greater tendency to have grown in the best environments. Thus, selection would increase the environmental sensitivity of the population as a whole with greater impact on the environmental variance. From the point of view of animal breeding, a commercial objective is usually to reduce the phenotypic variation between individuals, and it is possible to select for such homogeneity if there is genetic variance for the residual variation (Mulder et al., 2008; Hill and Mulder, 2011).

The existence of selection limits, as in the case of Goodale's experiment, is a common result, although in many cases no limits are observed even though the number of generations of selection is very high. One of the most notable is the so-called Illinois experiment (Dudley and Lambert, 2004), begun in 1896 to increase and decrease the percentage of protein and fat in the corn grain, with a continuous response for more than 100 generations, becoming the longest experiment carried out. Another is that of Yoo (1980), with

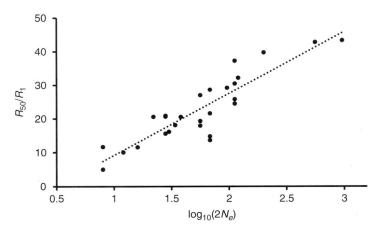

Figure 9.10 Linear relationship predicted by the infinitesimal model between the ratio of the responses in the 50th and first generations and doubling of the effective size in various selection experiments. (Data from Weber and Diggins, 1990.)

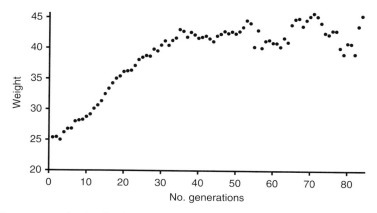

Figure 9.11 Response to selection for 60-day weight in mice from Goodale's (1938) experiment. Data of Wilson et al. (1971).

a continuous response of the number of abdominal bristles in *D. melanogaster* for almost 90 generations.

The limits to selection response can be explained by various circumstances. The most obvious is the depletion of genetic variation by the fixation of the selected alleles (see, for example, Gallego and López-Fanjul, 1983), since the increase in variance by mutation is small and could be undetectable if the number of generations is small (Figure 9.9). In addition, it is possible that at least part of the new mutations are deleterious, and are removed by natural selection. But there are several reasons why a limit to selection can be reached without an exhaustion of the genetic variation. For example, if favourable genes are dominant, selection will lead them to high frequencies but fixation will be difficult to achieve (see Figure 2.8, Chapter 2). Likewise, if there were overdominant genes, when they reached their equilibrium frequency, they would contribute null additive variance (Chapter 3) and, therefore, there would be a lack of selection response.

However, the most common cause of the selection limits without depletion of genetic variation is that natural selection is opposed to the change produced by artificial selection (Nicholas and Robertson, 1980). In fact, it is very common that there are lethal or highly deleterious genes with pleiotropic effect on the selected character and that, therefore, increase in frequency with selection. This was observed, for example, in Yoo's (1980) experiment, where some sudden jumps in the mean were due to lethal alleles with effect on the trait of up to five phenotypic standard deviations. Another example is the selection experiment for abdominal bristle number starting from a completely homozygous population of *Drosophila melanogaster* (López and López-Fanjul, 1993a, 1993b), where 1/3 of spontaneous mutations responsible for the response were lethals with pleiotropic effects on bristle number. We saw earlier that one way to detect this effect is by comparing the selection differential weighted by the number of offspring of each parent with the corresponding unweighted one. Another way is to relax the selection and see if part of the change is lost by natural selection. Figure 9.12 illustrates this procedure. In order to verify the existence or not of genetic variation in the limit, reverse selection can be carried out, that is, selection in the opposite direction to that of the response. If there is no change in the mean it is assumed that there is no additive genetic variation in the population (Figure 9.12a) (although the presence of overdominant loci in equilibrium could falsify this test). Otherwise, the absence or presence of change with the relaxation of selection would indicate that the remaining variation is neutral or not, respectively (Figures 9.12b and 9.12c). In the case of Goodale's experiment, these tests were not carried out, so there is no clear evidence about the absence or presence of genetic variation in the limit (Wilson et al., 1971).

9.4 Family, Within-Family and BLUP Selection

Equation (9.2) and all previous predictions refer to the case in which selection is carried out on the phenotype of individuals, what is called individual phenotypic selection or, simply, individual selection. Sometimes it is not appropriate to select single individuals, for example in some cases in aquaculture, and it is more convenient to select whole families, which is called family selection. On other occasions it is convenient, or only possible, to carry out selection within families. In other instances, for example, when the heritability of the trait is very low, it is convenient to use information of the individual's relatives, which is essential when only the individuals of the other sex can be evaluated (as in the case of milk production or fertility). In all these cases, the selection criterion is not the phenotype but the family mean, the deviation of the value of the individual to the family mean, or a combination of the phenotypic value of the individual and those of its relatives. Equation (9.2) can be applied in all cases when it is generalized to

$$R = i r_{CA} \sigma_A, \qquad (9.17)$$

where r_{CA}, which is called the accuracy of the selection criterion, is the correlation between the applied selection criterion (C) and the additive value (A) of the individual. In the case of individual selection, where the criterion is the phenotype of the individual (P), the correlation is $r_{PA} = \text{cov}(P,A)/(\sigma_P \sigma_A)$. As $P = A + E$ (Chapter 3), where E is the environmental deviation, which is independent of A, we have that $\text{cov}(P, A) = \text{cov}(A, A + E) = \sigma_A^2$. Thus, for individual phenotypic selection the selection accuracy is $r_{PA} = \sigma_A^2/(\sigma_P \sigma_A) = \sigma_A/\sigma_P = h$, the square root of the heritability.

In the case of within-family selection, the best male and the best female of each family are selected, and the selection criterion (C) is the phenotypic value of the individual in relation to the values of the other individuals of the same sex in its family (P_W). Note that an alternative type of selection is that of within-family deviations, where each individual is selected based on the deviation of its phenotypic value from the average of its family and, therefore, more than one individual of each sex can be selected per family and some families may not contribute selected individuals. A comparison between within-family selection and selection of within-family deviations was given by Hill et al. (1996). Here we refer to the first case. The phenotypic value of each individual can be decomposed in a portion due to the additive affects and another to the environmental ones, $P_W = A_W + E_W$. In turn, the additive value of the individual can be partitioned into the mentioned within-family component and a between-family component, A_B. Therefore, $\text{cov}(P_W, A) = \text{cov}(A_W + E_W, A_W + A_B)$, from which we obtain that $\text{cov}(P_W, A) = \text{cov}(A_W, A_W) = \sigma_{AW}^2 = (1 - r)\sigma_A^2$, where r is the genetic correlation between the individuals of the family. On the other hand, the variance of

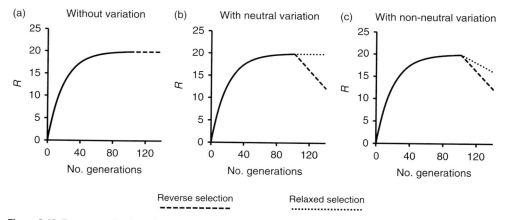

Figure 9.12 Reverse selection, that is, in the opposite direction to the selection made previously, indicates the existence or not of genetic variation in the population at the selection limit. If there is variation, the relaxation of selection (the stopping of the selective process) allows us to check if the variation in the limit is neutral or not.

the selection criterion is $\sigma_{PW}^2 = \sigma_{AW}^2 + \sigma_E^2 = (1-r)\sigma_A^2 + \sigma_E^2$. The accuracy of selection within families is, then, $r_{P_WA} = \text{cov}(P_W, A)/(\sigma_{P_W}\sigma_A)$. Substituting and noting that $\sigma_E^2/(\sigma_A^2 + \sigma_E^2) = 1 - h^2 - c^2$, and that the phenotypic correlation between sibs is $t = h^2 r + c^2$ (Chapter 6) if dominance is ignored, where c^2 represents the proportion of the phenotypic variance due to the common environment within families, we arrive at

$$r_{P_WA} = \frac{h(1-r)}{\sqrt{1-t}}.\qquad(9.18)$$

In the case of full-sib families, $r = 1/2$, so the accuracy of within-family selection will be approximately half that of individual selection, since it only makes use of the additive variance within families, which is half of the total additive variance in the case of full-sib families. When t is high (high h^2 and/or high c^2) the accuracy of within-family selection will increase, and it will be advised if there is substantial environmental variation common to families (c^2), although the accuracy of within-family selection will always be less than that of individual selection unless t is close to one.

Consider now the case of selection of complete families and assume families with n individuals. The family phenotypic mean is $P_F = (P_1 + P_2 + \ldots + P_n)/n$, where P_i is the phenotype of each member of the family. The covariance between the selection criterion and the additive value of a given individual, for example individual 1, is $\text{cov}(P_F, A_1) = \text{cov}[(P_1 + P_2 + \ldots + P_n)/n, A_1] = (1/n)[\text{cov}(P_1, A_1) + \text{cov}(P_2, A_1) + \ldots] = (1/n)[\text{cov}(A_1, A_1) + (n-1)\text{cov}(A_i, A_1)]$, where $i \neq 1$. The additive covariance between two sibs (ignoring dominance) is $\text{cov}(A_i, A_j) = r\sigma_A^2$, then $\text{cov}(P_F, A) = (1/n)[\sigma_A^2 + (n-1)r\sigma_A^2] = \sigma_A^2[1 + (n-1)r]/n$. On the other hand, the variance of the family means is $\sigma_{P_F}^2 = V[(P_1 + P_2 + \ldots + P_n)/n] = [nV(P_i) + n(n-1)\text{cov}(P_i, P_j)]/n^2$, where $i \neq j$. As $\text{cov}(P_i, P_j)/\sigma_P^2 = t$, we have that $\sigma_{P_F}^2 = [\sigma_P^2 + (n-1)t\sigma_P^2]/n = \sigma_P^2[1 + (n-1)t]/n$. Therefore, substituting in the equation of the accuracy of family selection, $r_{P_FA} = \text{cov}(P_F, A)/(\sigma_{P_F}\sigma_A)$ and, operating, we arrive at

$$r_{P_FA} = \frac{h[1 + (n-1)r]}{\sqrt{n[1 + (n-1)t]}}. \tag{9.19}$$

Considering full-sib families ($r = 1/2$) as before, $r_{P_FA} = (h/2)(n+1)/\sqrt{n[1 + (n-1)t]}$, that is, the accuracy of family selection will be, again, half that of individual selection by a factor that will increase it when n is high and t is low, that is, when h^2 and c^2 are both low (the noise due to low heritability will be compensated in family selection with the average of the values of the individuals in the family), but it will always be less than that of individual selection unless t is close to zero. Thus, family selection will be convenient when both the heritability and the common environmental variation are low.

Using the corresponding accuracy it is then possible to predict the response to within-family and family selection by means of equation (9.17). Note, however, that the selection intensity applied in each case may also be different. For example, with within-family selection, if the families are of few individuals, the intensity of selection may be less than that of other types of selection.

The selection response may be increased when a combination of within- and between-family information is used in the so-called selection indices, where the information of the individual and its siblings is used. For example, an index can be set as $I = b_W P_W + b_F P_F$, where P_W is, as before, the deviation of the phenotypic value of the individual to the family mean and P_F the deviation of the family mean to the general mean, and the corresponding coefficients b are the optimal weights that must be found with a regression procedure. To obtain the coefficients that optimize the response, the selection accuracy must be maximized, which is achieved by minimizing the sum of squares of the linear regression of the index (I) on the additive value (A) of individuals, that is, minimizing $\Sigma(I - A)^2$, whose result is the partial regression coefficients of I on A. The obtained index is

$$I = \left[\frac{1-r}{1-t}\right]P_W + \left[\frac{1 + (n-1)r}{1 + (n-1)t}\right]P_F. \tag{9.20}$$

From equation (9.20) it is deduced that family selection will be advised when $t \ll r$, within-family selection will be more convenient when $t \gg r$ and individual selection will be appropriate when $t \sim r$, and familial information is not relevant. The optimal index, however, is obtained using the information not only of the sibs but of all the relatives of the individual, carrying out BLUP evaluations (Chapter 6), where the information contained in the matrix of additive relationships of individuals of all generations is used. This takes into account the genetic changes that occur across generations and the fixed effects. It should be remembered, however, that BLUP evaluation requires estimates of the additive and common environmental components of variances of the base population, which can be estimated using REML (Chapter 6).

Figure 9.13 compares the responses to individual selection, family selection, within-family selection and optimized selection, the latter approximated by using a combination of information from the individual and its sibs with the selection index of equation (9.20). In the short term (say the first 10 generations), the optimized selection always provides the maximal response. Family selection response may be slightly higher than individual selection response initially if $t = h^2/2 + c^2$ is low, but then it is greatly reduced (panel a). Within-family selection produces a greater response than family selection with high t (panel b).

The long-term response, as shown in equation (9.12), depends not only on the accuracy and additive variance, but also on the effective population size. Family selection implies a remarkable decrease of N_e, since entire families are discarded for reproduction. In contrast, within-family selection implies the equalization of family contributions, so that the effective size is approximately equal to twice the number of breeding individuals (Chapter 5). Individual selection and index or BLUP selection will lead to effective sizes with intermediate values between the previous ones. Therefore, in general, it is expected that the long-term response of within-family selection will be greater than that of the other types of selection (Dempfle, 1975) and that of family selection will be much lower. In the example illustrated in Figure 9.13, the effective sizes were 37, 78, 8 and 19 (panel a) and 19, 79, 8 and 22 (panel b) for individual, within-family, familial and optimized selection, respectively. These effective sizes can be roughly predicted using expression (9.9) (see also Chapter 5) and the appropriate C^2 and Q values provided by Wray et al. (1994) for index selection, and by Bijma et al. (2001) for BLUP selection.

Selection indices can also be used to select more than one trait (Hazel, 1943; Hazel and Lush, 1943). Suppose that we intend to use an index of type $I = b_1P_1 + b_2P_2 + b_3P_3 + \ldots$ with information of several traits ($P_1 + P_2 + P_3 + \ldots$), where the interest is to improve the composite additive value, $A_C = a_1A_1 + a_2A_2 + a_3A_3 + \ldots$, the a_i terms being the weighting that is intended for each trait, for example, their economic values. The estimation of the b coefficients will be done, again minimizing $\Sigma(I - A_C)^2$ (see Falconer and Mackay, 1996, chapter 19). Obviously, the index can include individual and family information for each of the traits considered. The optimization of this type of selection, however, will be more precise if it is done by BLUP evaluation, considering the matrix of genetic covariances between traits.

9.5 Use of Molecular Markers

As we will see in Chapter 11, it is possible to determine the genomic position and the effect of some of the loci that affect quantitative traits, generally those with the greatest effect, or molecular genetic markers associated by linkage to them. From the numerous linkage studies carried out, it has been possible to detect hundreds of loci with effect on quantitative characters in domestic animal species (Weller, 2009) and cultivated plants (Collard and Mackill, 2008; Hospital, 2009). The information of molecular genetic markers associated with traits of interest has enabled the development of the so-called marker-assisted selection and, more recently, genomic selection.

The original idea of marker-assisted selection is that the additive effects of the markers associated with the quantitative trait loci are condensed into a molecular value for each individual, which is treated as a character correlated with the trait of interest in a selection index (Lande and Thompson, 1990). The method can be applied to BLUP selection (marker-assisted BLUP; Fernando and Grossman, 1989), and is particularly useful when phenotypic information about some individuals is not available. This occurs, for example, because the trait can only be evaluated in one sex or it cannot be evaluated until the post-productive stage,

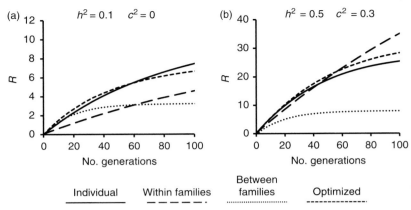

Figure 9.13 Simulated selection response with different criteria. In each generation, 20 individuals of each sex were reproduced ($N = 40$), evaluating 5 individuals of each sex per family (selected proportion ϕ = 0.2); h^2 and c^2 are the initial heritability and the proportion of the phenotypic variance due to the common environment among individuals of each family, constant throughout generations. The selection methods compared are individual phenotypic selection (the best 20 individuals of 100 are chosen in each sex), within-family selection (the best individual of 5 is chosen in each family and sex), family selection (the four families with the highest average are chosen) and index selection (the 20 best individuals of each sex are chosen using as a criterion an index that uses the information of the individual and its sibs; equation (9.20)).

as in the case of longevity, or even *post-mortem*, as the quantity and quality of meat. The success of marker-assisted selection has, however, been modest (Misztal, 2006; Collard and Mackill, 2008) for different reasons. First, the association found between markers and loci with an effect on the trait is sometimes spurious and, when it is real, it is gradually lost by recombination. Second, the extra response due to markers is reduced with the generations of selection, since the variance of marker-assisted loci decreases as their allele frequency increases until fixation (Meuwissen and Goddard, 1996). The long-term response can also be reduced if the selection on markers decreases the effective population size and, with it, the long-term response on polygenic variation. For example, Nomura (2000) found that N_e was reduced with marker-assisted selection relative to the corresponding value with phenotypic selection, if the within-family phenotypic correlation (t) was smaller than the genetic correlation (r), and that it was increased otherwise.

More recently, the availability of tens or hundreds of thousands of SNPs in many domestic species has allowed the so-called genomic selection to be applied (Meuwissen et al., 2001) that we will study in detail in Chapter 11. This makes use of all available molecular information of genetic markers without the need to know the specific genes that affect the trait of interest.

Problems

9.1 The number of eggs laid in one hour by 85 females of *Drosophila melanogaster* has been evaluated, a trait with an approximate heritability of $h^2 = 0.2$. If the offspring of 20% of the females with the greatest egg laying were selected to form the next generation, what would be the expected mean of the population in that generation?

Number of eggs	11	12	13	14	15	16	17	18	19	20	21	22	23
Number of females	2	3	6	4	8	9	12	13	11	9	5	2	1

9.2 In a selection experiment, 60 families of 10 sibs (5 of each sex) are evaluated in each generation for a trait with phenotypic variance 24 and heritability 0.3. Of the 300 individuals evaluated for each sex, the 60 with the highest phenotypic value are selected as parents for the next generation. What is the expected selection response in the first generation and the accumulated one in the initial five generations?

9.3 In the selection experiment described in the previous problem, what would be the effective size of the selected population and the standard error of the accumulated response?

9.4 Continuing with Problems 9.2 and 9.3, what would be the cumulative response up to generation 100, considering or not the contribution of mutation, if the mutational variance for the trait is $V_M = 0.002$?

9.5 In the selection experiment enunciated in Problem 9.2, individual selection was carried out. What would be the expected response to one generation of selection if within-family selection or family selection is carried out?

Self-Assessment Questions

1 For a given heritability and phenotypic variance, the response to artificial selection will be greater the greater the dominance variance.

2 To estimate the realized heritability it is necessary to carry out two or more generations of selection.

3 The scale effects on the phenotypic variance may be responsible for the asymmetric responses.

4 Since artificially selected parents belong to one extreme of the population phenotypic distribution, a negative linkage disequilibrium is generated in them that reduces the genetic variance of the offspring.

5 The Bulmer effect only affects the between-family component of the variance.

6 If, given a fixed number of individuals evaluated, the selection intensity is increased, the response to long-term selection will increase.

7 If artificial selection is applied to a population without genetic variability, the cumulative response will be due only to mutation and it will have an initially convex and then linear form.

8 The deleterious pleiotropic effects of selected loci may be one of the reasons for the existence of limits in the response to selection.

9 Selection within families is recommended when the heritability is low and the environmental variation common to the members of the families is scarce.

10 Marker-assisted selection increases the short-term response but may reduce the long-term one.

10 Natural Selection

Concepts to Study

- Fundamental Theorem of Natural Selection
- Robertson–Price equation
- Selection gradients
- Directional, stabilizing and diversifying selection
- Convergent selection
- Apparent stabilizing selection
- Background or purifying selection
- Selective sweep (hitchhiking alleles or sequences)

Objectives for Learning

- To understand the theorems of natural selection and their implications
- To know how to calculate selection gradients
- To learn the concepts of directional, stabilizing, diversifying and convergent selection
- To understand the meaning of apparent stabilizing selection
- To know the implications of background selection and selective sweeps

10.1 Quantitative Traits and Fitness

As we saw in Chapter 2, natural selection is the main force for the change of allele frequencies and the driving factor behind the evolution of living beings and their adaptation to the incessant environmental changes. Although there are other evolutionary agents that modify the allele frequencies, only selection produces changes that promote adaptation. We also indicated that natural selection acts directly on a single character, fitness, and the changes that occur on the different quantitative traits of an individual depend on the genetic correlation between them and fitness. Probably, all quantitative traits (and also qualitative ones) are more or less related to fitness, since selection acts at all levels, from the cell to the population (Endler, 1986). In previous chapters we indicated that some traits, the so-called main components of fitness, have a strong relationship with fitness. In fact, the empirical evaluation of this is carried out by its components, mainly viability, fecundity and mating success. But these, in turn, depend on other attributes. For example, the fertility of a mammal will depend on the number of litters produced throughout its life and the number of progeny per litter, and these characters, in turn, depend on others, such as body size, the rate of ovulation, embryonic survival or hormonal levels. In general, quantitative traits can be represented in a pyramid at the top of which is fitness, with its main components located immediately below, followed by a large number of interrelated traits through multiple and overlapping connections (Figure 10.1). At the base of the pyramid are the

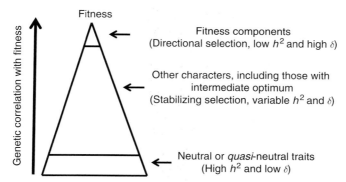

Figure 10.1 Illustrative scheme of the different types of quantitative traits according to their relationship with fitness and the type of selection to which they are subjected. The correlation between the traits and fitness increases as the pyramid is climbed. h^2: heritability. δ: rate of inbreeding depression. The area of the classes is not proportional to the number of traits included in them.

traits that are loosely related to fitness, which can be considered neutral for practical purposes. The traits at the tip of the pyramid show strong inbreeding depression and tend to have low heritabilities due to the great contribution of the environment and, to a lesser extent, to the existence of non-additive genetic components in their genetic variance. On the contrary, the characters in the base tend to have high heritabilities and low or no non-additive genetic variance and inbreeding depression.

10.2 The Response to Natural Selection

The action of natural selection can be quantified, analogously to the artificial selection we studied in the previous chapter, in terms of selection differential and response to selection. Consider a population with census size N, in which the contribution to the progeny of individual i is W_i, with $\overline{W}$ being the mean fitness of the population. The selection differential can be obtained as the average superiority of the selected parents weighted by their relative contribution to the next generation,

$$\sum_{i=1}^{N}\left(W_i - \overline{W}\right)W_i/N = \left(\sum_{i=1}^{N} W_i^2 - \overline{W}\sum_{i=1}^{N} W_i\right)/N = \left(\sum_{i=1}^{N} W_i^2/N\right) - \overline{W}^2 = \sigma_{P,W}^2,$$

that is, the selection differential is equal to the phenotypic variance for fitness. Since the selection response is $R = h^2 S$ (equation (9.1)), we have that $R_W = (\sigma_{A,W}^2/\sigma_{P,W}^2)\sigma_{P,W}^2$ and

$$R_W = \sigma_{A,W}^2. \tag{10.1}$$

Thus, the response to natural selection is equal to the additive variance for fitness, which is known as the Fundamental Theorem of Natural Selection (Fisher, 1930).

According to this theorem, natural selection produces response whenever there is additive variance for fitness. Once the population reaches the maximal fitness, the additive variance is zero and the response disappears. The incessant appearance of mutations with effect on fitness, mostly deleterious (Chapter 7), implies that additive variance is continuously generated for fitness, so that selection acts, also continuously, eliminating that variation. If the average fitness is rescaled as a function of the available resources, it is not necessary for the population census size to increase indefinitely as a result of selection, nor for it to be reduced to extinction, but rather populations can be found in a dynamic equilibrium where selection operates continuously to maintain the *statu quo*. However, new equilibria can be achieved sporadically depending on the appearance of new beneficial mutations, interaction with the external environment and relations with other species. This type of selection, in which the census size remains constant and fitness is dependent on the density and frequency of individuals, is called soft selection (Wallace, 1975). When the operation of selection implies an increase or reduction of the census size due to the birth and mortality of individuals and the limitation of the resources, it is called hard selection.

However, the following considerations should be made. First, the lack of additive variance for fitness implies absence of selection response, but there may be non-additive variance (dominance or epistatic), whose magnitude is not negligible for fitness components. Second, even if there is no additive variance for fitness, there could be additive variance for its main components even when the population fitness is maximum. In fact, it is common to observe the existence of antagonistic pleiotropy between fitness components, which generates negative correlations between them. For example, an increase in the fecundity of birds or mammals may imply a reduction in the viability of the progeny, due to limited resources or parental care, which may generate additive variance for the components. Therefore, although the fitness components tend to have low heritabilities (Chapter 3), their additive variance is considerable, which is demonstrated by scaling it to the mean (CV_A).

As we have already said, the action of natural selection on quantitative traits depends, in general, on the relation of these with fitness. Analogous to the correlated response to artificial selection experienced by a trait when another is selected (Chapter 9), the correlated selection differential for a trait X as a result of selection on fitness is $S_{CX} = \text{cov}_P(X, W)$, and the correlated response (equation (9.4)) would be $R_{CX} = b_{A(X.W)}R_W = [\text{cov}_A(X, W)/\sigma_{A,W}^2]\sigma_{A,W}^2$

$$R_{CX} = \text{cov}_A(X, W), \tag{10.2}$$

that is, equal to the genetic covariance between the trait and fitness, which is known as the Second Theorem of Natural Selection, Price equation or Robertson–Price equation (Robertson, 1966; Price, 1970; Queller, 2017; Walsh and Lynch, 2018, chapter 6). Note that a purely neutral trait is that whose genetic covariance with fitness is effectively nil (although the covariance can also be zero for non-neutral traits if the relationship between the trait and fitness is not linear). The ratio between the correlated selection differential and the phenotypic variance is equal to the linear regression of fitness on the trait, $b_{P(W.X)} = S_{CX}/\sigma_{PX}^2$, which is called selection gradient (Lande, 1979).

Let us now consider two characters X and Y, and express fitness relative to its mean in the population in terms of both as $W = b_{P(W.X)} X + b_{P(W.Y)}Y + e$, where the b terms are the coefficients of partial regression of fitness on the characters (the selection gradients) and e is the residue unexplained by the regression. Substituting in expression (10.2), we have $R_{CX} = \text{cov}_A(X, W) = \text{cov}_A\left(X, b_{P(W.X)}X + b_{P(W.Y)}Y\right)$, from which we get

$$R_{CX} = b_{P(W.X)}\sigma^2_{A,X} + b_{P(W.Y)}\text{cov}_A(X, Y).$$ (10.3)

Expression (10.3) was generalized by Lande and Arnold (1983) to predict the response to natural selection of multiple quantitative traits $X_1, X_2, \ldots, X_n$. Thus, the change in the mean of trait X_1 taking into account its correlation with fitness and with the other traits, is $R_{CX_1} = b_{P(W.X_1)}\sigma^2_{A,X_1} + b_{P(W.X_2)}\text{cov}_A(X_1, X_2) + \ldots + b_{P(W.X_n)}\text{cov}_A(X_1, X_n)$. In general, for trait i,

$$R_{CX_i} = b_{P(W.X_i)}\sigma^2_{A,X_i} + \sum_{j \neq i}^{n} b_{P(W.X_j)}\text{cov}_A(X_i, X_j),$$ (10.4)

or, in matrix notation,

$$\mathbf{R} = \mathbf{Gb},$$ (10.5)

where, considering n traits, $\mathbf{G}$ is the $n \times n$ matrix of additive variances–covariances between the traits and $\mathbf{b}$ the vector of the n selection gradients. For example, suppose that traits X_1 and X_2 are not correlated and that they have genetic variances 1 and 2 and selection gradients 0.5 and 0.1, respectively. Expression (10.5) would then be,

$$\begin{pmatrix} 1 & 0 \\ 0 & 2 \end{pmatrix}\begin{pmatrix} 0.5 \\ 0.1 \end{pmatrix} = \begin{pmatrix} 0.5 \\ 0.2 \end{pmatrix},$$

that is, both traits will undergo selective change, being greater for X_1 than for X_2. If the genetic covariance between both traits is negative, $\text{cov}_A(X_1, X_2) = -0.4$, the result is

$$\begin{pmatrix} 1 & -0.4 \\ -0.4 & 2 \end{pmatrix}\begin{pmatrix} 0.5 \\ 0.1 \end{pmatrix} = \begin{pmatrix} 0.1 \\ 0 \end{pmatrix},$$

which implies a lower response of trait X_1 and a lack of response of X_2. This illustrates in a very simple way how genetic correlations between characters can limit the ability of populations to respond to selection (Kirkpatrick, 2009).

Selection gradients can be obtained by estimating the partial regressions of relative fitness on the measures of the trait (Arnold and Wade, 1984a, 1984b). Consider, as an example, one of the studies carried out by Grant and Grant (1989) on the action of natural selection for the length and thickness of the beak in one of the species of finches of the Galápagos Islands, Darwin's finches (*Geospiza conirostris*). This species has a beak of variable thickness and length and, in the driest season, it feeds mainly on *Opuntia* cactus seeds. Some individuals with particularly thick beaks are also able to break the bark of trees and shrubs to search for arthropods. During the strong period of drought that took place after 1984, after the El Niño storm of the previous year, the flowers and fruits of *Opuntia* practically disappeared and, therefore, the hypothesis of a strong action of natural selection in favour of individuals with thicker beaks was proposed.

Individuals were sampled on Genovesa Island (Galápagos) before the drought period (Table 10.1), which were evaluated for different morphological traits and released after being tagged for later identification. In the first sample, 54% of the survivors were found and in the second, 39%, ascribing a fitness value of one for the survivors and zero for those missing. The measurements of morphological traits were transformed to logarithmic scale and

Table 10.1 *Selection gradients obtained by Grant and Grant (1989) in the analysis of morphological traits in the species* Geospiza conirostris *from Genovesa Island (Galápagos)*

Year	n (% survivors)	Trait	Selection gradient (b)
1983–4	137 (54%)	Beak length	**−0.22***
		Beak thickness	−0.05
		Body length	−0.02
1984–5	74 (39%)	Beak length	−0.05
		Beak thickness	**0.37***
		Body length	0.14

Note: n is the number of individuals analysed. * indicates statistical significance.

standardized to present mean zero and variance one. The selection gradient analysis obtained by partial regression of the relative fitness values of the individuals on the measures of the traits confirmed the hypothesis proposed (Table 10.1). In the first period, the intensity of selection was exercised for a reduction of the length of the beak, and in the second for the increase of its thickness, which indicates a clear selective pressure towards short and thick beaks motivated by the necessary change of feeding.

10.3 Directional, Stabilizing and Diversifying Selection

As we have said, natural selection acts continuously eliminating deleterious mutations for the main components of fitness, which is called purifying, negative or background selection, also increasing the frequency of beneficial mutations in parallel until their fixation, generating evolutionary novelties (positive selection). The form of selection that acts on fitness and its main components is directional selection (Figure 10.2), which would be analogous to artificial selection by truncation, although it is not done exactly that way, and entails an increase in the mean fitness of the population.

For characters less related with fitness, it is common to find the so-called stabilizing selection, whereby individuals with an intermediate phenotype are favoured and selection eliminates the most extreme (Figure 10.2). The attributes on which this type of selection acts are called traits with intermediate optimum. For example, the weight of newborns in humans is a classic example of stabilizing selection (Cavalli-Sforza and Bodmer, 1971), with newborns with intermediate weight having a lower mortality than that of more extreme individuals. The example of the number of eggs in the laying of birds in the wild, mentioned above, is another case.

There are several causes that can promote this type of selection. One, very frequent, is due to the pleiotropy of trait genes with respect to two or more components of fitness that act antagonistically. For example, body weight in mammals is a typical character with intermediate optimum in nature. Although artificial selection can increase or reduce the weight to a large extent (see, for example, Figure 9.11, Chapter 9), showing the presence of additive genetic variance for the trait, the fact is that the weight or size remains approximately constant in nature for long

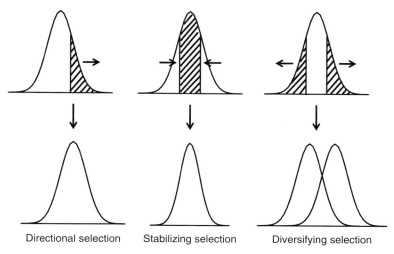

Directional selection   Stabilizing selection   Diversifying selection

Figure 10.2 Types of natural selection. The striped portion of the distribution of phenotypic values is favoured by selection. Directional selection produces a shift of the mean in the direction of selection. Stabilizing selection reduces the variance, keeping the mean invariable when the maximum fitness coincides with it. Diversifying selection increases the variation and can produce a bimodal distribution.

periods of time. The cause of the greater fitness of individuals with intermediate weight can be complex. It may be thought that females of greater weight will have greater litters and, therefore, greater fecundity. However, the smaller individuals will be faster and will have greater ability to hide, being able to escape more easily from predators and, therefore, show greater viability in nature. Only as a result of the combination of these two antagonistic effects for weight could individuals of intermediate weight be favoured, although many other factors could have a role.

This type of selection implies the invariable maintenance of the mean (hence its name), provided that it coincides with the optimum of fitness, and a reduction of the phenotypic variance that we will study later in this chapter. Obviously, if the optimum or the phenotypic distribution of the population is displaced for some reason, selection will be, in addition to stabilizing, directional, tending to take the mean of the distribution to the optimum. Note that although stabilizing selection implies the advantage of individuals with intermediate phenotype for the trait, this does not necessarily imply the existence of overdominance for fitness. However, marginal overdominance could be generated for loci with effect on the trait and antagonistic pleiotropic effects on fitness components, as explained in Section 2.8.1 of Chapter 2.

Sometimes stabilizing selection can be inferred even though the character is not subject to this type of selection. This occurs when there are mutations with effect on the trait and, in addition, deleterious pleiotropic effects. Such is the case of bristle number in *Drosophila*, generally considered a neutral trait. If mutations of great effect on the character appear that are also deleterious or even lethal, the extreme individuals that carry them will be removed generating an apparent stabilizing selection (Kearsey and Barnes, 1970).

It is possible that selection favours the most extreme individuals of the distribution, eliminating the intermediate ones, what is called diversifying, disruptive or divergent selection (Figure 10.2). This may occur, for example, when the population is subjected to two (or more) environments that involve the adaptation of different phenotypes to each of them, and implies an increase in the phenotypic variance and a bimodality (or multimodality) of the distribution. The selection may have two components, one directional and another divergent, if the areas selected at each side of the distribution are different. An example of diversifying selection will be given below.

Finally, when two or more populations are subject to stabilizing selection for the same optimum, the genetic differentiation between them for the trait is expected to be smaller than that expected with a neutral model, and then it is said that there is convergent selection.

See Roff (2001) for an in-depth study of the action of natural selection on the different types of quantitative traits in natural populations.

10.3.1 The Intensity of Real and Apparent Stabilizing Selection

It has been indicated that stabilizing selection implies an advantage of the individuals with intermediate phenotype. This is illustrated in Figure 10.3, where the phenotypic distribution of the character is, to simplify, normal with mean zero, variance V_P and normal density function $f(y) = (1/\sqrt{2\pi V_P})\exp(-y^2/2V_p)$. To model stabilizing selection, a fitness function is used

$$w(y) = \exp[(y - \theta)^2/2V_s] \tag{10.6}$$

(dashed line), which determines that fitness is maximum for an optimum θ which, in this example, coincides with the mean of the phenotypic distribution ($\theta = 0$). To understand this type of selection, it is useful to interpret that the fitness function gives the probability of survival conditioned to the phenotype. This function, although of Gaussian type, is not a density function, but its amplitude up to the point of inflection is equivalent to the standard deviation of the normal distribution and is expressed by $\sqrt{V_s}$, where V_s is an inverse measure of the intensity of stabilizing selection. The smaller the value of V_s, the greater the selection intensity.

The mean fitness of the population can be obtained by integrating the fitness function weighted by the trait density function,

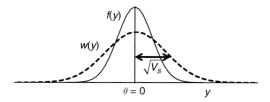

Figure 10.3 Illustration of stabilizing selection. The density function of the phenotypic values of the character is $f(y)$ (continuous line). The probability of survival is given by $w(y)$ (broken line), whose amplitude is the square root of V_s, which is an inverse measure of the selection intensity. The fitness optimum (θ) coincides, in this case, with the mean of the phenotypic distribution, which is assumed to be zero.

$$\overline{W} = \int_{-\infty}^{\infty} f(y)w(y)dy = \int_{-\infty}^{\infty} \frac{1}{\sqrt{2\pi V_P}} \exp\left[\frac{-y^2}{2}\left(\frac{V_P + V_s}{V_P V_s}\right)\right] = \sqrt{\frac{V_P V_s}{V_P + V_s}}. \qquad (10.7)$$

On the other hand, the trait function after selection will be

$$f^*(y) = \frac{f(y)w(y)}{\int_{-\infty}^{\infty} f(y)w(y)dy} = K\exp\left[\frac{-y^2}{2V_P}\right]\exp\left[\frac{-y^2}{2V_s}\right] = K\exp\left[\frac{-y^2}{2}\left(\frac{V_P + V_s}{V_P V_s}\right)\right], \qquad (10.8)$$

where K is a constant. From expression (10.8) it follows that the variance of the distribution after selection is

$$V_P^* = \frac{V_P V_s}{V_P + V_s} = V_P - \frac{V_P^2}{V_P + V_s} \qquad (10.9)$$

(Bulmer, 1985, p. 151). Equation (10.9) shows, first, that if selection acts on viability, the variance after selection is equal to the square of the mean survival (equation (10.7)) and, second, that stabilizing selection reduces the phenotypic variance of the trait, so that its impact can be measured as the proportional reduction in variance,

$$\Delta V_P = \frac{V_P^* - V_P}{V_P} = -\frac{V_P}{V_P + V_s}. \qquad (10.10)$$

Note that if stabilizing selection is not very strong, equation (10.6) can be expressed by the quadratic approximation $w(y) \approx 1 - (y^2/2V_s)$, which suggests that the selection intensity can be estimated by the regression of $w(y)$ on the square of the phenotypic values of the trait, that is, $b_{w.y^2}$, which is the stabilizing selection gradient, so that

$$V_s = -\frac{1}{2b_{w.y^2}}. \qquad (10.11)$$

For example, if $V_P = 1$ and $V_s = 5$, there is a decline in fitness of 10% in relation to the square of the phenotypic values ($b_{w.y^2} = -0.1$; equation (10.11)), and the phenotypic variance is reduced by 17% after selection ($\Delta V_P = -(1/6) = -0.17$; equation (10.10)). The experimental estimates of the intensity of stabilizing selection usually take values between $V_s = 20$ (strong intensity) and $V_s = 100$ (weak) (Kingsolver et al., 2001; Mackay, 2009).

It was previously mentioned that stabilizing selection may be apparent, if the character itself is not subject to this type of selection but the loci that affect it have pleiotropic effects on fitness, which requires that the correlation (ρ) between the character and fitness is high (Caballero and Keightley, 1994). For simplicity, suppose that the pleiotropic effect on fitness in heterozygosis is a fixed value sh. In Chapter 7 (Section 7.2.2) we saw that the genetic variance of a trait at the mutation–selection balance is approximately $V_G = V_M/sh$, where V_M is the mutational variance (Barton, 1990). From this it follows that the selection intensity with the pleiotropic model is $sh = V_M/V_G$. We also saw in Chapter 7 that the typical value of the mutational variance is $V_M = 0.001V_E$, where V_E is the environmental variance. Therefore, for a character with heritability 0.5 ($V_G = V_E$), the estimated value of sh would be only 0.001, which would indicate that the intensity of the apparent stabilizing selection by pleiotropy should be, in general, weak. For a deeper understanding of the mathematical models of multilocus evolution in quantitative traits under different forms of selection, see Barton and Turelli (1991) and Kirkpatrick et al. (2002).

10.3.2 Selection in Heterogeneous Environments

We have mentioned that natural selection can be diversifying when the selective advantage corresponds to the extreme individuals, being able to give rise to bimodal or multimodal distributions. When a population resides in areas with environmental heterogeneity, whether it is gradual (clines) or not, the adaptation of individuals to different habitats may entail differentiation for quantitative traits. If reproductive isolation is generated among the populations, either by geographic or genetic barriers, the differentiation can be accentuated to the point of promoting an ecological speciation process (Schluter, 2001).

An illustrative example of diversifying selection by environmental heterogeneity is presented in Figure 10.4 with reference to the species *Littorina saxatilis* (Rolán-Alvarez et al., 2015), a gastropod mollusc from the rocky intertidal area of the North Atlantic coasts. Dispersive capacity of the species is reduced, usually only a few metres although occasionally it can be greater, both due to its low mobility and because the females are ovoviviparous (carrying the embryos with direct development inside an embryonic pouch inside the shell). In the coasts exposed to the waves, there are two phenotypic forms (ecotypes) adapted to

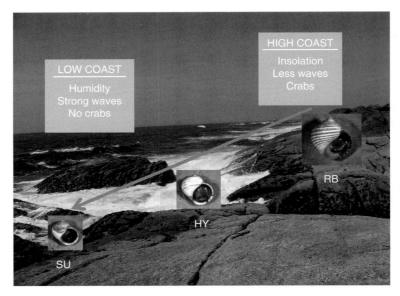

Figure 10.4 Illustrative example of selection in heterogeneous environments in the species *Littorina saxatilis*. The ecotypes RB and SU live adapted to different environmental conditions in the high and low part of the coast, respectively, and coexist and mate in the middle zone, where hybrids (HY) with intermediate characteristics are produced.

different environmental conditions: the RB (Ribbed Banded) ecotype (also called Crab ecotype), larger and with ribs and colour bands in the shell, usually lives between barnacles in the upper intertidal zone, while the SU (Smooth Unbanded) ecotype (also called Wave ecotype), smaller and without ribs or bands, is adapted to live among mussels in the lower area. In the middle intertidal zone both habitats overlap and it is possible to find both ecotypes and, in addition, a variable percentage of individuals with intermediate characteristics (hybrids). Despite the short distance between the high and low zones, the environmental characteristics of these are very different. In the high zone, there is a high level of insolation, the presence of predatory crabs and relatively weak waves. In the lower area, however, there are strong waves, reduced insolation and absence of crabs.

The existence of a strong diversification between the two ecotypes has been demonstrated for different qualitative and quantitative traits (morphological, physiological and behavioural). Among the most important, the RB ecotype has a thicker shell with ribs, which minimizes predation by crabs, and has a proportionally small shell aperture to prevent desiccation in the upper zone. The SU form, in contrast, has a thin shell with smaller size, which allows it to easily enter the rock cracks. This, together with the presence of a large aperture in the shell to release its powerful foot, allows it to protect itself from the strong waves. The differences between the ecotypes in size and aperture of the shell are illustrated in Figure 10.5. The measurements are obtained using the geometric morphometrics technique, which allows an estimate of size and shape as independent variables (Bookstein, 1991). The so-called centroid size is a global measure of the shell size and the first relative warp ($RW1$) is a measure of the relative shell aperture (Carvajal-Rodríguez et al., 2005). As indicated above, RB individuals have larger shell size and smaller aperture than SU ones. The figure shows the bimodal distribution typical of diversifying selection (Figure 10.2).

Although the individuals of each ecotype can migrate to the opposite zone, their survival in the hostile environment is very limited and they only coexist in the middle intertidal zone where they cross, producing hybrids with intermediate characteristics. The existence of a strong reproductive isolation determined by positive assortative mating for shell

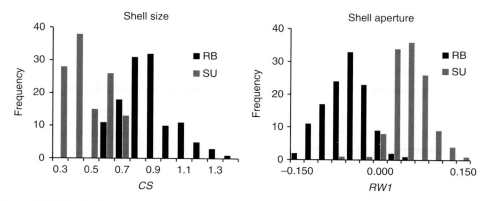

Figure 10.5 Distribution for centroid size (CS), a measure of the shell size, and the first partial component of shape, the relative warp $RW1$, a measure of the relative shell aperture, of individuals of ecotype RB and SU. (Data from Galindo, 2007.)

Table 10.2 *Estimates of heritability (h², genetic differentiation for morphological traits (Q$_{ST}$) and differentiation in allele frequencies for molecular markers (F$_{ST}$) in populations of the RB and SU ecotypes of* Littorina saxatilis

	h^2		Q_{ST} vs F_{ST}	
	RB	SU	Between transects (same ecotype)[a]	Between ecotypes[b]
Shell size[c]	0.60 ± 0.05	0.70 ± 0.05	0.01 ± 0.01	0.19 ± 0.03
Shell aperture[c]	0.45 ± 0.09	0.61 ± 0.09	0.09 ± 0.04	0.59 ± 0.05
Molecular markers[d]	–	–	0.02 ± 0.01	0.09 ± 0.01

Note: Data are obtained from two transects per locality in three localities of Galicia (Corrubedo, Silleiro and La Cetárea).
[a] Differentiation between populations of the same ecotype in different transects of the coast separated by 15–25 m, which is approximately the distance between the high and low zones of the coast.
[b] Differentiation between populations of distinct ecotypes (RB and SU) belonging to the same longitudinal transect of the coast.
[c] Heritabilities and values of Q_{ST} for morphological measures obtained by geometric morphometrics.
[d] Estimates of F_{ST} obtained with microsatellite loci and allozymes.
Source: Data from Conde-Padín et al. (2007).

size has been demonstrated. Individuals of the same size tend to mate with each other because of the physical difficulty involved in crossing between individuals of very different sizes and because males tend to follow the mucus of females of similar size. The isolation generated suggests that it is a case of possible incipient speciation due to ecological causes (Quesada et al., 2007; Butlin et al., 2014).

Since the embryos of the same female are kept in the embryo pouch for some time before laying and their shell is already formed, it is possible to measure their morphological characteristics and carry out estimates of genetic parameters such as heritability (Chapter 6) and the quantitative differentiation index (Q_{ST}; Chapter 8), with a sibling design (Conde-Padín et al., 2007). The comparison of Q_{ST} with F_{ST} for neutral molecular markers allows the existence of diversification selection for the trait to be deduced. Table 10.2 shows the results of a study where samples were taken from individuals of the RB and SU ecotypes in two longitudinal transects of the coast.

The heritabilities of the shell size and its aperture were similar in both ecotypes, in agreement with the similar effective population sizes estimated in both ecotypes (Fernández et al., 2005b). On the other hand, the genetic differentiation between ecotypes for the two characters was much greater than that obtained between populations of the same ecotype in different transects, the distance between transects and between the high and low intertidal zones being similar. Likewise, the allelic frequency differentiation for neutral molecular markers (F_{ST}) was similar to the Q_{ST} values between transects (within ecotype) but much lower than Q_{ST} between ecotypes, which indicates that diversifying selection promotes the divergence between the ecotypes for these characters. Finally, the F_{ST} values between ecotypes were greater than those corresponding to estimates between transects within ecotype, which is explained by the partial reproductive isolation between ecotypes.

10.3.3 Genetic Variance and Natural Selection

Except in the case of diversifying selection, natural selection tends to decrease genetic variation. This reducing effect, together with that caused by genetic drift, is balanced by the continuous contribution of variation by mutation. In the case of fitness components, we saw in Chapter 7 that the genetic variance at the deleterious mutation–selection balance can be approximated by $\hat{V}_A \approx 2U\overline{sh}$ (equation (7.13)). If we consider the average values of U and s in Figures 7.3 and 7.7 for eukaryotes (Chapter 7) and a value of $h = 0.2$ (Figure 7.4), we obtain predictions of $\hat{V}_A$ that range between 0.002 and 0.012 according to the point estimates used. These predictions can be compared with the estimates of additive variance obtained by Mukai and collaborators for the viability of *Drosophila* chromosome II, which are based on a large number of data. For example, for three different populations (Aomori, Raleigh and Orlando) values of $V_A = 0.006$, 0.021 and 0.044 were obtained (extrapolated for the complete *Drosophila* genome), respectively (Mukai, 1985). The first estimate falls within the range of previous theoretical predictions but the latter exceed these predictions. In fact, Mukai (1988) suggested that while the first estimate could be perfectly explained by mutation–selection balance, at least the last one could incorporate an excess of variance due to genotype–environment interaction. Fernández et al. (2005) suggested that an alternative explanation for the excess genetic variance could be the antagonistic pleiotropic effects of some mutations on fitness components different from viability. For a thorough in-depth treatment of the impact of natural selection on the genetic variability and evolutionary processes that occur in populations see Charlesworth and Charlesworth (2010).

At the opposite extreme to the components of fitness in the pyramid mentioned at the beginning of the chapter are the purely or nearly neutral characters. We also saw in Chapter 7 that the genetic variance of equilibrium for these characters is given by $\hat{V}_A \approx 2N_e V_M$ (equation (7.7)), that is, it is a direct function of the effective population size and the mutational variance. Therefore, for neutral traits, the genetic variance would increase indefinitely with the effective population size. With this premise, it should be assumed that species with higher census sizes should have greater neutral genetic variation. However, it has been repeatedly observed that the neutral genetic diversity in most prokaryotic and eukaryotic species ranges by an order of magnitude, while the population census sizes of the species can differ by more than three orders of magnitude (Lynch and Conery, 2003; Leffler et al., 2012), which is called the Lewontin paradox, as it was first noted by Lewontin (1974).

One possible explanation for this paradox is that species effective sizes do not increase proportionally to population census sizes, that is, species with very large census sizes have effective sizes proportionally smaller than species with small census sizes. The reasons for this deviation from linearity are less clear. Two main explanations have been proposed, which are likely to be complementary. The first is that species with high census sizes experience large demographic disturbances, such as a high variation in reproductive success and large oscillations in census size, which greatly reduces the effective size of the species in relation to its census size (Chapter 5). There are genomic data that support this hypothesis (Romiguier et al., 2014). The second is that the effective size can be reduced due to the constriction of genetic variation as a consequence of the action of natural selection on the genome, in the form of purifying selection of deleterious mutations (background selection) or selection of beneficial variants that, particularly in regions of the genome with low

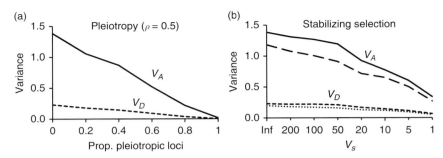

Figure 10.6 Magnitude of the additive (V_A) and dominance (V_D) variances in a population of census size $N = 1000$ at mutation–selection–drift equilibrium, where mutations appear on a quantitative trait with rate $U = 0.1$ per generation, mutational effects with exponential distribution, mean $a = 0.1$ and average dominance coefficient $h = 0.2$ (model of Figure 7.5, Chapter 7). (a) A proportion of the trait loci has deleterious effects on fitness. These pleiotropic effects are obtained with the same parameters as the effects for the trait, and the correlation between both is $\rho = 0.5$. (b) The trait is subject to stabilizing selection with intensity V_s. The second and fourth lines starting from the top indicate a combination of stabilizing selection and pleiotropic effects on fitness in 20% of the trait loci with correlation $\rho = 0.5$ between both types of effects.

recombination, would greatly reduce neutral variation (see next section). There are also multiple genomic data (for example, Corbett-Detig et al., 2015), as well as theoretical deductions (Santiago and Caballero, 2016) that support this explanation. Therefore, although neutral genetic variation increases indefinitely with effective size, the constriction of this by demographic or selective causes implies that the differences in genetic variation between species with small or large numbers of individuals do not have to be so great.

In addition to the above arguments, it should be noted, as explained above, that even characters that are considered neutral may have deleterious pleiotropic effects on fitness. When this model is considered, the genetic variance does not grow indefinitely in proportion to the effective size, but an asymptotic limit is reached for effective sizes of the order of 10^4–10^5 that depends fundamentally on the joint distribution of effects for the trait and fitness and the correlation between pleiotropic effects (Keightley and Hill, 1990; Caballero and Keightley, 1994). Figure 10.6a illustrates this phenomenon by simulations of a population in which a variable proportion of the loci that affect the trait have deleterious pleiotropic effects on fitness. As can be seen, the greater the proportion of pleiotropic loci, the smaller the genetic variance of equilibrium for the trait.

Consider, finally, the traits subjected to stabilizing selection. As we saw in a previous section, this type of selection reduces the genetic variation of the population continuously, and in the equilibrium this reduction must be compensated with the increase in variation by mutation. An illustration of the decline in additive variance with the intensity of selection is presented in Figure 10.6b. There are basically two predictive models of the magnitude of the genetic variance of equilibrium for infinite populations. The Kimura–Lande model (Kimura, 1965; Lande, 1975) assumes loci with a high mutation rate per locus and an infinite number of alleles per locus whose effects are small in relation to the existing variation and present normal distribution. Therefore, the model applies to situations where $a^2 << uV_s$, where a is the allelic effect on the trait and u the mutation rate per locus. At equilibrium, the genetic variance predicted by this model is $\hat{V}_G = \sqrt{2nV_MV_s}$, where n is the number of loci. Turelli's (1984) model assumes low mutation rates and allelic effects so great in relation to the intensity of stabilizing selection that the mutations remain at low frequency, so that stabilizing selection on the trait results in selection against rare alleles. This model is called the House of Cards model, since each mutation disturbs the equilibrium reached under selection. The predictive formula in this case is $\hat{V}_G = 4nuV_s = 4UV_s$, that must be applied when $a^2 \gg 2uV_s$, and there are modifications to include genetic drift (Stochastic House of Cards; Turelli, 1984). Stabilizing selection implies that genetic variation grows with the effective size until reaching a limiting value determined by the previous predictions based on the mutational parameters. Evidently, a trait subject to stabilizing selection may also exhibit deleterious pleiotropic mutational effects on fitness in a proportion of the relevant loci, in which case the restriction of genetic variation would be greater than that which occurs only with stabilizing selection. The results of this scenario are shown in Figure 10.6b.

An analysis of the impact of stabilizing selection on a quantitative trait assuming an epistatic model of variation was carried out by Ávila et al. (2013). Strong pervasive epistasis did not substantially modify the genetic properties of the trait. The impact of epistasis on the changes in genetic variance components when large populations were subjected to successive bottlenecks of different sizes was investigated. An initial increase of the different components of the genetic variance, as well as a dramatic acceleration of the between-line divergence, were always associated with synergistic epistasis but were strongly constrained by selection.

10.4 Genomic Footprint of Natural Selection

We have seen that natural selection produces, except in cases of balancing selection, a reduction of genetic variability. In Chapters 5 and 9 we studied the impact of selection on the effective population size and we saw that the reduction of this is more intense when recombination is low (Figure 5.1). At the genomic level, the impact of selection can be studied by evaluating the diversity in different regions of the genome. Purifying or background selection, resulting in the elimination of deleterious mutations, reduces the neutral variation in a relatively homogeneous manner, since there are usually a multitude of deleterious mutations dispersed throughout the genome (Figure 10.7a). Selection of beneficial alleles usually produces more conspicuous results, since the rapid increase in frequency of a favourable allele produces a selective sweep (Maynard Smith and Haigh, 1974), where nearby neutral variants, or deleterious alleles (the so-called hitchhiking alleles or sequences) will spread to fixation along with the beneficial allele (Figure 10.7b). The lower the recombination rate in the region surrounding the selected locus, the greater the dragged portion and the more extreme its effects, producing drastic reductions in diversity in the areas close to the aforementioned locus (Figure 10.8). In Chapter 11 we will study some of the procedures that are used to detect the genome-wide selection footprint.

Problems

10.1 In a field experiment, the partial regression of the individuals' fitness (W, scaled to the population mean) on the phenotypic values of two traits (X and Y) was estimated,

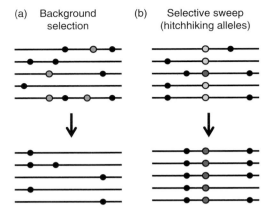

(a)   Background
        selection

(b)   Selective sweep
        (hitchhiking alleles)

Figure 10.7 Effect of selection on neutral genetic variation. (a) For deleterious background selection, natural selection eliminates deleterious mutations (grey circles) and also part of the neutral mutations linked to them (black circles). (b) For selective sweep, selection is considered for a locus with a beneficial allele (dark grey circles) versus another allele that is not (light grey circles). The rapid increase in frequency of the beneficial allele drags the alleles from other linked neutral loci (black circles). In both cases, the smaller the recombination, the greater the area of the genome with reduced variation due to the selective effect.

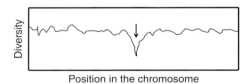

Position in the chromosome

Figure 10.8 Selective sweeps due to the rapid increase in frequency of a beneficial allele generate a reduction of genetic diversity in its vicinity (arrow), which will be higher depending on the recombination rate in the region and the effect of the selected allele.

obtaining selection gradients $b_{P(W.X)} = 0.8$ and $b_{P(W.Y)} = 0.1$. There are also estimates of the additive genetic variance of the two traits and their genetic correlation: $\sigma^2_{A,X} = 20$, $\sigma^2_{A,Y} = 30$ and $r_A = 0.41$. (a) What are the expected correlated responses in the traits by the action of natural selection? (b) What would be those responses if the genetic correlation between the two traits were of the same magnitude but negative?

10.2 The phenotypic variance for a character is $V_P = 30$ and that character is subject to stabilizing selection with intensity $V_s = 20$. By how much will the phenotypic variance be reduced by the effect of selection? What decline in fitness is expected for each unit of increase in the square of the phenotypic values?

Self-Assessment Questions

1 According to the Fundamental Theorem of Natural Selection, the genetic variance for the components of fitness is zero.
2 Only neutral traits show a null genetic covariance with fitness.

3 Stabilizing selection, by favouring individuals with intermediate phenotype, necessarily implies overdominance gene action for fitness.

4 Convergent selection implies stabilizing selection for the same optimum in two or more populations.

5 The greater the intensity of stabilizing selection, the greater the value of V_s, a measure of the width of the Gaussian curve that determines fitness.

6 The intensity of apparent stabilizing selection on a trait, due to pleiotropic effects on fitness, depends on the deleterious effect of mutations in heterozygosis.

7 A value of quantitative genetic differentiation, Q_{ST}, greater than the corresponding value of differentiation in allele frequencies for neutral markers, F_{ST}, allows diversifying selection to be inferred.

8 For a purely neutral character, it is predicted that the genetic variance increases linearly with the effective population size.

9 All types of natural selection reduce genetic variance.

10 Selective sweeps of beneficial alleles reduce the neutral genetic diversity near the selective locus, but background selection can increase it.

11 Genomic Analysis of Quantitative Traits

Concepts to Study

- QTL (quantitative trait locus)
- Interval mapping
- Genome-wide association studies (GWAS)
- Extended haplotype homozygosity (EHH)
- Genomic selection

Objectives for Learning

- To know the nature of Quantitative Trait Loci (QTLs)
- To understand the method of interval mapping of QTLs
- To learn the procedure to carry out genome-wide association studies (GWAS)
- To understand the bases of the detection of the footprint of selection by the extended haplotype homozygosity method and its variants
- To know the principles of genomic selection

The development of genotyping platforms with high density of SNP markers makes possible today a detailed analysis of the variation along the genome. Advances in genomic sequencing techniques have allowed access to more than 80 million SNPs in humans (Auton et al., 2015: 1000 Genomes Project Consortium) and several million in model species (Mackay et al., 2012). In parallel, there has been a great development of statistical methods for the estimation of genetic effects and mapping of quantitative trait loci (QTLs). Among the new applications of genomic information are the search and mapping of QTLs, the detection of the selection footprint and genomic selection, methods that we will outline throughout the chapter. Their development has allowed us to obtain information, previously very limited, on the number, genomic position and effects of QTLs (Mackay et al., 2009; Hill, 2012), with wide applications in animal and plant breeding and conservation (Ibáñez-Escriche and Simianer, 2016), and the prediction of risk to diseases (Wray et al., 2007; de los Campos et al., 2010).

11.1 Mapping of Quantitative Trait Loci

The location of quantitative trait loci can be determined through the identification of candidate genes or through their association by linkage to molecular markers. The procedure of the candidate gene consists of assuming that a given gene could have an effect on a trait, sequencing part of the gene or the complete gene in different individuals, and checking the possible association between the variants found and the phenotypic value of the trait. With this method it was possible to map the position of many QTLs in a variety of cultivated plants

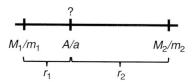

Figure 11.1 Model considered for the mapping of QTLs in intervals (interval mapping). The locus that affects the quantitative character (QTL) with alleles A and a can be located between two biallelic markers at a distance of r_1 and r_2 cM, respectively.

(Tanksley, 1993) and domestic animals (Andersson and Georges, 2004). This method, however, is of limited use, given the large number of loci that can affect a trait and the fact that many QTLs can be found in regulatory genomic regions far from the coding genes considered as candidates.

For the methods of association by linkage to a marker, the central idea is that, if a QTL is linked to an identifiable marker, an association can be found between the allelic segregations of both. The first analyses were performed with a few markers, such as blood groups (Neimann-Sorensen and Robertson, 1961). With the development of techniques to detect a greater number of markers in the genome, such as RFLPs or microsatellites, statistical methods of maximum likelihood and regression were developed. In Chapter 6 (Section 6.3.5) a method based on the animal model was outlined, consisting of comparing the likelihood of the model considering or not a random effect produced by the inclusion of a single molecular marker (George et al., 2000). A more powerful method is the so-called interval mapping (Lander and Botstein, 1989). When there are pairs of markers separated by a small distance and there is a QTL between them, it is possible to estimate the relative position of the QTL and its effect on the character. The model is illustrated in Figure 11.1. The two biallelic markers (M_1 and M_2) can be found flanking a QTL (locus A) at a distance such that the recombination frequencies between the markers and the QTL are r_1 and r_2, and it is assumed that there is no interference.

To carry out the mapping, the most suitable designs start from the crossing of two highly inbred lines that differ for a character (Figure 11.2) and the analysis is carried out either in the F_2, that is, in the panmictic descendants of the hybrid, or in the backcross of the hybrid to one of the parents, which is the case illustrated in the figure. Suppose that the parental lines are homozygous for both markers and the QTL, as indicated in the scheme. In the backcross, the parental line will only produce M_1AM_2 gametes, whereas the hybrid will produce eight types of gametes whose frequencies will depend on the frequency of recombination in the interval between the two markers and the QTL. The frequencies of the resulting genotypes in the offspring are indicated in the figure, where half of the individuals will have genotype AA and the other half genotype aa. Suppose that the locus is additive, with a being the average effect of allelic substitution (half of the effect of allele a in this case). The phenotypic value (P) of individual i would then be

$$P_i = M + \alpha g_i + e, \qquad (11.1)$$

where M is the general mean, e is a normal deviation with mean zero and variance σ^2, and g_i is the number of copies of the allele a (0 or 1 in this case).

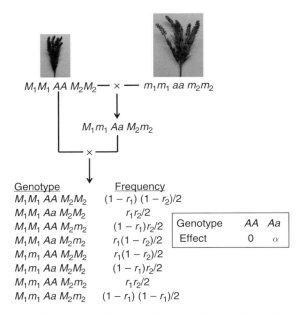

Figure 11.2 Design to determine the location of a QTL using the technique of interval mapping. Two inbred lines are crossed that differ for the trait of interest, assuming that a locus affecting the trait (QTL) is located between two genetic markers of known position (see Figure 11.1). By backcrossing the hybrid with one of its parents, offspring are obtained whose genotypes and expected frequencies are a function of the recombination frequencies (r_1 and r_2) between the markers and the QTL.

The estimation of the effect and position of the QTL can be obtained by maximum likelihood (Lander and Botstein, 1989), where the likelihood function is

$$L(M, \alpha, \sigma^2) = \prod_i [G_i(0)L_i(0) + G_i(1)L_i(1)], \tag{11.2}$$

where $L_i(x)$ is a normal deviation with mean $P_i - (M + \alpha x)$ and variance σ^2, and $G_i(x)$ is the probability that the genotype for the QTL is $x = 0$ for AA and 1 for Aa conditioned to the markers observed in the individual and their frequencies (Figure 11.2). For example, the probability that the individual had genotype Aa given that it was of genotype $M_1M_1M_2 M_2$ for the markers would be $r_1r_2/ 2$.

The decimal logarithm of the ratio between the likelihood function (11.2) and that of the function that assumes that there is no QTL in the interval studied gives rise to the so-called LOD (*log odds*) value, similar to the test explained in Chapter 6,

$$\text{LOD} = \frac{\log_{10}[L(M, \alpha, \sigma^2 | with \text{ QTL})]}{\log_{10}[L(M, \alpha, \sigma^2 | without \text{ QTL})]}, \tag{11.3}$$

which in large samples is distributed as a χ^2. LOD values higher than approximately two would indicate the existence of a QTL in the interval and the method would provide the most probable position and its effect (Figure 11.3).

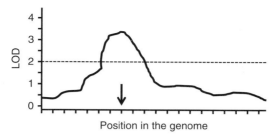

Figure 11.3 Illustrative graph of the LOD value in an interval mapping analysis. A value of LOD greater than 2 indicates significance, and the most probable position of the QTL is indicated by the arrow.

The above procedure can also be carried out by means of a regression analysis, both for crosses of inbred (Haley and Knott, 1992) and non-inbred (Knott et al., 1996) lines, which is computationally simpler and faster than maximum likelihood. The estimation of genetic components can also be incorporated into the method using the NOIA model (from Natural and Orthogonal InterActions; Alvarez-Castro et al., 2008).

11.2 Genome-Wide Association Studies

The development of massive genotyping techniques, particularly of next generation sequencing technologies, has allowed the study of the association between hundreds of thousands of SNPs and the phenotypic variation for many quantitative traits. These studies are called genome-wide association studies (GWAS) and, in humans, have enabled the identification of thousands of SNPs associated with hundreds of diseases, including cancer, obesity, coronary heart disease, stroke and Crohn's disease (Welter et al., 2014), as well as common traits, such as height and body mass index (Visscher et al., 2012, 2017). These studies are of great interest because they allow the identification of the positions or regions of the genome most related to any character, whether of medical or production interest (for example, traits of economic interest in domestic animals and plants), providing important information on the location of the QTLs, their effects and frequencies.

The GWAS on individual markers is simply a regression analysis of the type

$$\mathbf{y} = \mathbf{1}\mu + \mathbf{Mg} + \mathbf{e}, \tag{11.4}$$

where $\mathbf{y}$ is the vector of the phenotypic observations for the character, $\mathbf{1}$ is a vector of ones, μ is the population mean, $\mathbf{M}$ is the vector of the number of copies of an allele for each marker (with values 0, 1 and 2 for the three possible genotypes), $\mathbf{g}$ is the vector of effects of the markers, which are considered fixed effects, and $\mathbf{e}$ is a vector of random residual deviations with distribution $N(0, \sigma_e^2)$, where σ_e^2 is the residual variance. An illustrative example is presented in Figure 11.4. The underlying basic idea is that some SNPs physically close to a given QTL will be in strong linkage disequilibrium with it. Recall that linkage disequilibrium is reduced by recombination (Chapter 2) and, therefore, the greater the distance between two given positions, the lower the expected value of the disequilibrium between the two. This is illustrated by an example in Figure 11.5, which shows the rapid decline in linkage disequilibrium (measured

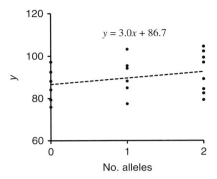

Figure 11.4 Example of regression of the phenotypic value of the individuals for a character (y) on the genotype (number of alleles) of a SNP in GWAS. The slope of the regression is an estimate of the average effect of allelic substitution associated with the SNP, in this case, three units of the measure of the character.

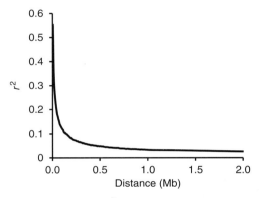

Figure 11.5 Decline in linkage disequilibrium (r^2; equation (2.7), Chapter 2) with genomic distance (in megabases). Data of 164,722 SNPs genotyped in 109 individuals of Atlantic salmon of the River Ulla (Galicia, Spain). (Data provided by María Saura, personal communication.)

as r^2; equation (2.7), Chapter 2) with salmon data from the River Ulla (M. Saura, personal communication). Therefore, only SNPs in positions physically close to a QTL will have a strong linkage disequilibrium (say $r^2 > 0.2$) with this. If this is the case, the position of the QTL may be approximated by that of the marker with a certain margin of error.

The estimated effects for the markers are statistically evaluated and those with probability below the critical genomic value are considered associated with QTLs. Normally, tens of thousands of SNPs are needed to detect QTLs with statistical significance. For example, it has been calculated that r^2 values of at least 0.2 are required between a SNP and a QTL to detect the second with a probability of 80% if it contributes 0.05 to heritability and 1000 individuals are analysed. A disequilibrium value of the order of $r^2 = 0.2$ is usually obtained at a distance of about 100–200 kb. For a 3 Gb genome, therefore, at least 15,000–30,000 SNPs would be needed to ensure the presence of one marker per 100 kb on average. Figure 11.6 shows an example of GWAS analysis for furcal length in 109 salmon from the

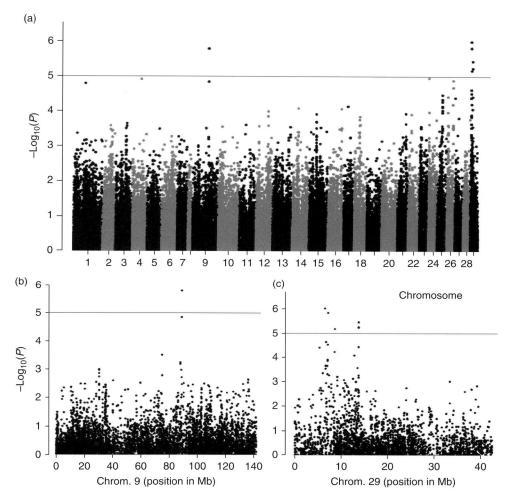

Figure 11.6 Probabilities associated with each of the 163,750 SNPs analysed by GWAS for the trait furcal length in 109 individuals of Atlantic salmon from the River Ulla (Galicia, Spain). The horizontal line indicates the genomic statistical significance. (a) Analysis of the 29 salmon chromosomes. (b, c) Detail of chromosomes 9 and 29. (Data provided by María Saura, personal communication.)

River Ulla (Galicia, Spain), genotyped for more than 160,000 SNPs. In this case, significant SNPs were detected on chromosomes 9 and 29.

The hundreds of GWAS carried out to date have allowed information on the distribution of effects of the QTLs to be obtained. The character for which the greatest number of results have been obtained is human height. The combined analysis of more than 250,000 people has allowed the detection of a total of 679 SNPs associated with 426 genomic loci for height (Wood et al., 2014), whose effects are presented in Figure 11.7a. The mean effect observed is 0.06 phenotypic standard deviations, with a deviation approximately equal to 7 cm, that is, the average effect of variants affecting height is about 4 mm. However, it can be observed that the distribution of effects is leptokurtic, with many more effects of small size

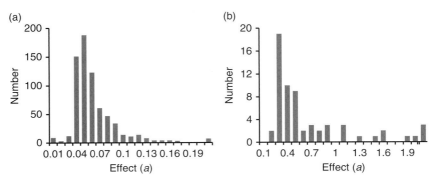

Figure 11.7 Distribution of effects in homozygosis of QTLs (a, in phenotypic standard deviations for the trait) obtained by GWAS. (a) Data of 679 SNP effects for height in humans associated with 426 loci, with an average effect of 0.06 standard phenotypic deviations (1 deviation is approximately equal to 7 cm). (Data from Wood et al., 2014.) (b) Data of effects for different quantitative traits associated with human diseases. (UK10K Consortium, 2015.)

than of large size since, presumably, many QTLs with effects lower than 0.03 standard deviations have not yet been detected due to lack of statistical power.

The existence of a majority of QTLs with small effect and a few genes of large effect is consistent with previous predictions and results (Shrimpton and Robertson, 1988), with the data of the distribution of mutational effects that we studied in Chapter 7, and with the results of artificial selection experiments (Chapter 9) where jumps in the response and heterogeneity between lines are often seen due to major genes. Hayes and Goddard (2001) obtained maximum likelihood estimates of the effects of QTLs in pigs and dairy cattle, finding a gamma distribution of effects with estimates of the shape parameter $\beta = 1.5$ in pigs and 0.4 in cows (see Chapter 7) and mean effects of 0.4 and 0.3 phenotypic standard deviations. The analysis showed that 35% and 17% of the QTLs accounted for 90% of the genetic variance in these species, respectively.

Panel (b) of Figure 11.7 illustrates the effects found for QTLs associated with different human diseases in a study using the sequenced genome of 10,000 people (UK10K Consortium, 2015). The effects point to the same leptokurtosis as the height data but the mean effect of the QTLs is an order of magnitude higher.

The GWAS technique applied to individual SNPs is the most used. However, there are other procedures that use the probability value generated by groups of SNPs in a specific segment of the genome (Wu et al., 2011; Uemoto et al., 2013). Although these methods are theoretically more powerful for the detection of QTLs, the studies carried out are inconclusive in this regard. On the other hand, the use of complete sequences of the genome allows QTLs to be detected with frequencies and effects smaller than those of traditional chips (Thornton et al., 2013; Caballero et al., 2015).

Genome-wide association studies provide estimates of the effect and frequency of each SNP associated with QTLs, so it is possible to quantify the contribution of each one to the genetic variance or heritability of the trait. As discussed in Chapter 6 (Section 6.3.6), a surprising result is that the total contribution obtained by summing those of the QTLs detected by GWAS is usually much smaller than the heritability of the estimated character with the traditional methods of resemblance between relatives (Chapter 6), what has been called 'missing heritability' (Maher, 2008; Manolio et al., 2009). For example, the heritability of height in humans referred to above is of the order of 0.6–0.8 (Zaitlen et al., 2013; Yang et al., 2015), but the 426 loci found by Wood et al. (2014) (Figure 11.7a) only explain 15% of the heritability obtained by familial (usually twins) data. Even when all the available SNPs are incorporated into the estimates, the estimation percentage only amounts to about 60%, and something similar occurs for other traits (Vinkhuyzen et al., 2013; Zaitlen et al., 2013).

There are several explanations for the problem of missing heritability (Eichler et al., 2010; Nolte et al., 2017) but we will discuss the two main ones, which are currently subject to intense debate. The first is that genomic association studies are not able to detect all the nucleotide variants responsible for the variation of the trait, as suggested by the fact that the number of QTLs found increases continuously as the sample sizes increase (Yang et al., 2015), and as inferred from theoretical studies (Eyre-Walker, 2010). The second one suggests that the gap between GWAS heritabilities and familial ones may occur because the latter (which are, in fact, usually estimates of broad-sense rather than narrow-sense heritability) are inflated by the contribution of environmental or non-additive genetic components of variance (Zuk et al., 2012; Hemani et al., 2013). It is well known that many quantitative trait loci are involved in networks where interactions are frequent, for example those whose products intervene in complex metabolic pathways or those that are responsible for the regulation of the expression of other genes (Mackay, 2014). From the point of view of quantitative genetics, these interactions are reflected in the fact that an unknown fraction of the genetic variation can be epistatic (V_I; Chapter 3). This variance may contribute to the estimates of broad-sense heritability obtained by resemblance between twins in humans (Chapter 6) possibly creating part of the difference with the heritability deduced from GWAS. However, the data from multiple analyses indicate that the magnitude of the epistatic variance is almost always small and much smaller than the additive variance (Chapter 6). A theoretical study carried out by Hill et al. (2008) concludes that if the distribution of locus effects is such that most of the alleles have extreme frequencies (close to 0 or 1), which is a reasonable assumption in many cases when talking about populations of high census size, most of the genetic variation is additive and there would be almost no epistatic variance, as suggested by the above data. Logically, if alleles are rare, it is difficult for possible epistatic associations with other alleles to manifest, since it would be necessary for the interacting alleles to be segregating in the population with a certain non-negligible probability. In addition, a meta-analysis carried out by Polderman et al. (2015) of the heritability of human traits based on 50 years of twin studies including nearly 18,000 traits found that genetic variation for about 69% of the traits is inconsistent with a substantial influence from shared environment or non-additive genetic sources. Yet, in the remaining 31% there was a significant deviation, what would imply some contribution from non-additive or environmental effects in twin heritability estimates. Therefore, the contribution of non-additive genetic components to the broad-sense familial heritabilities is a possibility, at least for some traits, to explain the gap between GWAS heritability estimates and those based on twin studies (López-Cortegano and Caballero, 2019).

11.3 Detection of the Footprint of Selection

The analysis of genetic diversity in genomic sequences allows the formulation of hypotheses about the action of natural selection. There are several methods available for this, based on the comparison between nucleotide diversity and polymorphism within and between species (Hart and Clark, 2007). We also saw in Chapter 10 that the action of selection produces temporary footprints at the genomic level that are characterized by the loss of genetic diversity, in a diffuse way, in the case of background selection, or more localized, in the case of selective sweeps of beneficial mutations (Figures 10.7 and 10.8). Different procedures have been developed that exploit the linkage disequilibrium between neutral molecular markers and selected genes to detect such footprints. A problem common to most methods is the difficulty in distinguishing between the effects of selection and those produced by historical demographic changes in the population. The foundations of these methods are explained below. Further details are given by Walsh and Lynch (2018, chapters 9 and 10).

11.3.1 Principles of Methods Based on Polymorphism and Divergence

There are two types of methods for detecting the footprint of natural selection based on intraspecific polymorphism and divergence between species. The first includes the procedures that detect the recent footprint of selection, using only polymorphism or combining polymorphism with divergence. The second covers those that are capable of detecting selective effects that have occurred in remote times using phylogenetic techniques on divergence data. While methods of the first type entail the difficulty of distinguishing between demographic and selective effects, the latter are quite robust with respect to this problem.

A procedure of the first type based on using intraspecific polymorphism is based on the following idea. In Chapter 7 we saw that the expected heterozygosity at a locus under the neutral model is equal to $4N_eu$, which is usually represented by the symbol θ, where N_e is the effective population size and u the mutation rate. To estimate this population parameter there are two types of estimators. One is the nucleotide diversity, θ_π, which is defined as the average number of nucleotide differences between two given genomic sequences (if expressed as a rate per nucleotide is usually referred to as π). The expected value of θ_π is $4N_eu$. On the other hand, the genetic diversity in a sample of n nucleotide sequences can also be estimated using the Watterson (1975) estimator (θ_w), which is obtained by counting the number of polymorphic sites (S; also called nucleotide polymorphism, P, if expressed as a rate per nucleotide) corrected for sample size as S/a, where $a = 1 + 1/2 + 1/3 + \ldots + 1/(n-1)$. Therefore, under the neutral model, we expect $\theta_\pi = \theta_w$. Both estimators behave differently when deviations from the neutral model occur, which is the basis of Tajima's D statistic to test neutrality (Tajima, 1989). Selective sweeps of beneficial alleles lead to a reduction in θ_π proportionally greater than that of θ_w. In contrast, balancing selection, such as that induced by overdominance or frequency-dependent selection, implies a larger increase for θ_π than for θ_w. Therefore, the comparison between these two measures of variability (θ_π and θ_w) gives an idea of the type of selection in action.

There are also methods to detect the recent footprint of natural selection that use the comparison between the interspecific divergence (number of nucleotide differences between the sequences of two species) and the intraspecific polymorphism (HKA-Hudson-Kreitman-Aguadé test, McDonald-Kreitman test, etc.). For example, under neutrality, the relationship between the number of non-synonymous/synonymous polymorphic changes within species must be equal to the relationship between the number of non-synonymous/synonymous changes fixed between species, since the rate of change (for polymorphism and divergence) would be a function of the neutral mutation rate. An excess of interspecific non-synonymous changes would indicate the presence of diversifying selection.

Among the methods of the second type, the *dn/ds* test stands out, capable of detecting selection in phylogenetic branches that lead to ancestral forms that could have been extinct for thousands or millions of years. This has been used for the analysis of many species including domestic ones, for example, to study the pig domestication process (Groenen et al., 2012). The *dn/ds* test is applied to coding regions and compares the rate of substitutions for non-synonymous (*dn*) and synonymous (*ds*) positions, using the divergence between species, that is, it does not take into account the intraspecific polymorphism. Under neutrality the two categories of positions diverge at the same speed (*dn/ds* = 1). A ratio *dn/ds* > 1 would imply an acceleration of the rate of fixation of non-synonymous mutations due to the action of diversifying selection, whereas *dn/ds* < 1 would imply a decreased fixation rate for non-synonymous mutations due to the action of purifying selection. Since the synonymous and non-synonymous positions analysed are intermingled throughout the coding genes, the demographic changes that affect the genome as a whole must affect both categories of positions equally. Therefore, a difference in the speed at which substitutions between species accumulate for synonymous and non-synonymous positions is not explained by demographic effects, but by selection.

11.3.2 Linkage Disequilibrium Methods

One of the most frequently used methods for the detection of the footprint of selection is based on the relationship between the frequency of an allele and the linkage disequilibrium in its surrounding regions. Processes such as genetic drift or selection generate linkage disequilibrium between loci that will decay by recombination as a function of the distance between them, more slowly the lower the recombination rate (Figure 2.3, Chapter 2). When there is selection and the frequency of a beneficial allele increases rapidly by dragging the physically linked neutral allelic variants (Figure 10.7), the disequilibrium generated will be greater than that expected with a neutral model and its decay with distance will also be smaller. This will be manifested as a greater homozygosity (lower diversity) in the vicinity of the selected locus (haplotype homozygosity). If selection is strong enough, this process will occur faster than recombination or mutation can act to break up the haplotype, and thus a signal of high haplotype homozygosity will be observed surrounding the adaptive locus. The extended haplotype homozygosity (*EHH*) method proposed by Sabeti et al. (2002), is based on this principle.

The idea is to look for segments of the genome (cores) with a high density of SNPs that allow the identification of different haplotypes (alleles; usually one ancestral and others derived). In Figure 11.8a, one of these cores is indicated with six possible alleles present in the

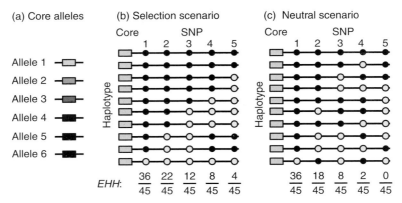

Figure 11.8 Illustration of an example of application of the extended haplotype homozygosity (EHH) method. It is considered a (a) haplotypic core determined by a dense series of SNPs, where there may be different alleles (haplotypes 1–6) with different frequencies in the population. If (b, c) one of the core alleles is considered, homozygotes can be analysed for different sequences in a series of SNPs (the black and grey circles indicate the two alleles of each SNP) of an extended region adjacent to the core (only the sequence to the right of the core is considered for simplicity). In a model in which an allele of the core has been subjected to selection (panel b), it is expected that the *EHH* value (probability of homozygosis of two randomly chosen sequences) decays with distance more slowly than expected in a neutral model (panel c).

population. Next, a broader (extended) region on both sides of it is considered, with much less density of SNPs than the core, in order to evaluate the decay of linkage disequilibrium with distance. In Figures 11.8b and 11.8c, this process is illustrated with 10 samples from the population, and the extended genomic region with a series of SNPs (the genomic region to the left of the core is ignored for simplicity). Panel (b) assumes a case in which the allele of the core is under selection, which implies a high homozygosity (probability that two sequences taken at random from the 10 available are homozygous for the SNPs in question) in the extended region. Thus, the homozygosity of SNP-1 closest to the core is $EHH = 36/45$, that of SNP-1 and SNP-2 is $EHH = 22/45$, that of SNP-1, SNP-2 and SNP-3 is $EHH = 12/45$, and so on. Panel (b) illustrates a neutral allele of the core, where homozygosity decays more rapidly with the distance to the core than in the previous scenario.

The decay of homozygosity with distance to the core in the extended region (*EHH*) is shown in Figure 11.9a for a selective and a neutral scenario. The average value of *EHH* in a given extended region can be calculated for all core alleles (six in the case of the example in Figure 11.8a). These values will depend on the allele frequency in the population, so that the lower the frequency, the higher the expected *EHH* value. The observed *EHH* values of the different core alleles can then be compared to their expected value under the neutral model (Figure 11.9b). Those alleles that show an *EHH* value significantly greater than this expected value will indicate a selective effect.

There are some variations of this method. For example, given that there may be local differences in the recombination rate, the *EHH* value of each core can be scaled to the combined *EHH* value considering all the cores analysed, obtaining a relative value, *rEHH* (Sabeti et al., 2002). An alternative method to reduce the effects of the heterogeneity of recombination rates and the complication introduced by the demographic processes consists

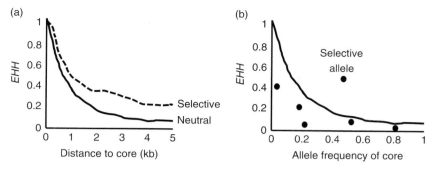

Figure 11.9 (a) Decay of extended haplotype homozygosity (EHH) of a genomic region adjacent to a given haplotypic core. In a selective scenario, the observed homozygosity decay is lower than that occurring in a neutral scenario. (b) The average *EHH* value of the six alleles of the core (circles) of Figure 11.8a can be compared to their maximum expected value under the neutral model (line) as a function of their frequencies. An *EHH* value greater than its expected neutral value will be indicative of selection for this allele of the core.

in comparing the decay of the *EHH* values occurring for the ancestral allele and the derived alleles of the core. For this, it will be necessary to find out which of the haplotypes of the core is the ancestral one by means of a phylogenetic analysis. The *EHH* decay is added (integrated) over the extended region on both sides of the core, which constitutes the *iHH* value (integrated haplotype homozygosity), and is calculated for the ancestral allele (iHH_A) and the derived one (iHH_D). The statistic used is $iHS = \ln(iHH_A/iHH_D)$ (integrated haplotype statistic; Voight et al., 2006), which is standardized to present mean zero and variance one. Equal decay rates for the ancestral and derived alleles ($iHS = 0$) will imply a neutral scenario, while large negative/positive iHS values will indicate abnormally long haplotypes with the derived/ancestral allele, which is indicative of a selective scenario. Finally, the method has also been extended to make cross-population comparisons (XPEHH; Sabeti et al., 2007). The extended haplotype homozygosity method has also been widely applied for the detection of selection in domestic species (see, for example, Qanbari et al., 2011).

11.3.3 Detection of Diversifying and Convergent Selection

The methods of detection of the selection footprint due to linkage disequilibrium explained in the previous section apply to a single population. If we consider diversifying selection among populations (Chapter 10), different methods must be used. This type of selection can produce a quantitative genetic differentiation between populations and the effects of selection on the character can be quantified by comparing the value of Q_{ST} for the trait (Chapter 8) and that of F_{ST} for neutral markers, which will indicate the level of divergence among populations due to causes different from selection (see the example in Section 10.3.2 of Chapter 10). Likewise, convergent selection could occur when stabilizing selection leads the populations to reach the same optimum of fitness, in which case a $Q_{ST} < F_{ST}$ relation would be expected. Although the comparison $Q_{ST} - F_{ST}$ allows us to detect the existence of selection on the character, it is not possible to infer which are the loci involved in the process. This deduction can be carried out when a sufficiently large number of neutral genetic markers is available so that some are in

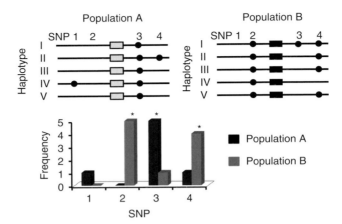

Figure 11.10 Illustrative example of the detection of the footprint of diversifying selection between two populations. The different alleles of the QTL (marked with a rectangle) are selected differentially in populations A and B, which implies the dragging of some allelic variants of SNPs in linkage disequilibrium with the QTL alleles. Some SNPs will show a differentiation in allele frequencies between populations greater than what is expected by chance, being a footprint of selection on the QTL. The asterisks indicate significant differences in frequencies between populations.

linkage disequilibrium with the QTLs. Figure 11.10 illustrates this procedure. Let us assume that there is selection for the locus indicated with a rectangle, with one allele favourably selected in population A and another one in population B. Selection produces the drag of neutral variants in strong disequilibrium with the respective selected alleles. As a result, the value of differentiation in allele frequencies between populations for some markers will be higher than expected with a neutral model. Thus, by examining the distribution of F_{ST} values for neutral markers and comparing it with the expected one under a neutral model it is possible to detect the markers linked to selected QTLs. The neutral distribution of F_{ST} values can be approximated by a χ^2 distribution, in particular, the value $(n-1)F_{ST}/\overline{F}_{ST}$ has χ^2 distribution with $n-1$ degrees of freedom and variance $2\overline{F}_{ST}^2/(n-1)$, where n is the number of populations and $\overline{F}_{ST}$ is the average of F_{ST} values (Lewontin and Krakauer, 1973).

The methodology, originally proposed by Lewontin and Krakauer (1973), was subsequently developed by other authors using frequentist (Beaumont and Nichols, 1996) or Bayesian (Foll and Gaggioti, 2008) procedures. Figure 11.11 illustrates an analysis of the first type with *Littorina saxatilis* AFLP data corresponding to the ecotype RB and SU populations presented in Figure 10.4 (Chapter 10). The distribution of the combined values of F_{ST} and H (expected heterozygosity) for the markers is compared with the simulated distribution for neutral markers with an average differentiation equal to that observed. Markers whose allele frequency differentiation values (F_{ST}) are higher than the 95% statistical significance threshold are considered associated by linkage disequilibrium to QTLs subjected to diversifying selection for the character. Those with a level of differentiation lower than the low significant threshold for the neutral model are indicative of convergent selection among populations.

Sometimes markers with high differentiation are grouped into more or less large regions of the genome due to linkage, forming what are called 'islands of differentiation',

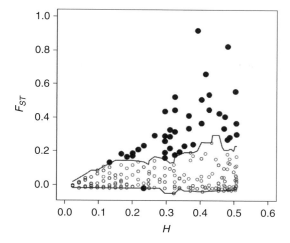

Figure 11.11 Illustration of the detection of AFLP markers associated with selected loci in populations of the RB and SU ecotypes of *Littorina saxatilis* from the locality of Burela (Galicia, Spain) using the DFDIST programme (Beaumont and Nichols, 1996). Each circle indicates the value of F_{ST} and heterozygosity (H) of a marker. The lines indicate the 95% limits of the expected values under a neutral model. The black circles indicate the footprint of diversifying selection between ecotypes (above the upper limit) or convergent selection (below the lower limit). (Data from Butlin et al., 2014, provided by Juan Galindo.)

which facilitate the identification of selective loci (Nosil and Feder, 2012). The main limitation of the method lies in the difficulty of distinguishing between the differentiation generated by diversifying selection and that due to demographic causes or other genetic factors (Pérez-Figueroa et al., 2010; Bierne et al., 2011). However, despite the high type I error of the method (Vilas et al., 2012), it has been empirically verified that the detected markers are located at an average distance of about 10 cM from the QTLs (Via, 2012), which shows its usefulness.

11.4 Genomic Selection

We saw in Chapter 9 that the use of genetic markers can improve the response to artificial selection through marker-assisted selection (MAS). The possibility of having tens or hundreds of thousands of SNPs in many domestic species makes it possible to apply genomic selection (Meuwissen et al., 2001; Goddard, 2008) that incorporates all the available molecular information and not only the markers associated with loci of great effect, as with MAS. Genomic selection, already widely applied in animals (Goddard et al., 2010) and plants (Desta and Ortiz, 2014), exploits the linkage disequilibrium between nearby loci generated by genetic drift or selection, for which it is necessary to have a density of markers high enough to ensure that all loci with effect on the trait are in disequilibrium with a marker. With genomic selection it is possible to predict the additive values of individuals with high precision (of the order of 80%) and estimates can be made at an early age, which is of great practical use by shortening the generational interval. It is also very useful in scenarios where it is difficult to obtain genealogical relationships, such as in aquaculture. The expectations of genomic selection are high and it is expected that the rate of genetic progress in domestic species could be doubled with this method. It is important to emphasize that to carry out genomic selection it is not necessary to know the genomic position of the markers. However, if this is known, genomic selection also provides a method of mapping the loci that affect the character.

The methodology necessary to carry out genomic selection is varied and there is no consensus on which is the most appropriate. It can be applied on individual markers or on haplotypes and there are several statistical methods, of which we will mention the most used. Genomic selection is conceptually carried out in two steps. In the first, the effects of genetic markers are estimated in a reference population where the animals are genotyped for the markers and evaluated for the trait of interest. In the second, the additive values of the individuals of the population to be selected (evaluated population) are estimated using the previously estimated effects, where individuals are genotyped for the markers without requiring the phenotypic values for the trait. The second step is simple because it only requires adding the estimated effects of the markers. The complication arises in the first step, mainly because the number of markers whose effect is to be estimated is generally much greater than the number of individuals evaluated.

To understand the procedure, suppose first that the additive value of an individual (A) is obtained as the sum of the average effects (α) of all loci that affect the trait, that is, $A = \boldsymbol{\alpha}'\mathbf{x}$, where $\boldsymbol{\alpha}$ is the vector of average effects and $\mathbf{x}$ the vector that indicates the number of copies of the allele with effect on the trait (with terms $x = 0$, 1 and 2 for the three possible genotypes). The idea is to estimate A using the genetic markers associated with the trait loci, that is, $E[A] = \mathbf{b}'\mathbf{m}$, where $\mathbf{b}$ is the vector of estimated effects and $\mathbf{m}$ that of the number of copies of each marker. This is the first step to which we referred earlier. To estimate the effects associated with the markers, these are considered random in the model $P = \hat{\mathbf{b}}'\mathbf{m} + e$, where P is the phenotypic value of the individual and e is a normal deviation with mean zero and variance σ^2, which includes the environmental effects and the rest of effects not considered by the markers. The estimation of the coefficients can be carried out by different methods (Gianola, 2013), such as Bayesian analysis, ridge regression (SNP-BLUP or RR-BLUP) or GBLUP (Chapter 6).

With the Bayesian method, initially proposed by Meuwissen et al. (2001), the coefficients are estimated as

$$\hat{b}' = \frac{\int b f(b) \Pr(P|b)\,db}{\int f(b) \Pr(P|b)\,db}, \tag{11.5}$$

where the integration is applied to the distribution of b values assumed a priori [$f(b)$], and $\Pr(P|b)$ is the probability of obtaining the phenotypic value given the effect of the marker. Equation (11.5) is a Bayesian probability but can also be considered a conditioned mean from the frequentist point of view. Meuwissen et al. (2001) proposed two Bayesian models: Bayes A, in which the a priori distribution of values of b follows a t distribution (with tails longer than normal), and Bayes B, in which it is also assumed that a proportion of the markers has no effect. Subsequently, other alternative models have been proposed (Gianola, 2013).

Now consider the ridge regression method, so called because the least squares equations are reduced according to the degree of genomic relationship, and that is technically

Table 11.1 *Example to illustrate the estimation of effects associated with eight molecular markers (SNP) in eight individuals (A–H) using the ridge regression method or SNP-BLUP*

Individual	SNP								Phenotype *(y)*
Reference population									
A	0	1	1	0	1	0	1	1	92.54
B	2	1	2	1	0	0	1	0	104.51
C	0	2	0	0	2	1	0	0	108.89
D	2	1	0	2	1	0	1	0	80.99
Evaluated population									
E	1	1	0	1	1	1	1	1	
F	2	1	0	0	0	2	0	1	
G	0	2	0	0	0	0	2	1	
H	1	1	0	2	0	2	1	0	

Note: 0, 1 and 2 indicate the SNP genotype (the number of copies of one allele).

denominated SNP-BLUP or RR-BLUP from Ridge Regression-BLUP. With this method, the coefficients are obtained by solving the equation analogous to (6.27) of Chapter 6,

$$\hat{\mathbf{b}} = [\mathbf{M}'\mathbf{M} + \mathbf{I}\lambda]^{-1}\mathbf{M}'\mathbf{y}, \tag{11.6}$$

where $\mathbf{M}$ is the matrix that indicates the number of alleles for each marker and individual (with terms 0, 1 and 2 for the three possible genotypes), $\mathbf{I}$ is the identity matrix, $\hat{\mathbf{b}}$ is the vector of estimated effects for the markers, $\mathbf{y}$ is the vector of phenotypic observations for the trait, and $\lambda = \sigma_E^2/(\sigma_A^2/n) = n(1-h^2)/h^2$, where σ_A^2, σ_E^2 and h^2 are the additive and environmental variances, respectively, and the heritability of the trait, that must be previously known, and n the number of markers.

Consider a small example as an illustration. Suppose we have molecular data from eight SNP markers for eight individuals, four of which constitute the reference population and are measured for the character, and the other four from the evaluated population, which are not measured and of which we want to predict their genomic value (Table 11.1). Suppose also that the heritability of the trait is $h^2 = 0.5$.

In the reference population, the $\mathbf{M}$ matrix and the $\mathbf{y}$ vector will be, therefore,

$$\mathbf{M} = \begin{pmatrix} 0 & 1 & 1 & 0 & 1 & 0 & 1 & 1 \\ 2 & 1 & 2 & 1 & 0 & 0 & 1 & 0 \\ 0 & 2 & 0 & 0 & 2 & 1 & 0 & 0 \\ 2 & 1 & 0 & 2 & 1 & 0 & 1 & 0 \end{pmatrix}, \quad \mathbf{y} = \begin{pmatrix} 92.54 \\ 104.51 \\ 108.89 \\ 80.99 \end{pmatrix},$$

and $\lambda = 8(1 - 0.5)/0.5 = 8$. We then have that

$$\mathbf{M'M} + \mathbf{I}\lambda = \begin{pmatrix} 16 & 4 & 4 & 6 & 2 & 0 & 4 & 0 \\ 4 & 15 & 3 & 3 & 6 & 2 & 3 & 1 \\ 4 & 3 & 13 & 2 & 1 & 0 & 3 & 1 \\ 6 & 3 & 2 & 13 & 2 & 0 & 3 & 0 \\ 2 & 6 & 1 & 2 & 14 & 2 & 2 & 1 \\ 0 & 2 & 0 & 0 & 2 & 9 & 0 & 0 \\ 4 & 3 & 3 & 3 & 2 & 0 & 11 & 1 \\ 0 & 1 & 1 & 0 & 1 & 0 & 1 & 9 \end{pmatrix}, \mathbf{M'y} = \begin{pmatrix} 0 & 2 & 0 & 2 \\ 1 & 1 & 2 & 1 \\ 1 & 2 & 0 & 0 \\ 0 & 1 & 0 & 2 \\ 1 & 0 & 2 & 1 \\ 0 & 0 & 1 & 0 \\ 1 & 1 & 0 & 1 \\ 1 & 0 & 0 & 0 \end{pmatrix} \begin{pmatrix} 92.54 \\ 104.51 \\ 108.89 \\ 80.99 \end{pmatrix} = \begin{pmatrix} 371 \\ 495.82 \\ 301.56 \\ 266.49 \\ 391.51 \\ 108.89 \\ 278.04 \\ 92.54 \end{pmatrix}.$$

$$(\mathbf{M'M} + \mathbf{I}\lambda)^{-1} = \begin{pmatrix} 0.0844 & -0.0106 & -0.0156 & -0.0304 & -0.0005 & 0.0025 & -0.0156 & 0.0047 \\ -0.0106 & 0.0897 & -0.0114 & -0.0070 & -0.0316 & -0.0129 & -0.0095 & -0.0041 \\ -0.0156 & -0.0114 & 0.0882 & -0.0008 & 0.0033 & 0.0018 & -0.0150 & -0.0072 \\ -0.0304 & -0.0070 & -0.0008 & 0.0964 & -0.0052 & 0.0027 & -0.0124 & 0.0028 \\ -0.0005 & -0.0316 & 0.0033 & -0.0052 & 0.0887 & -0.0127 & -0.0063 & -0.0060 \\ 0.0025 & -0.0129 & 0.0018 & 0.0027 & -0.0127 & 0.1168 & 0.0035 & 0.0023 \\ -0.0156 & -0.0095 & -0.0150 & -0.0124 & -0.0063 & 0.0035 & 0.1086 & -0.0086 \\ 0.0047 & -0.0041 & -0.0072 & 0.0028 & -0.0060 & 0.0023 & -0.0086 & 0.1140 \end{pmatrix}.$$

Substituting in (11.6) we obtain the vector of estimated effects for the markers $(\hat{\mathbf{b}})$ and, with them, the predicted genomic additive values for the four individuals of the reference population $(\hat{\mathbf{a}}_R)$,

$$\hat{\mathbf{b}} = \begin{pmatrix} 9.41 \\ 18.44 \\ 11.61 \\ 5.76 \\ 14.78 \\ 4.72 \\ 9.01 \\ 4.30 \end{pmatrix}; \hat{\mathbf{a}}_R = \mathbf{M}\hat{\mathbf{b}} = \begin{pmatrix} 58.14 \\ 75.26 \\ 71.16 \\ 72.58 \end{pmatrix}. \tag{11.7}$$

The same values would also be obtained with the Bayesian method (equation (11.5)). Thus, for the four individuals of the evaluated population, the required estimated additive values would be

$$\hat{\mathbf{a}}_E = \mathbf{M}_E\hat{\mathbf{b}} = \begin{pmatrix} 1 & 1 & 0 & 1 & 1 & 1 & 1 & 1 \\ 2 & 1 & 0 & 0 & 0 & 2 & 0 & 1 \\ 0 & 2 & 0 & 0 & 0 & 0 & 2 & 1 \\ 1 & 1 & 0 & 2 & 0 & 2 & 1 & 0 \end{pmatrix} \begin{pmatrix} 9.41 \\ 18.44 \\ 11.61 \\ 5.76 \\ 14.78 \\ 4.72 \\ 9.01 \\ 4.30 \end{pmatrix} = \begin{pmatrix} 66.42 \\ 51.00 \\ 59.19 \\ 57.81 \end{pmatrix}. \tag{11.8}$$

An alternative form of estimation is to use GBLUP evaluation, in which BLUP is carried out using the genomic relationship matrix rather than the genealogical one (see Chapter 6). By having multiple markers with a high density in the genome, the estimation of coancestries between individuals is performed with a precision much higher than that obtained with the genealogical matrix, since the former allows the proportion of genes that have two specific individuals in common to be obtained instead of resorting to its expected values. This can be used to carry out selection by GBLUP or to combine the genealogical and molecular matrices with this purpose (Legarra et al., 2009). The molecular coancestry matrix uses the identity in state of alleles while the genealogical one uses the identity by descent and, therefore, refers to different base populations (Toro et al., 2002), although it is possible to correct the second so that they can be combined (Powell et al., 2010). The equation to predict the additive values with GBLUP, analogous to equation (6.26) of Chapter 6 is

$$\hat{\mathbf{a}} = [\mathbf{Z'Z} + \mathbf{G}^{-1}\lambda/n]^{-1}\mathbf{Z'y}, \tag{11.9}$$

where $\mathbf{Z}$ is the incidence matrix of the data (Chapter 6), $\mathbf{G} = \mathbf{MM'}/n$ is the molecular genomic matrix, n is the number of markers and $\lambda = n(1 - h^2)/h^2$. The estimation of the markers effects would be obtained through

$$\hat{\mathbf{b}} = \mathbf{M'}(\mathbf{G} \times n)^{-1}\hat{\mathbf{a}}. \tag{11.10}$$

In the previous example, the matrices for the reference population are

$$\mathbf{Z'y} = \begin{pmatrix} 1 & & & \\ & 1 & & \\ & & 1 & \\ & & & 1 \end{pmatrix} \begin{pmatrix} 92.54 \\ 104.51 \\ 108.89 \\ 80.99 \end{pmatrix} = \begin{pmatrix} 92.54 \\ 104.51 \\ 108.89 \\ 80.99 \end{pmatrix}; \quad \mathbf{Z'Z} = \mathbf{Z} \ ; \ \lambda/n = 1;$$

$$\mathbf{G} = \mathbf{MM'}/8 = \begin{pmatrix} 0.625 & 0.500 & 0.500 & 0.375 \\ 0.500 & 1.375 & 0.250 & 1.000 \\ 0.500 & 0.250 & 1.125 & 0.500 \\ 0.375 & 1.000 & 0.500 & 1.375 \end{pmatrix};$$

$$\mathbf{G}^{-1} = \begin{pmatrix} 3.751 & -1.604 & -1.639 & 0.739 \\ -1.604 & 2.263 & 0.891 & -1.532 \\ -1.639 & 0.891 & 1.800 & -0.855 \\ 0.739 & -1.532 & -0.855 & 1.951 \end{pmatrix};$$

$$(\mathbf{G} \times 8)^{-1} = \begin{pmatrix} 0.469 & -0.200 & -0.205 & 0.092 \\ -0.200 & 0.283 & 0.111 & -0.191 \\ -0.205 & 0.111 & 0.225 & -0.107 \\ 0.092 & -0.191 & -0.107 & 0.244 \end{pmatrix};$$

$$[\mathbf{Z'Z} + \mathbf{G}^{-1}\lambda/n]^{-1} = \begin{pmatrix} 0.296 & 0.120 & 0.144 & 0.030 \\ 0.120 & 0.467 & -0.015 & 0.208 \\ 0.144 & -0.015 & 0.475 & 0.094 \\ 0.030 & 0.208 & 0.094 & 0.467 \end{pmatrix}.$$

Substituting these expressions in equations (11.9) and (11.10), we obtain the same predictions of the additive values and marker effects obtained with the regression method (equations (11.7)), and the same would occur with the data of the evaluated population (equation (11.8)).

The accuracy of GBLUP depends basically on the effective population size, the genetic length of the genome and the number of markers. The variance of the genomic relationships of individuals around their genealogical value is approximately equal to $\log(2N_eL)/(2N_eLn_c)$ (Goddard et al., 2010), where L is the length of a chromosome in Morgans and n_c is the number of haploid chromosomes. The smaller that variation and the greater the number of markers used (n), the greater the precision, which is then a function of $n/(N_eLn_c)$ (Goddard et al., 2010).

Problems

11.1 The following table presents the 16 polymorphic nucleotides in five sequences of 500 bp (the other nucleotides are monomorphic). Calculate the nucleotide diversity (θ_π). What can be deduced from the comparison between the number of polymorphic sites and the nucleotide diversity?

1	2	3	4	5	6	7	8	9	10	11	12	13	14	15	16
T	C	T	A	C	C	T	C	C	T	C	G	G	T	T	A
T	C	C	T	A	C	C	T	C	C	T	G	G	T	T	T
C	T	C	C	C	C	C	T	C	T	T	T	G	C	T	A
C	T	C	C	C	C	C	T	T	C	T	G	A	C	T	T
C	T	C	C	C	T	C	T	T	T	T	G	G	C	C	A

11.2 The polymorphism and divergence between synonymous nucleotide positions (they produce no change in amino acid) and replacement positions (non-synonymous: they produce a change) have been studied for the alcohol dehydrogenase (*Adh*) gene in samples from four *Drosophila* species. In these samples, 71 fixed and 106 polymorphic synonymous positions were found, as well as 23 fixed and 13 polymorphic replacement positions. Do these data indicate any kind of selection?

Self-Assessment Questions

1 The procedure of QTL mapping by candidate genes requires partial sequencing of the relevant gene.

2 Interval mapping of QTLs using genetic markers can only be carried out by crossing highly inbred lines.

3 For the detection of a QTL by GWAS it is necessary that the magnitude of the linkage disequilibrium (r^2) between the markers and the QTL is greater than 0.2.

4 The complete sequencing of genomes allows the detection of QTLs of smaller effect and frequency than those found by SNP chips.

5 Epistatic interactions are a possible explanation of the observation that the total contribution to heritability of the QTLs found by GWAS is often lower than the heritability of the trait estimated by resemblance among relatives.

6 If the value of the nucleotide diversity (θ_π) for a set of sequences of the same species is greater than the ratio S/a, where S is the number of polymorphic nucleotides in the sequences, this suggests the selective sweep of a beneficial allele.

7 A ratio $dn/ds > 1$ implies diversifying selection between species.

8 Genomic selection allows the generational interval to be shorten.

9 Genomic selection can also provide a method for mapping QTLs.

10 The accuracy of GBLUP increases with the increase in the variance of the genomic relationships of individuals around their genealogical value.

Solutions to Problems and Self-Assessment Questions

Chapter 1

Problem 1.1

(a) With $n = 20$ loci, the number of genotypic classes would be $2n + 1 = 41$. (b) The frequency with which heterozygous individuals would be found for the 20 loci would be $(1/2)^{20} = 9.5 \times 10^{-7}$. (c) The probability of finding individuals with a phenotypic value equal to 10 would be:

$$\binom{40}{10}\left(\frac{1}{2}\right)^{10}\left(\frac{1}{2}\right)^{30} = 7.7 \times 10^{-4}.$$

Problem 1.2

The sums of the 50 phenotypic values and their squares are $\Sigma X_i = 1134$ and $\Sigma X_i^2 = 26,276$, respectively, so the mean is $M = 1134/50 = 22.68$ and the phenotypic variance $V_P = [\Sigma X_i^2 - (\Sigma X_i)^2/n]/(n-1) = [26,276-(1,134)^2/50]/49 = 11.36$. Since the genetic variance is $V_G = 4$, the estimate of the environmental variance is $V_E = V_P - V_G = 11.36 - 4 = 7.36$. The phenotypic coefficient of variation is $CV_P = \sqrt{V_P}/M = \sqrt{11.36}/22.68 = 0.145$, and the genotypic coefficient of variation is $CV_G = \sqrt{V_G}/M = \sqrt{4}/22.68 = 0.088$.

Self-Assessment Questions

1-F. 2-T. 3-F. 4-F. 5-T. 6-T. 7-F. 8-F. 9-T. 10-T.

Chapter 2

Problem 2.1

(a) The allele frequencies (p_1, p_2 and p_3 corresponding to the A_1, A_2 and A_3 alleles, respectively) are calculated from the genotype frequencies as follows:

$$p_1 = [23 + (61/2) + (28/2)]/200 = 0.338,$$

$$p_2 = [39 + (61/2) + (41/2)]/200 = 0.450,$$

$$p_3 = [8 + (28/2) + (41/2)]/200 = 0.213 \text{ or simply } p_3 = 1 - p_1 - p_2.$$

Heterozygosity is the expected global frequency of heterozygotes and is calculated by means of equation (2.2), that is, $H = 1 - (0.338)^2 - (0.450)^2 - (0.213)^2 = 0.638$.

(b) To determine if the population is in Hardy–Weinberg equilibrium for the locus, the observed genotype frequencies should be compared with those expected in Hardy–Weinberg equilibrium (HW), which are indicated in the following table. The expected absolute frequencies (E) are obtained by multiplying the frequencies at equilibrium by the total number of individuals, that is, 200.

Genotypes	A_1A_1	A_1A_2	A_1A_3	A_2A_2	A_2A_3	A_3A_3	Total
Observed (O)	23	61	28	39	41	8	200
Frequencies HW	p_1^2	$2p_1p_2$	$2p_1p_3$	p_2^2	$2p_2p_3$	p_3^2	
Expected (E)	22.8	60.7	28.7	40.5	38.2	9.0	200

The fit between observed (O) and expected (E) frequencies is statistically contrasted by a χ^2 test, since the summation $\Sigma(O - E)^2/E$ is distributed as a χ^2 distribution with a number of degrees of freedom equal to the number of genotypic classes that are compared less the number of parameters that are needed to obtain the expected values. In this case there are six classes and to obtain the expected values two of the three allele frequencies are needed (the third is determined once the first two are known) and the total number of individuals in the sample. Therefore, the number of degrees of freedom is d.f. = 6 − 2 − 1 = 3. The statistic is

$$\sum_{i=1}^{6} \frac{(O_i - E_i)^2}{E_i} =$$
$$= \frac{(23 - 22.8)^2}{22.8} + \frac{(61 - 60.7)^2}{60.7} + \frac{(28 - 28.7)^2}{28.7} + \frac{(39 - 40.5)^2}{40.5} + \frac{(41 - 38.2)^2}{38.2} + \frac{(8 - 9.0)^2}{9.0} = 0.39.$$

The critical value from which the statistic would have a significant value (with a probability of 95%) and, therefore, the null hypothesis would be rejected, is obtained from the tables of the χ^2 distribution that, with 3 degrees of freedom, is 7.81. Since the value found is lower than this critical value, we accept the hypothesis that the population is in Hardy–Weinberg equilibrium for the locus studied. Two important aspects of this test should be noted. First, it must be performed with absolute frequencies, not relative ones. Second, since the χ^2 distribution is continuous and approaches a scenario of discrete classes, the expected number of individuals in the classes should not be too small, say less than 5. If this is the case, it is possible to follow different procedures, one of which is to gather different classes so that they all have at least five expected individuals.

Problem 2.2

The relative gametic (haplotypic) frequencies are obtained from the absolute ones dividing between the total number of gametes (200). For example, the frequency of the AB haplotype is $p_{AB} = 29/200 = 0.145$, and analogously for the other gametic types (table below). The allele frequencies are $p_A = (29 + 37)/200 = 0.33, p_a = 1 - p_A = 0.67; p_B = (29 + 88)/200 = 0.585, p_b = 1 - p_B = 0.415$.

(a) Introducing the four gametic frequencies in equation (2.4) we have that the linkage disequilibrium is $D = (0.145 \times 0.230) - (0.185 \times 0.440) = -0.0481$, whose sign indicates a defect of AB and ab gametic types. The maximum disequilibrium, given the allele frequencies, is given by equation (2.6), $D' = -0.0481/(0.33 \times 0.585) = -0.249$. Finally, the squared correlation of allele frequencies (equation (2.7)) is $r^2 = (-0.0481)^2/(0.33 \times 0.67 \times 0.585 \times 0.415) = 0.043$.

Haplotypes	AB	Ab	aB	ab	Total
Observed (O)	29	37	88	46	200
Frequencies	0.145	0.185	0.440	0.230	1
Frequencies (Linkage equilibrium)	$p_A p_B$ 0.19	$p_A p_b$ 0.14	$p_a p_B$ 0.39	$p_a p_b$ 0.28	1
Expected (E)	38.6	27.4	78.4	55.6	200

(b) To check if the disequilibrium is significant, the observed (haplotype) frequencies should be compared with those expected under gametic equilibrium with a χ^2 test for independence with 4 (classes) $-$ 2 (frequencies p_A and p_B) $-$ 1 (total sample) $= 1$ degree of freedom. The expected values with linkage equilibrium are obtained by the product of the allele frequencies corresponding to each gametic type (see table). The observed and expected absolute values are used to obtain the statistic

$$\sum_{i=1}^{4} \frac{(O_i - E_i)^2}{E_i} = \frac{(29 - 38.6)^2}{38.6} + \frac{(37 - 27.4)^2}{27.4} + \frac{(88 - 78.4)^2}{78.4} + \frac{(46 - 55.6)^2}{55.6} = 8.6.$$

This value could have also been approximated as the product of $r^2 \times 200$. The critical value of the χ^2 distribution with a 95% probability and one degree of freedom is 3.84, so the hypothesis of gametic equilibrium is rejected.

(c) If the frequency of recombination between the two SNPs is $c = 0.05$, the number of generations needed for the disequilibrium to be reduced by a half is obtained with equation (2.5), $t = [\ln(\tfrac{1}{2})]/[\ln(1 - 0.05)] = 13.5$ generations.

Problem 2.3

(a) Since it has not been indicated that there are other forces of change in allele frequencies affecting the locus in addition to genetic drift, the expected frequency of the allele at generation 50 will be equal to its initial frequency, that is, $E[q_{50}] = q_0 = 0.4$.

(b) The variance of the allele frequencies between lines is given by equation (2.9), that is,

$$\sigma^2_{q_{50}} = (1 - 0.4) \times 0.4 \times \left[1 - \left(1 - \frac{1}{2 \times 40}\right)^{50}\right] = 0.112.$$

(c) The initial expected frequency of heterozygotes is $H_0 = 2q_0(1 - q_0) = 2 \times 0.4 \times (1 - 0.4) = 0.48$. After 50 generations its expected value (equation (2.11)) will be reduced to a value $H_{50} = 0.48 - (2 \times 0.112) = 0.256$.

Problem 2.4

(a) In an island model, allele frequencies will tend to become uniform by migration, reaching the average allele frequency. The expected value of the allele frequency can be obtained by equation (2.16), where p_t is the frequency on a given island in generation t, p_0 its initial frequency and $P = 0.25$ the average frequency in the set of islands. Therefore, the expected frequencies at generation $t = 10$ will be 0.267, 0.233, 0.302 and 0.198, respectively. After

50 generations the frequencies on the four islands will be practically equal to the equilibrium value, 0.25.

(b) If on the first island, which had an initial allele frequency of 0.3, the frequency is 0.275 at generation 10, we can estimate the migration rate again using equation (2.16), where $p_0 = 0.3$, $p_{10} = 0.275$ and $P = 0.25$. Clearing, we obtain $m = 1 - [(0.275 - 0.25)/(0.3 - 0.25)]^{0.1} = 0.067$.

Problem 2.5

(a) The frequencies of the three genotypes (phenotypes) before and after selection are shown in the following table.

Genotypes	AA	Aa	aa	Total
Observed before selection	80	240	180	500
Frequencies	0.16	0.48	0.36	1
Observed after selection	72	130	33	235
Proportion of surviving individuals	0.9	0.542	0.183	
Relative fitness (w)	1	0.60	0.20	
General model	1	$1 - sh$	$1 - s$	
	$s = 0.8$		$h = 0.5$	

The genotype frequencies are in Hardy–Weinberg equilibrium, with allele frequencies $p = 0.4$ and $q = 0.6$ for the A and a alleles, respectively. The proportion of surviving individuals of each genotype is obtained by dividing the number observed after selection by the corresponding number before selection. These proportions of survivors scaled to the highest one (that of AA) provide the fitness (w) of each genotype. The selection coefficient against allele a is obtained by equalling $1 - s = 0.2$, that is, $s = 0.8$, and the dominance coefficient by equalling $1 - sh = 0.6$, that is, the locus is additive ($h = 0.5$).

(b) The mean fitness of the population is given by equation (2.17), $\overline{w} = 1 - sq(q + 2ph) = 1 - [0.80 \times 0.6 \times (0.6 + 2 \times 0.4 \times 0.5)] = 0.52$. The expected change in the allele frequency after one generation of selection is obtained by the expression of Table 2.7, $\Delta p = spq/(2\overline{w}) = 0.185$.

Problem 2.6

(a) Relative fitnesses are obtained as in the previous problem, and are shown in the following table. It is a model of overdominance or advantage of the heterozygotes.

Genotypes	AA	Aa	aa	Total
Observed before selection	35	144	146	325
Frequencies	0.11	0.44	0.45	1
Observed after selection	22	130	75	227
Proportion of surviving individuals	0.63	0.90	0.51	
Relative fitness (w)	0.69	1	0.57	
Model	$1 - s_A$	1	$1 - s_a$	
	$s_A = 0.31$		$s_a = 0.43$	

(b) The locus is in Hardy–Weinberg equilibrium, with allele frequencies $p = 0.33$ and $q = 0.67$. The mean fitness of the population is given by the expression in Table 2.8, $\overline{w} = 1 - p^2 s_A - q^2 s_a = 1 - (0.33)^2(0.31) - (0.67)^2(0.43) = 0.77$. The maximum possible fitness of the population with Hardy–Weinberg equilibrium would be reached with the allele frequencies of equilibrium, at which selection would lead. The equilibrium values are calculated with the expressions in Table 2.8, that is, $\hat{p} = s_a/(s_A + s_a) = 0.43/(0.31 + 0.43) = 0.58$ and $\hat{q} = s_A/(s_A + s_a) = 0.42$, and the maximum fitness is $\overline{w} = 1 - \hat{p}^2 s_A - \hat{q}^2 s_a = 0.819$. Therefore, the population is not at equilibrium and the observed fitness (0.77) is not the maximum possible one.

Self-Assessment Questions

1-T. 2-F. 3-T. 4-F. 5-F. 6-F. 7-T. 8-F. 9-T. 10-F.

Chapter 3

Problem 3.1

Denominating AA, Aa and aa to the genotypes for this locus, where allele a is that with the detected effect and allele A the standard one, the relative genotypic values of the three genotypes are zero, one and three bristles, respectively, which indicates a homozygous value of $a = 3$ and a dominance coefficient of $h = 1/3$. The allele frequencies of A and a are $p = 0.8$ and $q = 0.2$, respectively, and in the following calculations it is necessary to assume that the locus is in Hardy–Weinberg equilibrium.

(a) The dominance effect (equation (3.2)) is given by $d = a(h - \frac{1}{2}) = -0.5$, and the contribution of the locus to the population mean is (equation (3.3)) $M = aq + 2dpq = 0.44$ bristles. The average effect of allelic substitution (equation (3.6)) is $\alpha = ah - 2dq = 1.2$, and the average effects of the two alleles (equations (3.4) and (3.5)) are $\alpha_1 = -q\alpha = -0.24$ for allele A and $\alpha_2 = p\alpha = 0.96$ for allele a.

(b) The contributions of the locus to the additive and dominance variances (equations (3.9) and (3.10)) are $V_A = 2\alpha^2 pq = 0.46$ bristles^2, and $V_D = (2dpq)^2 = 0.03$ bristles^2. The genotypic variance is, therefore, $V_G = V_A + V_D = 0.49$ bristles^2.

Problem 3.2

(a) Since the mean height of the population is 170 cm, the phenotypic value of the man, deviated from the mean, is $P = 185 - 170 = 15$ cm. The value of $H^2 = 0.8$ indicates that 80% of the phenotypic value is genetic, and 20% environmental. Therefore, the expected genotypic value of the man is $E[G] = P \times H^2 = 15 \times 0.8 = 12$ cm, and the environmental deviation in the environment in which he grew is $E = 3$ cm. Since the environmental deviations are values deviated from zero, in the average environment $E = 0$, so that we would expect that the height of the man in this environment would have been $P = 170 + 12 = 182$ cm.

(b) To deduce the genotypic value of the child we must obtain an estimate of the additive value of his parents. The value of $h^2 = 0.6$ indicates that 60% of the phenotypic value is genetic additive. Therefore, the expected additive value of the man is $[A] = P \times h^2 = 15 \times 0.6 = 9$ cm. The phenotypic value of the woman is $P = 165 - 170 = -5$ cm, then her additive value is $E[A] = P \times h^2 = -5 \times 0.6 = -3$ cm. The expected genotypic value in the child is then $E[G_{child}] = [A_{father} + A_{mother}]/2 = (9-3)/2 = 3$ cm. If this grows in his father's environment (with $E = 3$), the expected phenotypic value, deviated from the population mean, is $E[P_{child}] = 3 + 3 = 6$, so that his expected height will be $170 + 6 = 176$ cm.

Problem 3.3

The expressions necessary to solve this problem can be found in Table 3.6. For locus A, the genotypic value of the homozygote ($a = 3$), the dominance coefficient ($h = 1/3$), and the allele frequencies of the alleles A ($p = 0.8$) and a ($q = 0.2$) were indicated in Problem 3.1. For the new locus B the corresponding values are $a' = 5$, $h' = 0.4$, $r = 0.9$ and $s = 0.1$, respectively.

(a) Since the genotypic value of the double homozygote *aabb* is four times greater than its expected additive value (32 versus 8), the synergistic epistatic factor is $k = 4$, and the excess of the double homozygote value with respect to its additive value is $c = (k-1)(a + a') = 24$. The dominance effects of the two genes are the same in this case, $d = ah - a/2 = d' = a'h' - a'/2 = -0.5$. The dominance epistatic effects are then $d_e = d - (c/2)s^2 = -0.62$ and $d_e' = d' - (c/2)q^2 = -0.98$, and the average effects of an allele substitution are $\alpha_e = ah - 2d_eq = 1.248$ and $\alpha_e' = a'h' - 2d_e's = 2.196$, respectively.

(b) The effects of epistatic interaction are $i(\alpha_e \times \alpha_e') = cqs = 0.48$, $i(\alpha_e \times d_e') = (c/2)q = 2.40$, $i(d_e \times \alpha_e') = (c/2)s = 1.20$, and $i(d_e \times d_e') = (c/4) = 6.0$.

(c) The contributions of the two loci to the additive, dominance and epistatic variances of the character are, respectively, $V_A = 2\alpha_e^2 pq + 2\alpha_e'^2 rs = 1.37$, $V_D = (2d_e pq)^2 + (2d_e'rs)^2 = 0.07$, and $V_I = V_{AA} + V_{AD} + V_{DD} = c^2 q^2 (1-q^2) s^2 (1-s^2) = 0.22$. In this case, the epistatic variance is of greater magnitude than the dominance variance.

Problem 3.4

As the pea is an autogamous species it is expected that the individuals of the same variety are homozygous for all loci of the genome and identical to each other, for example of genotype *AAbbccDDEEff* ... The individuals of another variety can be distinguished in a series of loci with respect to the first being, for example, of genotype *aaBBccddEEFF* ... When crossing two individuals of the two varieties, it is expected, therefore, that the F_1 is heterozygous for all loci in which the parental varieties differ and, again, all individuals of that generation are of identical genotype, in this case of genotype *AaBbccDdEEFf* ... As there is no genetic variation among F_1 individuals, the only cause of phenotypic variation for the character is the environmental variation, so the estimate of the phenotypic variance in this generation is an estimate of the environmental one, that is, $V_E = 0.12$ cm^2. When the F_1 is selfed to obtain the F_2, the heterozygous genes have Mendelian segregation, so that the F_2 will be genetically variable for such loci. Therefore, the estimate of the phenotypic variance in the F_2 will be composed of a component of environmental variation and another of genetic variation, being an estimate of the total phenotypic variance, $V_P = 0.25$ cm^2. Assuming that the environmental variance is the same in the two generations we can obtain an estimate of the genotypic variance as $V_G = V_P - V_E = 0.25 - 0.12 = 0.13 \text{ cm}^2$. Therefore, the broad-sense heritability or degree of genetic determination (equation (3.15)) will be $H^2 = V_G/V_P = 0.13/0.25 = 0.52$. The estimate includes only the genetic variation due to the loci in which the two varieties differ, and these are only a sample of all possible varieties of pea, so the estimate obtained has its limitations.

Problem 3.5

(a) The fact of obtaining two measures of the trait per individual allows us to differentiate the causes of environmental variation that affect the individual as a whole from those that differentially affect their different parts. The variance between measures of different individuals includes the genotypic variance (V_G) and the general environmental variance (V_{EG}), that is, $\sigma_B^2 = V_G + V_{EG} = 145$, while the variance between measurements of the same individual includes only the special environmental variance, $\sigma_W^2 = V_{ES} = 64$. The total phenotypic variance is $V_P = V_G + V_{EG} + V_{ES} = 209 \text{ seeds}^2$, and the repeatability is obtained by equation (3.23) as $r = (V_G + V_{EG})/V_P = 0.69$. Assuming that the two means taken in each individual have the same genetic base, this estimate is an upper limit of H^2, the degree of genetic determination.

(b) If measurements of four fruits per individual were obtained instead of two, the variance between measures within individuals would be expected to be reduced by a half, $\sigma_W^2 = (64 \times 2)/4 = 32$, so that the phenotypic variance would be reduced to $V_P = 177$, and the repeatability would increase to a value of $r = 0.82$.

Self-Assessment Questions

1-T. 2-T. 3-T. 4-T. 5-F. 6-F. 7-F. 8-T. 9-F. 10-F.

Chapter 4

Problem 4.1

In the initial generation, the expected heterozygosity at the locus is $H_0 = 2p_0q_0 = 0.42$. Since these are ideal populations, the inbreeding coefficient is given by equation (4.13), $F_t = 1 - (1 - 1/2N)^t$. Substituting $N = 20$, the expected inbreeding coefficients in generations 5, 20 and 100 are $F_5 = 0.119$, $F_{20} = 0.397$ and $F_{100} = 0.920$. The corresponding expected heterozygosities, using equation (4.16) or (4.19), are $H_5 = 0.370$, $H_{20} = 0.253$ and $H_{100} = 0.033$. Finally, the expected values of the variance of allele frequencies between populations (equation (4.14)) are $\sigma_{q_5}^2 = 0.025$, $\sigma_{q_{20}}^2 = 0.083$ and $\sigma_{q_{100}}^2 = 0.193$.

Problem 4.2

Since the species carries out autogamy with a proportion of 50%, the expected value of the equilibrium inbreeding coefficient in the population is, using equation (4.22), $\alpha = \beta/(2 - \beta) = 0.5/(2 - 0.5) = 0.333$. In the greenhouse, the line with census size $N = 20$ is maintained approximately in ideal conditions, so that a coefficient of inbreeding would be reached relative to the generation of its foundation which can be obtained by equation (4.13), $F_{10} = 1 - (1 - 1/40)^{10} = 0.224$. The total inbreeding including that of the species and that reached in the line due to its reduced census size is obtained by combining both inbreeding values by means of equation (4.36), that is, $F_{total} = 1 - [(1 - 0.333) \times (1 - 0.224)] = 0.482$.

Problem 4.3

The purged inbreeding coefficient at generation t (g_t) is obtained by means of equation (4.26), where F_t is the inbreeding coefficient without purging, defined by equations (4.11) or (4.13). Therefore, using equations (4.11) and (4.26) on a recurring basis we obtain the values of $F_{20} = 0.397$, as obtained in Problem 4.1, and $g_{20} = 0.147$. The expected heterozygosity, using equation (4.16), is $H_{20} = 0.358$, higher than the value without purging (0.253), and the expected value of the variance of allele frequencies between populations (equation (4.14)) is $\sigma^2_{q_{20}} = 0.031$, lower than that obtained without purging (0.083).

Problem 4.4

(a) In a finite island model, the equilibrium value of F_{ST} is given by equation (4.39). Substituting the known data in the equation gives a value of $F_{ST} = 0.168$. If mutation is ignored ($u = 0$), the result is the same up to the third decimal place. The differentiation index D, on the other hand, is predicted by the approximation (4.42) and the value obtained is 0.001, that is, D predicts much less differentiation than F_{ST}.

(b) Assuming a marker with a high mutation rate ($u = 10^{-3}$), the expected value of F_{ST} is slightly reduced with respect to that of section (a), being 0.157. However, the D index is 0.474, showing that this gives great relevance to the allelic diversity between sub-populations generated by mutation.

Problem 4.5

(a) The value of F_{ST} can be obtained by equation (4.33) approximately, since the number of sub-populations (tanks, $n = 10$) is sufficiently high. With it we obtain $F_{ST} = (0.32 - 0.21)/0.32 = 0.344$. To scale this estimate to its maximum value given the allele frequencies, equation (4.40) is used, which results in an estimate of $F_{ST}' = 0.445$. The estimate of D (equation (4.41)) is 0.155.

(b) In the case of microsatellites, the value of $F_{ST} = (0.89 - 0.83)/0.89 = 0.067$ is much lower than that for the SNPs. Given the high heterozygosity of microsatellites, this estimate is expected to be an underestimation. In fact, when the maximum value is scaled (equation (4.40)), an estimate of $F_{ST}' = 0.433$ is obtained, which presents a good concordance with the estimate of the SNPs. The D index in this case gives a value of 0.392, again indicating the importance of allelic diversity for this type of marker.

Self-Assessment Questions

1-T. 2-F. 3-F. 4-T. 5-T. 6-F. 7-T. 8-T. 9-T. 10-T.

Chapter 5

Problem 5.1

When the number of reproducers of each sex differs, the effective population size can be predicted by equation (5.3). In this case, with $N_m = 4$ males and $N_f = 16$ females, we have $N_e = (4 \times 4 \times 16)/(4 + 16) = 12.8$. Using, as in Problem 4.1, expressions (4.13), (4.14) and (4.16) substituting N for N_e, the expected inbreeding coefficient in generations 5, 20 and 100 are $F_5 = 0.181$, $F_{20} = 0.549$ and $F_{100} = 0.981$, considerably higher than those obtained in the ideal population scenario. The expected heterozygosities are $H_5 = 0.344$, $H_{20} = 0.189$ and $H_{100} = 0.008$, and the expected values of the variance of allele frequencies between populations $\sigma_{q_5}^2 = 0.038$, $\sigma_{q_{20}}^2 = 0.115$ and $\sigma_{q_{100}}^2 = 0.206$.

Problem 5.2

The inbreeding of the species is the same as that calculated in Problem 4.2, that is, the asymptotic value obtained by equation (4.22), $\alpha = \beta/(2 - \beta) = 0.5/(2 - 0.5) = 0.333$. As the line with census size $N = 20$ is now maintained with the natural reproduction of the species (50% selfing), the effective size of the line is given by equation (5.9). If we assume that the number of descendants by self-fertilization and cross-fertilization are distributed following independent Poisson distributions, the variance of parental contributions to offspring is $S_k^2 = 2 + 2\beta = 2 + (2 \times 0.5) = 3$. Substituting into equation (5.9) we obtain $N_e = 15$. The inbreeding coefficient reached in the line relative to the generation of its foundation is obtained, again, by means of equation (4.13), but replacing N by N_e, that is, $F_{10} = 1 - (1 - 1/30)^{10} = 0.288$. The total inbreeding, combining the inbreeding values by means of equation (4.36) will then be $F_{\text{total}} = 1 - [(1 - 0.333) \times (1 - 0.288)] = 0.525$.

Problem 5.3

With $N_m = 16$ males and $N_f = 32$ females, the effective size of autosomal loci is given by equation (5.3), $N_e = (4 \times 16 \times 32)/(16 + 32) = 42.67$. In the case of loci linked to the X chromosome it would be obtained by means of the expression in Table 5.2, that is, $N_e = 9N_m N_f/(4N_m + 2N_f) = (9 \times 16 \times 32)/[(4 \times 16) + (2 \times 32)] = 36$. In the case of loci in the Y chromosome, the scenario can be considered equal to that of a haploid species, where N is the number of males. Therefore, using the equation in Table 5.2, with a Poisson distribution of the number of descendants, the effective size is $N_e = N/2 = 8/2 = 4$.

Problem 5.4

The necessary expression to estimate the effective size in this situation is (5.13), where the cumulative term of selection (Q) is given by equation (5.14). With a deleterious mutation genomic rate of $U = 0.2$ and a mean effect of mutations of $s = 0.1$, the coefficient of genetic variation for fitness, that is, the genetic variance for fitness scaled to a mean of one, is $C^2 = Us = 0.02$ (see Section 5.3.2). Since the species self-fertilizes with frequency $\beta = 0.5$ we have, as in Problem 5.2, $\alpha = 0.333$. To obtain the value of Q we will assume, to simplify, that selection does not reduce the genetic variance, that is, $G = 1$ (the magnitude of this reduction will be studied in Chapter 9). On the other hand, the correlation between the selective advantages of parents (r) can be approximated by β in the case of selfing (see Section 5.3.1). Therefore, substituting in equation (5.14) we have $Q = 2/[2 - (1 + r)] = 2/(2 - 1 - 0.5) = 4$. As in Problem 5.2, the variance of the contributions of parents to offspring in the absence of selection is $S_k^2 = 2 + 2\beta = 3$. Substituting $N = 20$, $\alpha = 0.333$, $S_k^2 = 3$, $Q = 4$ and $C^2 = 0.02$ in equation (5.13) we obtain $N_e = 11.36$.

Problem 5.5

(a) The prediction of the effective population size can be carried out by means of equation (5.6), for which it is necessary to calculate the variance of the contributions of the couples to progeny. The sum of the number of descendants of the $n = 8$ pairs is $\Sigma k_i = 16$ and the sum of squares $\Sigma k_i^2 = 56$, with which the variance is $S_k^2 = [\Sigma k_i^2 - (\Sigma k_i)^2/n]/(n - 1) = 3.429$. Substituting this value along with $N = 16$ in equation (5.6) we obtain $N_e = 11.79$.

(b) If each pair contributes a male and a female to the offspring, $S_k^2 = 0$ and $N_e = 32$. If the contributions of the couples were of individuals of the same sex, maintaining an equal total number of descendants of each sex, the result would be the same, that is, $N_e = 32$.

Self-Assessment Questions

1-T. 2-T. 3-F. 4-F. 5-T. 6-T. 7-F. 8-T. 9-F. 10-F.

Chapter 6

Problem 6.1

Using equation (6.5) with $j = 1$ offspring per family, we have $\sigma_b^2 = k/n$. If only one parent is used ($k = 1$), $\sigma_b^2 = 1/n$ and $SE(h^2) = 2/\sqrt{n}$. To obtain a value of $SE(h^2) = 0.05$ then $n = 1600$ pairs are needed. In the case of using the two parents ($k = 2$), $\sigma_b^2 = 2/n$, $SE(h^2) = \sqrt{2/n}$ and the minimum number of pairs needed would be $n = 800$, half that in the case of a single parent.

Problem 6.2

(a) The estimate of heritability based on the correlation between groups of full sibs can be obtained with equation (6.7). If dominance and the environmental variation common to sibs are ignored, $h^2 = 2t_{FS} = 0.24$.

(b) If parents have been paired with assortative mating for the trait with correlation $\rho = 0.5$, the correction indicated in Section 6.1.3 should be applied. In the case of full-sib groups, the correlation in this scenario is $t_{FS} = 1/2\ h^2(1 + \rho h^2)$. Substituting the observed value of $t_{FS} = 0.12$ and that of ρ and clearing, we obtain an estimate of $h^2 = 0.217$, slightly lower than the previous one.

(c) If $\rho = 1$, the heritability value would be $h^2 = 0.20$.

Problem 6.3

Using the expressions in Table 6.3 we know that the intraclass correlation of monozygotic twins, $t_{MZ} = \sigma_B^2/\sigma_T^2 = 0.695$, is an estimate of $h^2 + d^2 + c^2$, and that of dizygotic twins, $t_{DZ} = \sigma_B^2/\sigma_T^2 = 0.439$, is an estimate of $1/2\,h^2 + 1/4\,d^2 + c^2$. Therefore, $2t_{DZ} - t_{MZ} = 0.182$ is an estimate of $c^2 - \frac{1}{2}\,d^2 \approx c^2$, the proportion of the phenotypic variance for the character due to the environmental variance common to sibs, and $2(t_{MZ} - t_{DZ}) = 0.513$ is an estimate of $h^2 + (3/2)d^2 \approx H^2$, the degree of genetic determination of the character or the proportion of the phenotypic variance due to hereditary factors.

Problem 6.4

It is possible to obtain an estimate of the heritability by means of equation (6.36). The phenotypic values of the individuals are $P_1 = 16.6$, $P_2 = 22.1$, $P_3 = 18$ and $P_4 = 12.4$, with phenotypic mean $\overline{P} = 17.275$ and variance $\sigma_P^2 = 16$. Calculating the similarities between the phenotypic values as the products of their squared deviations to the mean, standardized by the variance, $Z_{ij} = [(P_i - \overline{P})(P_j - \overline{P})]/\sigma_P^2$, we obtain the following values, which can be correlated with the corresponding molecular coancestries:

Pairs of individuals		Z_{ij}	$f_{M,ij}$
A	A	0.02846	1
A	B	−0.20344	0.25
A	C	−0.03057	0.5
A	D	0.205546	0.5
B	A	−0.20344	0.25
B	B	1.454206	1
B	C	0.218508	0.75
B	D	−1.46928	0.5
C	A	−0.03057	0.5
C	B	0.218508	0.75
C	C	0.032833	1.25
C	D	−0.22077	0.375
D	A	0.205546	0.5
D	B	−1.46928	0.5
D	C	−0.22077	0.375
D	D	1.484501	1

The covariance between the Z and f_M values is $\text{cov}(Z, f_M) = 0.1109$ and the variance of the coancestries, $\text{var}(f_M) = 0.08958$. Using expression (6.36), the estimate of heritability would be $h^2 = \text{cov}(Z, f_M)/[2\text{var}(f_M)] = 0.62$.

Problem 6.5

The standard error of heritability estimates using genomic relationships is given by the equation in Section 6.3.6, $\text{SE}[h^2] = \sqrt{2/[n \times \text{var}(r_M)]}$, where $\text{var}(r_M)$ is the variance of the additive relationships, which can be approximated by $\text{var}(r_M) = 1/(16L) - 1/(3L^2)$ where L is the length of the genome in morgans. For the human genome $L = 35$ and $\text{var}(r_M) = 0.0015$. Substituting in the expression of the expected standard error and assuming that this is 0.1, we obtain $n = 133{,}333$ pairs, which indicates that the method requires a very high number of pairs to obtain accurate estimates.

Self-Assessment Questions

1-F. 2-T. 3-F. 4-T. 5-F. 6-F. 7-T. 8-T. 9-T. 10-T.

Chapter 7

Problem 7.1

(a) The probability of fixation of an additive beneficial mutation is estimated by equation (7.1), $P_f = [1 - \exp(-N_e s/N)]/[1 - \exp(-2N_e s)] = [1 - \exp(-50 \times 0.1/100)]/[1 - \exp(-2 \times 50 \times 0.1)] = 0.049$.

(b) If the mutation were recessive, the probability of fixation can be approximated by the equation in Table 7.1, $P_f = \sqrt{2N_e s/\pi}/N = \sqrt{2 \times 50 \times 0.1/3.1416}/100 = 0.018$.

(c) If the mutation is deleterious and additive, since $N_e s = -5 \ll -1$, its fixation probability is $P_f \approx 0$. The average persistence time of the mutation in the population is $2[\ln(-1/sh) + 1 - 0.5772] = 6.84$ generations and the number of copies $-1/sh = 20$.

Problem 7.2

Since the mean viability of the lines was reduced from 0.8 to 0.35 in 100 generations, the average decline per generation is $\Delta M = (0.8 - 0.35)/100 = 0.0045$. On the other hand, the variance between the mean viabilities of the lines increased in the same period from 0.05 to 0.14, with which $\Delta V = (0.14 - 0.05)/100 = 0.0009$. If the distribution of effects is assumed to be exponential, the coefficient of variation of mutational effects is $C^2 = 1/\beta = 1$. Using equations (7.2) and (7.3) we then obtain the following estimates: $U = 2\Delta M^2/\Delta V = 0.045$, $\bar{s} = \Delta V/2\Delta M = 0.1$ and $V_M = \Delta V/2 = 0.00045$.

Problem 7.3

The variance between lines for the trait can be approximated by $V_{B,t} \approx 2tV_M$. Therefore, in 100 generations it will be $V_{B,100} \approx 2 \times 100 \times 0.0005 = 0.1$. At the end of that period, the additive variance is expected to be the mutation–drift equilibrium one, that is, $V_A \approx 2N_e V_M = 2 \times 20 \times 0.0005 = 0.02$. More precisely, the additive variance can be obtained by the recursive equation $V_{A(t)} = V_{A(t-1)}[1 - 1/(2N_e)] + V_M$, which at generation $t = 100$ provides an estimate of $V_A \approx 0.0184$.

Problem 7.4

At the deleterious mutation–selection balance the expected frequency of a recessive deleterious allele ($h = 0$) is given by equation (7.9). In the case of a lethal ($s = 1$), $q = \sqrt{u} = 0.00161$, which is an order of magnitude greater than the observed frequency (0.00014). Therefore, the estimates are not compatible with full recessivity of the lethal allele. If the allele is not recessive, the equilibrium frequency is given by equation (7.10) and, clearing, the most probable value of the dominance coefficient is $h = u/q = 0.0186$.

Problem 7.5

If we assume that the population is at the deleterious mutation–selection equilibrium, the expressions in Table 7.2 can be used to obtain estimates of the mutational parameters. We have the fitness data of panmictic (P) and selfed (S) individuals, as well as the values of the composite parameter $T = 4S - 2P$, from which we obtain the means, variances and covariances as indicated in the following table.

							Means	(Co)variances
P	0.78	0.75	0.81	0.79	0.83	0.85	$W_P = 0.802$	$V_{WP} = 0.00130$
S	0.63	0.61	0.70	0.69	0.72	0.74	$W_S = 0.682$	$V_{WS} = 0.00262$
T	0.96	0.94	1.18	1.18	1.22	1.26	$W_T = 1.123$	$V_{WT} = 0.01895$
								$\mathrm{cov}_{WP,T} = 0.00369$

Using the expressions in Table 7.2, it is possible to obtain estimates of the following mutational parameters: the harmonic and arithmetic means of the dominance coefficients of mutations weighted by their selection coefficient:

$$\hat{h}_b = \frac{\mathrm{cov}_{WP,T}}{V_{WT}} = \frac{0.00369}{0.01895} = 0.195 \text{ and } \hat{h}_{1/b} = \frac{V_{WP}}{\mathrm{cov}_{WP,T}} = \frac{0.00130}{0.00369} = 0.351, \text{ respectively;}$$

the inbreeding depression, $\delta \approx W_S - W_P = 0.682 - 0.802 = -0.12$;

the deleterious haploid genomic mutation rate, $\hat{U} \approx \frac{2\hat{h}_b \delta}{2\hat{h}_b - 1} = \frac{2 \times 0.195 \times (-0.12)}{(2 \times 0.195) - 1} = 0.077$;

and the mean selection coefficient of mutations, $\hat{s} \approx \frac{V_{WP}}{2\hat{U}\hat{h}_{1/b}} = \frac{0.00130}{2 \times 0.077 \times 0.351} = 0.024$.

Self-Assessment Questions

1-F. 2-T. 3-F. 4-T. 5-T. 6-F. 7-T. 8-F. 9-T. 10-F.

Chapter 8

Problem 8.1

(a) The expressed load in the large size panmictic population is given by equation (8.4), $A = -\ln(W_0) = -\ln(0.8) = 0.22$, which can also include environmental factors of decline in mean fitness. The load masked in heterozygosis that is revealed by inbreeding (inbreeding load) is $B = -\ln(W_F/W_0)/F = -\ln(0.5/0.8)/0.12 = 3.92$ lethal equivalents.
(b) The mean fitness of the small size population when a coefficient of inbreeding of 0.2 is reached is given by equation (8.5), $W_F = W_0 e^{(-BF)} = 0.8 \times e^{(-3.92 \times 0.2)} = 0.365$.

Problem 8.2

(a) In a panmictic population of large census size, the inbreeding load or inbreeding depression rate can be approximated by expression (8.8). Since we have an estimate for the harmonic mean of the dominance coefficients of 0.1, this implies $\overline{1/h} = 10$. Therefore, $B \approx U(\overline{1/h} - 2) = 0.2 \times (10 - 2) = 1.6$ lethal equivalents, which is also the rate of inbreeding depression.

(b) Since the mean fitness of the panmictic population is 0.62, we have $A = -\ln(W_0) = -\ln(0.62) = 0.478$. At mutation–selection equilibrium we know that the value $A/[2(A + B)]$ is an upper estimate of the harmonic mean dominance of mutations. Substituting, $0.478/[2 \times (0.478 + 1.6)] = 0.115$, which is compatible with the available value of the harmonic mean of h (0.1).

Problem 8.3

Since the lines are maintained with $N_m = 6$ males and $N_f = 24$ females each generation, the effective size can be approximated by equation (5.3), $N_e = 4N_mN_f/(N_m + N_f) = (4 \times 6 \times 24)/(6 + 24) = 19.2$. The expected inbreeding coefficient at generation $t = 10$ is given by equation (4.13) by replacing N with N_e, that is, $F_t = 1 - [1 - 1/(2N_e)]^t = 1 - [1 - 1/(38.4)]^{10} = 0.232$. The expected values of the genetic variance within lines (V_{GW}; equation (8.9)) and between lines (V_{GB}; equation (8.10)) at generation 10 would be $V_{GW} = V_{G0}(1 - F_t) = 3.5 (1 - 0.232) = 2.688$, and $V_{GB} = 2F_tV_{G0} = 2 \times 0.232 \times 3.5 = 1.623$.

Problem 8.4

(a) The observed components of variance are $\sigma_w^2 = 0.43$ within families, $\sigma_f^2 = 0.11$ between families and $\sigma_{GB}^2 = 0.45$ between localities. The heritability, ignoring dominance and common environment to the members of the families, can be estimated as $h^2 = 2\sigma_f^2/(\sigma_f^2 + \sigma_w^2) = (2 \times 0.11)/(0.11 + 0.43) = 0.407$. The genetic variance within localities is $\sigma_{GW}^2 = 2\sigma_f^2 = 0.22$. The estimate of Q_{ST} is obtained with equation (8.12), $Q_{ST} = \sigma_{GB}^2/(\sigma_{GB}^2 + 2\sigma_{GW}^2) = 0.45/[0.45 + (2 \times 0.22)] = 0.506$.

(b) The value of Q_{ST} (0.506) is considerably higher than that of F_{ST} (0.023) corresponding to neutral markers, which suggests diversifying selection between localities for the trait.

Problem 8.5

(a) In Problems 4.1 and 4.3 of Chapter 4, the standard (F_t) and purged (g_t) inbreeding coefficients were calculated for the case of a line of census size $N = 20$ maintained for $t = 20$ generations. The first is given by equation (4.13), $F_t = 1 - (1 - 1/2N)^t = 1 - (1 - 1/40)^{20} = 0.397$, and the second must be calculated with the recurrent equation (4.26), being $g_{20} = 0.147$. The mean fitness and the initial inbreeding load are $W_0 = 0.86$ and $B_0 = 2.4$, respectively. The corresponding values in the line after 20 generations are calculated by means of equations (8.14) and (8.15), that is, $B_{20} = B_0g_{20}(1 - F_{20})/F_{20} = 2.4 \times 0.147 \times (1 - 0.397)/0.397 = 0.536$, and $W_{20} = W_0\exp(-B_0g_{20}) = 0.86 \times \exp(-2.4 \times 0.147) = 0.604$.

(b) If there had been no purging, the corresponding values would be $B_{20} = B_0(1 - F_{20}) = 2.4 \times (1 - 0.397) = 1.447$, considerably higher than that for the purging scenario, and $W_{20} = W_0\exp(-B_0F_{20}) = 0.86 \times \exp(-2.4 \times 0.397) = 0.332$, which is approximately half that in the case where purging takes place.

Self-Assessment Questions

1-F. 2-T. 3-F. 4-T. 5-T. 6-F. 7-T. 8-F. 9-T. 10-T.

Chapter 9

Problem 9.1

The mean laying in the population is $\overline{X} = [(11 \times 2) + (12 \times 3) + \ldots + (23 \times 1)]/85 = 17.118$ eggs. If 20% of the females with the largest laying are selected, that is, $85 \times 0.2 = 17$ females at the end of the distribution, the mean laying of this group is $\overline{X}_S = [(20 \times 9) + (21 \times 5) + (22 \times 2) + (23 \times 1)]/17 = 20.706$ eggs. Therefore, the selection differential would be $S = \overline{X}_S - \overline{X} = 3.588$. Using equation (9.1) the expected response in the next generation is $R = h^2 S = 0.2 \times 3.588 = 0.718$ and, therefore, the expected mean in the next generation will be $17.118 + 0.718 = 17.835$ eggs.

Problem 9.2

Since 60 individuals are selected out of 300 in each sex, the selected proportion is $\phi = 0.2$, which corresponds to an approximate selection intensity of $i = 1.4$. The expected response in a generation is given by expression (9.2), $R = ih^2\sigma_P = 1.4 \times 0.3 \times \sqrt{24} = 2.06$. Assuming that this is linear during the initial five generations, the cumulative response would simply be $t \times R = 5 \times 2.06 = 10.3$.

Problem 9.3

To estimate the effective population size, we can use equation (9.9) where $Q = 2/(2 - G)$ is the term that represents the cumulative effect of selection, $G = 1 - kh^2$ is the genetic variance remaining after selection, where k is equal to 0.781 for the selection intensity applied, and $C^2 = i^2(h^2/2)$ is the variance of the selective advantages of individuals. Substituting, we have, $G = 1 - (0.781 \times 0.3) = 0.766$; $Q = 2/(2 - 0.766) = 1.62$; $C^2 = (1.4)^2 \times (0.3/2) = 0.294$; and the effective population size is $N_e = N/(1 + Q^2 C^2) = 120/[1 + (1.62^2 \times 0.294)] = 67.72$.

   The expected standard error of the cumulative response due to genetic drift and sampling can be approximated by equation (9.5), $E[V(R)] = V_P[(th^2/N_e) + (1/n)]$ where $n = 10$ is the number of individuals evaluated per family. Thus, substituting, $E[V(R)] = 24 \times [(5 \times 0.3/67.72) + (1/10)] = 2.93$, and the standard error of the response is $\sqrt{2.93} = 1.71$.

Problem 9.4

The long-term response assuming the infinitesimal model and considering the contribution of mutation is given by expression (9.16), $R_T = R_t + R_{M,t} = (2N_e i/\sigma_P) [\sigma_{A,0}^2 F_t + \sigma_M^2 (t - 2N_e F_t)]$. The expected inbreeding coefficient in generation 100 is $F_{100} = 1 - (1 - 1/2N_e)^{100} = 1 - [1 - 1/(2 \times 67.72)]^{100} = 0.523$ and the initial additive variance $\sigma_{A,0}^2 = V_P h^2 = 24 \times 0.3 = 7.2$. Substituting the available values in expression (9.16) we obtain $R_T = (2 \times 67.72 \times 1.4/\sqrt{24}) \times [(7.2 \times 0.523) + 0.002 \times (100 - 2 \times 67.72 \times 0.523)] = 148.12$. If mutation is ignored, $R_t = (2 \times 67.72 \times 1.4/\sqrt{24}) \times (7.2 \times 0.523) = 145.86$.

Problem 9.5

The response to one generation of selection is given, in general, by equation (9.17) with r_{CA} being the accuracy of selection. In the case of individual selection $r_{CA} = h$, the square root of the heritability, while with within-family and family selection r_{CA} is given by expressions (9.18) and (9.19), respectively. The necessary data are available from the above problems: heritability of the trait $h^2 = 0.3$, additive genetic variance $\sigma_A^2 = 7.2$, intraclass correlation of full sibs $t = h^2/2 = 0.15$, genetic correlation among full sibs $r = 0.5$, and number of individuals per family and sex $n = 5$. The accuracy of within-family selection is then $r_{P_WA} = h(1-r)/\sqrt{1-t} = \sqrt{0.3} \times (1-0.5)/\sqrt{1-0.15} = 0.297$, and that of family selection, $r_{P_FA} = h[1+(n-1)r]/\sqrt{n[1+(n-1)t]} = \sqrt{0.3} \times [1+(5-1)\times0.5]/\sqrt{5[1+(5-1)\times0.15]} = 0.581$. The responses to one generation of selection are: individual selection ($R = ih\sigma_A = 1.4 \times \sqrt{0.3} \times \sqrt{7.2} = 2.06$), within-family selection ($R = ir_{P_WA}\sigma_A = 1.4 \times 0.297 \times \sqrt{7.2} = 1.11$), and family selection ($R = ir_{P_FA}\sigma_A = 1.4 \times 0.581 \times \sqrt{7.2} = 2.18$).

Self-Assessment Questions

1-F. 2-F. 3-T. 4-T. 5-T. 6-F. 7-T. 8-T. 9-F. 10.T.

Chapter 10

Problem 10.1

(a) Applying equation (10.3) we can predict the response correlated to natural selection. For trait X it is: $R_{CX} = b_{P(W.X)}\sigma_{A,X}^2 + b_{P(W.Y)}\text{cov}_A(X,Y)$. Since the genetic correlation is $r_A = 0.41$, the genetic covariance is $\text{cov}_A(X,Y) = r_A\,\sigma_{A,X}\sigma_{A,Y} = 10$. Therefore, $R_{CX} = (0.8 \times 20) + (0.1 \times 10) = 17$. Analogously, for trait Y, $R_{CY} = b_{P(W.Y)}\sigma_{A,Y}^2 + b_{P(W.X)}\text{cov}_A(X,Y) = (0.1 \times 30) + (0.8 \times 10) = 11$. Trait X will have a greater correlated response given its greater positive correlation with fitness.

(b) If the genetic correlation between the two traits were negative, $\text{cov}_A(X,Y) = -10$, the responses would be $R_{CX} = (0.8 \times 20) + (0.1 \times -10) = 15$ and $R_{CY} = (0.1 \times 30) + (0.8 \times -10) = -5$, implying a negative correlated response for trait Y.

Problem 10.2

The phenotypic variance after stabilizing selection can be obtained by means of equation (10.9), that is, $V_{P*} = V_P V_s/(V_P + V_s) = (30 \times 20)/(30 + 20) = 12$. So that the proportional reduction in variance is (equation (10.10)), $\Delta V_P = -0.6$, or 60%. The decline in fitness for each unit of increment in the square of the phenotypic values is obtained by equation (10.11), $b_{w.y^2} = -1/(2V_s) = -1/40 = -0.025$, that is, 2.5%.

Self-Assessment Questions

1-F. 2-F. 3-F. 4-T. 5-F. 6-T. 7-T. 8-T. 9-F. 10-F.

Chapter 11

Problem 11.1

The number of polymorphic nucleotides is $S = 16$ (the polymorphism would be $P = 16/500 = 0.032$). To obtain the nucleotide diversity (θ_π) we calculate the average number of differences between all pairs formed by the n sequences and this is divided by the total number of comparisons. The number of comparisons per position is $n(n - 1)/2 = 5(5 - 1)/2 = 10$. Positions 1, 2, 9, 10, 14 and 16 imply six differences, positions 3, 5, 6, 7, 8, 11, 12, 13 and 15 imply four differences and position 4 implies seven differences. Therefore, the average nucleotide diversity is $\theta_\pi = [(6 \times 6) + (4 \times 9) + (7 \times 1)]/10 = 7.9$. The nucleotide diversity per position would be $\pi = 7.9/500 = 0.0158$.

Under the neutral model it is predicted that $E[\theta_\pi] = 4N_e u$ and also $E[\theta_w] = S/a = 4N_e u$, where $a = 1 + 1/2 + 1/3 + 1/4 = 2.083$. In this case, $\theta_w = 16/2.083 = 7.68$, which is approximately equal to θ_π (7.9), suggesting sequence neutrality. A relation $\theta_\pi < \theta_w$ indicates a recent selective sweep and $\theta_\pi > \theta_w$ balancing selection (for example, overdominance).

Problem 11.2

The observed data can be presented as a table and the corresponding expected values obtained assuming independence between the two combinations of position classes:

Observed	Fixed	Polymorphic	Total
Synonymous	71	106	177
Non-synonymous	23	13	36
Total	94	119	213

Expected	Fixed	Polymorphic	Total
Synonymous	$177 \times 94/213 = 78.11$	$177 \times 119/213 = 98.89$	177
Non-synonymous	$36 \times 94/213 = 15.89$	$36 \times 119/213 = 20.11$	36
Total	94	119	213

To demonstrate whether or not independence exists, a χ^2 test is carried out:

$$\sum_{i=1}^{4} \frac{(O_i - E_i)^2}{E_i} = \frac{(71 - 78.11)^2}{78.11} + \frac{(106 - 98.89)^2}{98.89} + \frac{(23 - 15.89)^2}{15.89} + \frac{(13 - 20.11)^2}{20.11}$$

$$= 6.86.$$

The value obtained is higher than the critical value with a probability of 95% for a χ^2 distribution with 1 degree of freedom (3.84), so the hypothesis of independence and, therefore, of neutrality is rejected. There is an excess of non-synonymous positions fixed in different species, which suggests diversifying selection.

Self-Assessment Questions

1-T. 2-F. 3-T. 4-T. 5-T. 6-F. 7-T. 8-T. 9-T. 10-F.

Glossary

A

Accuracy of selection (r_{CA}). It is the correlation between the selection criterion used for selection (phenotype, family mean, deviation of the value of the individual from the family mean, additive value estimated by a selection index or by BLUP, etc.) and the expected additive value of the individual. In the case of selection on the individual phenotype, the accuracy is equal to h, the square root of the heritability.

Additive genetic variance. It is the variance of the additive values of the individuals.

Additive genic variance. In a multilocus model, it is the magnitude of the additive genetic variance obtained with the sum of the additive variances contributed by each locus. Therefore, it does not consider gametic phase disequilibrium or other sources of multilocus genetic variation.

Additive value or breeding value (A). It is the value of the individual obtained as the sum of the average effects on the trait of the alleles it carries and is responsible for the resemblance between relatives. The additive value of an individual can be estimated as twice the deviation of the mean value of its progeny (with the other parent taken at random from the population) from the population mean (see 'Progeny test').

Additivity or additive gene action. It occurs when the genotypic value of an individual is equal to the sum of the effects on the trait of the alleles carried by the individual and, therefore, the genotypic value of the heterozygote is intermediate between that of the corresponding homozygotes.

Allele, gamete and genotype frequencies. The genetic composition of a population is described in terms of the relative frequencies of (1) individuals with different genotypes (genotype frequencies), (2) the different alleles of each locus (allele frequencies), and (3) gametes differing in their allelic constitution (gametic frequencies). When more than one locus is considered, the gametic frequencies become important in the analysis of the physical and genetic association between loci.

Allelic diversity. It is the average number of alleles per locus that segregate in the population.

Animal model. It is a mixed model applicable to the phenotypic data of an individual for a quantitative trait. It includes at least the population mean, the additive effect of the individual and a residual effect. It may also include other fixed and random effects ascribed to dominance, common environment, and so on.

Antagonistic pleiotropy. It is the type of pleiotropy in which the effects of a locus on two traits occur in opposite directions, that is, the alleles increasing the value of a trait diminish the value of the other trait.

Apparent stabilizing selection. It is the selection that appears for a trait that is not subjected to stabilizing selection but there are deleterious mutations of great effect, so that the extreme individuals that carry them are eliminated.

Assortative mating. It is the tendency for mating to occur between individuals of the same (positive) or different (negative) genotype or phenotype. Negative assortative mating is often called disassortative.

Automatic selective advantage hypothesis of self-fertilization. Self-fertilization implies a 50% selective advantage over allogamy, which can be compensated by inbreeding depression. If the inbreeding load of an allogamous population is equal to or lower than two lethal equivalents, a mutation allowing self-fertilization could be fixed in the population, leading to a state of complete autogamy.

Average allelic excess (e). It is the effect of an allele on a quantitative trait obtained as the mean of the character for the individuals that carry a copy of the allele in question. It coincides with the average effect of allelic substitution in the case of a panmictic population but may differ in the case of a non-panmictic one.

Average effect of an allele (a_i). It is the trait effect associated with each allele (i) of a given locus. It depends on the average effect of allelic substitution (α) and the allele frequencies. The sum of the average effects of the alleles carried by an individual constitutes the additive value or breeding value of the individual.

Average effect of allelic substitution (a). It is the common term appearing in the expression to calculate the average allelic effects for a quantitative trait, and it is obtained as the regression coefficient of the genotypic values of the individuals on the allelic dose. It coincides with the average allelic excess in the case of a panmictic population but may differ in the case of a non-panmictic one.

B

Back-mutation. It is the mutation that restores the DNA sequence modified by a previous mutation, recovering the wild genotype and phenotype. When the new mutation occurs in a different position from the first one, so that the genotype is not recovered, but the wild phenotype is recovered, it is usually referred to as a suppressor mutation.

Background, purifying or negative selection. It is the selection that produces the elimination of deleterious mutations. Its effect is a reduction in diversity, particularly in the regions of the genome with low recombination rate.

Balanced chromosomes. They are chromosomes that carry multiple inversions that prevent or reduce genetic recombination and carry visible markers for their identification. They allow useful chromosomal manipulation in various procedures: mutation accumulation experiments, obtaining lines devoid of genetic variability, estimation of the mutation rate of lethal alleles, estimation of the genetic variance and the average coefficient of dominance of mutations, and so on.

Balancing selection. It is the type of selection that maintains a polymorphism in the allele frequencies. Overdominance for fitness and frequency-dependent selection are examples of this type of selection.

Bateman–Mukai method. It is a method to obtain estimates of the deleterious genomic mutation rate and the mean effect of mutations in mutation accumulation experiments. Estimates are obtained from the mean rate of decline in fitness of the lines and the rate of increase in variance of fitness between them.

Beneficial mutation. It is the mutation that increases the fitness of the carrier individual.

BLUE and BLUP. The best linear unbiased estimation (BLUE) and prediction (BLUP) techniques are the most commonly used methods in the estimation of fixed effects and the prediction of additive values, respectively, with the animal model.

Breeder's equation. It is the equation $R = h^2 S$, due to Jay L. Lush, that predicts the response to selection (R) from the selection differential (S) and the heritability (h^2).

Broad-sense heritability or degree of genetic determination (H^2). It is the proportion of the phenotypic variance of a character due to genetic causes for a given population and under specific environmental conditions.

Bulmer effect. It is the reduction of the genetic variance between families with selection due to the fact that the selected parents are less variable than the whole population.

C

Castle–Wright method. It is a method to roughly estimate the number of loci that affect a quantitative trait from the selection in opposite directions of the cross between two highly inbred lines that differ for that trait. The estimate is obtained from the ratio of the square of the total response for divergent selection and eight times the additive variance of the trait. The estimate is for the effective number of loci, which is the number of loci that there would be if all had the same effect.

Circular mating. It is a type of mating in which individuals are numbered and each individual is crossed with the next one consecutively to contribute with a descendant, giving rise to continuous mating between half sibs. In the case of being carried out with pairs of individuals, each pair contributes with a male and a female to the offspring, and the male of each pair is crossed with the female of the next pair consecutively.

Coancestry or kinship coefficient (f) between two individuals. It is the probability that taking at random an allele of a locus from each of the individuals, these are identical by descent, that is, both are copies of an allele of an ancestor common to the parents of the individual.

Coefficient of additive relationships (r). It is the numerator of the correlation between the additive values of two individuals, and it is equal to twice the coefficient of coancestry between them. In the allocation of the covariance between relatives to causal components of variance, it is the weighting factor attached to the additive variance components.

Coefficient of dominance relationships (u). It is the probability that, at a given locus, the alleles of an individual are identical by descent to those of the other individual. In the allocation of the covariance between relatives to causal components of variance, it is the weighting factor attached to the dominance variance components.

Coefficient of inbreeding (F). It is the probability that the two alleles carried by an individual in a given locus are identical by descent, that is, both are copies of an allele of an ancestor common to the parents of the individual.

Compensatory mating. It consists of crossing individuals from large families with others from small families, with the purpose of producing negative correlations between the drift caused by selection and that caused by the finite population size.

Compensatory mutation. It is the mutation that fully or partially compensates for the deleterious effect of another occurring in a different locus.

Complementarity. It is the combination of heritable traits of commercial interest when crossing takes place between two lines or strains of plants or domestic animals.

Concordance between characters. It is a way to evaluate the phenotypic differences between monozygotic and dizygotic twins, and it is defined as the proportion of pairs of twins in which the same character is manifested.

Conditional lethal. It is a lethal allele that is expressed only in certain environmental conditions.

Convergent selection. It is the type of selection for which a trait is subject to stabilizing selection for the same optimum in two or more populations, so that the genetic differentiation between populations for the trait is expected to be lower than that corresponding to a neutral model.

Correlated response. It is the response that occurs in a character Y, which presents genetic correlation r_A with the character subjected to selection (X). The term $r_A h_X h_Y$ is called coheritability, and the correlated response is a function of it, where h_X and h_Y are the square root of the heritability for each character.

Cumulative effect of selection on the effective population size (Q). It is the factor by which the selective value of an individual is transmitted to its offspring and is halved each generation by the effect of segregation and recombination. This factor increases the variance of parental contributions to offspring, reducing the magnitude of the effective population size.

D

Deleterious mutation. It is the mutation that reduces the fitness of the individual which carries it.

Deleterious mutation–selection balance. It is the equilibrium reached by the allele frequencies as a consequence of the continuous appearance of deleterious mutations and their elimination by natural selection.

Demographic estimates of the effective population size. They are the estimates of effective size that are obtained by using the predictive formulas of this parameter when the necessary demographic data are available, such as the number of breeders of each sex or the variance of their offspring contributions.

Directional and non-directional dominance. Directional dominance occurs when for most loci that affect a trait the recessive allele is the one that increases (or reduces) the value of the trait. If this trend does not exist, we speak of non-directional dominance. For a character to have inbreeding depression, there must be directional dominance.

Directional selection. It is the type of selection that produces a directional change in the population mean and occurs for the main components of fitness or during adaptation to environmental changes. It is also the type of selection induced by artificial selection.

Discrete generations equivalent. For the calculation of the effective population size from genealogies in populations with overlapping generations, it is required to obtain an equivalent to discrete generations for each individual, which is calculated from the number of generations that separate the individual from its ancestors.

Diversifying, disruptive or divergent selection. It is the type of selection in which the most extreme individuals of the phenotypic distribution are favoured, which can produce the bimodality of the distribution. It is common in heterogeneous environments.

Dominance. It occurs when the effect of an allele masks, partially or totally, the effect of others.

Dominance and overdominance hypotheses of inbreeding depression. Inbreeding depression is generally due to the expression of deleterious recessive or partially recessive alleles in homozygosis (dominance hypothesis), but on some occasions it could also be due to overdominance alleles in homozygosis (overdominance hypothesis).

Dominance coefficient (h). It is the coefficient that modulates the effect associated with a given allele in heterozygosis by reference to its effect in homozygosis (a), this effect being ah. If $h = 0.5$, the value of the heterozygote is intermediate between that of the two homozygotes, and we talk about additive gene action or additivity. If h varies between 0 and 1 (except 0.5), there is dominance of one allele and recessiveness of the other. Values of h lower than 0 or greater than 1 indicate that the value of the heterozygote is the highest (overdominance) or the lowest (underdominance).

Dominance effect (d). It is the deviation of the genotypic value from its additive prediction indicating the existence of dominance for a locus with effect on a quantitative trait. In the case of deleterious mutations, it coincides with the purging coefficient.

Drift load. It is the reduction of the average fitness of the population due to changes in the allele frequencies that occur in populations of reduced census size by genetic drift.

E

Effective number of alleles (n$_e$). It is the number of alleles that there would be in a population if all of them were segregating at the same frequency. It can be obtained as the inverse of the sum of squares of the allele frequencies $1/\sum q_i^2$.

Effective number of loci or factors. It is an estimate of the number of loci of a quantitative trait obtained by the Castle–Wright method from the selection in opposite directions of the cross between two highly inbred lines that differ for that trait. The estimate assumes that all loci have equal effect.

Effective population size (N$_e$). It is the size of an ideal population that would give rise to the genetic drift or inbreeding observed in the real population, which leads to the so-called drift and inbreeding effective sizes. There are also other procedures for its prediction, giving rise to eigenvalue, mutation and coalescence effective sizes.

Environmental correlation (r$_E$). Strictly, it is the correlation between the environmental deviations of the phenotypic value of individuals for two quantitative traits. In practice, it includes not only environmental deviations but also non-additive genetic values.

Environmental deviation (E). It is the contribution of the environment to the phenotypic value of an individual. It is usually considered to have positive or negative sign with equal probability.

Environmental variance (V$_E$). It is the proportion of the phenotypic variance due to environmental factors. It can be decomposed into general environmental variance (V_{EG}), due to non-localized or permanent causes, and special environmental variance (V_{ES}), due to localized or temporary causes.

Epistasis or epistatic interaction. It is the lack of additivity between the effects of different loci on the trait, that is, the joint genotypic effects are not the simple sum of the effects of each locus separately.

Epistatic variance (V$_I$). It is the part of the genotypic variance that is due to the epistatic interactions between loci.

Equalization of contributions. The usual procedure followed in conservation programmes whereby couples contribute an equal number of individuals to the offspring. With this method the rates of genetic drift and inbreeding are half those which would be obtained if the parental contributions were random.

Extended haplotype homozygosity (EHH). It is one of the most commonly used methods for detecting the footprint of selection with high-density marker data. It is based on the differential decay of the linkage disequilibrium and, therefore, of the homozygosity of markers with the distance to a selected locus.

Expected heterozygosity (H). It is the expected frequency of heterozygotes with Hardy–Weinberg equilibrium and is the most common measure to quantify genetic diversity. It is also known as gene diversity.

F

Family selection. It is the type of artificial selection in which the best families are selected as reproducers of the next generation.

Fitness. It is the capacity of individuals to contribute offspring to the next generation, and it is the trait on which natural selection acts directly. It is constituted by the combination of several reproductive traits that are called main components of fitness.

Fixation index (F$_{ST}$). It is the statistic that quantifies the variation of allele frequencies between sub-populations. It can also be defined as the inbreeding coefficient of individuals of a structured population due to subdivision, or the correlation between two alleles taken at random from a sub-population relative to that of two alleles taken at random from the global population.

Founder effect. It refers to populations founded with very few individuals, with consequences in genetic drift, inbreeding and linkage disequilibrium.

Fundamental Theorem of Natural Selection. It is the theorem proposed by R. A. Fisher that indicates that the response to natural selection for fitness is equal to its additive variance. The so-called Second Theorem of Natural Selection, proposed by A. Robertson and G. R. Price, states that the correlated response to natural selection for a quantitative trait is equal to the genetic covariance between that trait and fitness.

G

Gametic or linkage disequilibrium (D). It is the state of association in allele frequencies between two or more loci, which is reduced each generation by a fraction that depends on the recombination rate between them.

Gene. A functional genomic element that generally encodes a protein or RNA.

Gene diversity. It is the expected heterozygosity or expected frequency of heterozygotes at Hardy–Weinberg equilibrium.

General (GCA) and specific (SCA) combining ability. The general combining ability of a line is the mean of the trait of all crosses in which this line is involved, while the specific combining ability of a cross is the deviation of the mean of the trait in that cross with respect to its expected value.

Genetic correlation (r_A). It is the correlation between the additive genetic values of individuals for two quantitative traits.

Genetic draft. It is the effect by which the frequency of a neutral allele can change during generations to a greater extent than expected by simple genetic drift, as a result of its drag by selected genes. See 'Hitchhiking alleles or sequences'.

Genetic drift. It is the random change of the allele frequencies due to the finite population census size.

Genetic estimates of the effective population size. They are those obtained using information from molecular genetic markers. There are several methods based on the excess of heterozygosity, the linkage disequilibrium, the temporal change in allele frequencies, the increase in coancestry or the combination of various statistics.

Genome-wide association studies (GWAS). It is the most used method to detect the position and effect of QTLs. It is based on the regression of the phenotypic value of the individuals on the genotype of the marker. The method has allowed the detection of hundreds of loci with effect on different traits of humans, cultivated plants and domestic animals.

Genomic selection. It is a type of selection in which information of high-density molecular markers is used. It consists in the estimation of the average effects associated with all markers by a Bayesian, regression or GBLUP procedure in a reference population, and then used to estimate the additive value of each individual in the evaluated population.

Genotype. It is the genetic constitution of an individual.

Genotype–environment covariance. It occurs when the genotypic values are not independent of the environmental deviations.

Genotype–environment interaction. It occurs when the phenotypic value of an individual deviates from the sum of the genotypic value and the environmental deviation.

Genotypic value (G). It is the value associated with a specific genotype for the quantitative trait. It is decomposed into additive value (A), equal to the sum of the effects of the alleles carried by the individual, dominance deviation (D), or the minimum quadratic deviation of the additive value from the genotypic value, and epistatic interaction (I), which is the deviation of the multilocus genotypic value with respect to the sum of the genotypic values of the loci separately.

Genotypic variance (V_G). It is the proportion of the phenotypic variance due to the genotypic values of the individuals. That contributed by each locus is decomposed into additive variance (that of additive values) and dominance variance (that of dominance deviations), while in multilocus genotypes, there may be contributions due to gametic disequilibrium and epistatic interactions.

GREML and GBLUP. They are the methods of estimation of variance components and prediction of additive values similar to REML and BLUP but that use the matrix of additive and dominance molecular relationships, obtained from genetic markers.

H

Haploid genomic mutation rate (U). It is the frequency with which mutations arise per haploid genome each generation in the germ line. For quantitative traits, it is estimated, together with other parameters such as the average effect of the mutation and its average allelic dominance, from mutation accumulation experiments, from estimates of genomic data, or from data of means and variances in natural populations assuming the deleterious mutation–selection balance.

Hard selection. It is the type of selection that operates independently of the densities and frequencies of the individuals, and implies a reduction of the census size due to the mortality of the individuals and the limitation of resources (see 'Soft selection').

Hardy–Weinberg equilibrium. It is the principle that indicates that, with random pairing of gametes, the genotype frequencies for a locus are determined by the allele frequencies. If p and q are the frequencies of the alleles A and a of a locus, the frequencies of the three genotypes AA, Aa and aa are given, respectively, by p^2, $2pq$ and q^2.

Heritability or narrow-sense heritability (h^2). It is the proportion of the phenotypic variance of a character explained by the additive genetic variance. It is a fundamental parameter for the prediction of the response to selection, both for natural selection, which is the main objective of evolutionary genetics, and for the artificial selection practised for the genetic improvement of plants and animals.

Heterosis. It is the increase of the mean of a character in the offspring of a cross between inbred lines.

Heterostyly. It is a floral polymorphism in individuals with different lengths of style and stamens in order to avoid self-fertilization. When there are only two floral types, we speak of distyly.

Heterozygote. Individual carrying a genotype with two different alleles in a locus.

Hill–Robertson effect. It is the process of interference between alleles of different loci, particularly if they are linked, which reduces their probability of fixation.

Hitchhiking alleles or sequences. They are the alleles of neutral loci whose frequency changes as a consequence of the change in the allele frequency of a selective locus to which they are intimately linked in the genome.

Homozygote. Individual carrying a genotype with two identical alleles in a locus.

Homozygote by descent. It is the homozygote in which the two alleles are copies of another allele carried by a common ancestor of the parents of the individual in question.

Homozygote in state. It is the homozygote in which the two alleles come from alleles carried by ancestors not common to the parents of the individual in question.

House of Cards model. Model proposed by M. Turelli to explain the genetic variation at equilibrium with stabilizing selection, which assumes low rates of mutation and allelic effects so great in relation to the intensity of the stabilizing selection that mutations remain at a low frequency. It is so named because each new mutation can destroy the equilibrium previously reached under selection.

I

Ideal population of Wright–Fisher. It is the population in which very simplified characteristics are assumed (diploid species of hermaphroditic individuals with constant census size, panmixia, discrete generations and absence of mutation, migration and selection) in order to carry out the prediction of the genetic drift process.

Inbreeding depression (δ). It is the change in the population mean for a quantitative trait as a consequence of the expression in homozygosis of recessive or overdominant alleles due to inbreeding.

Index of sub-population differentiation for quantitative traits (Q_{ST}). It is the index that indicates the proportion of genetic variation among sub-populations relative to the total variation in a subdivided population. The comparison between the value of Q_{ST} for an additive trait and that of F_{ST} for neutral molecular markers can provide information about the existence of convergent or divergent selection between populations for the character and, in general, $Q_{ST} > F_{ST}$ suggests divergent selection.

Indirect genetic effect. It is the genetic effect of an individual that influences the phenotypic value of other individuals. An example is the maternal effects.

Infinitesimal model. It is the basic model of quantitative variation formulated by R. Fisher, in which a character is controlled by a very large number of independent loci, of infinitesimal and summable effect. The phenotypic value of an individual (P) is the sum of its genotypic value (G) and an environmental deviation (E). The distribution of phenotypic effects follows a normal distribution with phenotypic mean equal to the genotypic mean and variance equal to the sum of the genotypic and environmental variances. Genotypic values and environmental deviations are also normally distributed, the latter with mean zero.

Interval mapping. It is one of the most common methods to map QTLs (quantitative trait loci), consisting of using pairs of known and nearby located markers, to find out the presence of a QTL in the interval between the markers, either by Bayesian or regression methods.

Intraclass correlation coefficient (t). It is the phenotypic correlation between relatives, or the ratio between the phenotypic variance of the means of the character for groups of relatives and the total variance. This correlation is a function of the heritability.

Islands of differentiation. These are regions of the genome in which the allele frequencies differ among populations due to the drag by linkage to loci subjected to diversifying selection.

L

Lethal. Allele that produces the death of the individual in homozygosis.

Lewontin's paradox. Observation noted for the first time by R. C. Lewontin that neutral genetic diversity in most prokaryotic and eukaryotic species differs by an order of magnitude, while population census sizes of species may differ by more than three orders of magnitude. The possible explanation for this paradox is that the effective population sizes of the species do not increase proportionally to the population census sizes but, due to demographic or genetic causes, the species with high census sizes have proportionally lower effective population sizes than the species with low census sizes.

Linkage disequilibrium variance. It is the contribution to the variance of the multilocus genotypic values when the allele frequencies of different loci are not independent.

Locus. Genomic element located in a fixed position of the genome, which may or may not have an impact on the phenotype of the individual and may have different variants (alleles).

M

Main components of fitness. They are the reproductive characters whose combination constitutes the essence of fitness. The most important are viability, fecundity and mating success.

Major gene. The number of genes controlling a quantitative trait is usually unknown and the different genotypic classes cannot be distinguished. However, sometimes there are genes of great effect on the character that are called major genes, for which it is possible to identify the genotypes of individuals.

Marginal or apparent overdominance. It is the form of overdominance that occurs due to the combined effect of several non-overdominance loci. It can occur by antagonistic pleiotropy or by the repulsion effect of different closely linked loci.

Marker-assisted selection. It is the type of selection which incorporates information from genetic markers for which an association with the character to be selected has been found.

Mass selection. It is the type of artificial selection in which the selected individuals are mated as a group, without establishing specific pairs.

Maternal effect. It is the environmental effect produced by the maternal phenotype or genotype on the offspring phenotype which can be extended for more than one generation.

Maximum avoidance of inbreeding. It is a system of matings designed by S. Wright in which crossings between relatives are avoided to a maximum. If the population census size is $N = 2^n$ individuals, crossing between individuals with a common ancestor in the last n generations can be avoided.

Meristic traits. They are the quantitative characters whose observable expression occurs in discrete classes, for example, the number of descendants or the number of elements of a given morphological structure.

Migration–drift equilibrium. It is the equilibrium value of the inbreeding coefficient achieved in a subdivided population as a consequence of the balance between the genetic drift that takes place in the sub-populations and the migration of individuals between them.

Minimum coancestry contributions. Procedure for the control of inbreeding followed in conservation programmes and based on establishing the contributions that individuals must make to their offspring so that the average coancestry between them is minimal.

Minimum effective size of a viable population. It is the minimum effective size that can guarantee the viability of a population. Its definition depends on whether the concept is applied in the short or long term.

Molecular coancestry coefficient or molecular similarity (f_M). It is the coancestry coefficient obtained from genetic marker data. It not only includes identity by descent but also identity in state.

Molecular coefficient of inbreeding or molecular homozygosity (F_M). It is the coefficient of inbreeding obtained from genetic marker data. It not only includes identity by descent but also identity in state.

Muller's Ratchet. It is the process that occurs in asexual species by which the minimum number of mutations carried by each genome increases irreversibly with the passage of generation due to the absence of recombination.

Mutation. It is any change that occurs in the DNA sequence during cell reproduction. Germ line mutations are the only ones that are transmitted to offspring.

Mutation load (L). It is the reduction of the average fitness of a population due to the segregation of deleterious mutations. In a high census size population, where the frequencies of mutations are at the deleterious mutation–selection balance, the mutation load only depends on the rate of mutation. In a panmictic population, there is an expressed load and a masked one in the heterozygotes that is revealed by inbreeding (inbreeding load).

Mutational heritability (h_M^2). It is the ratio between the mutational variance, V_M, and the environmental variance, V_E, of a trait.

Mutational meltdown. It is the recurrent cycle of fixation of deleterious mutations, reduction of the average fitness and decrease of the census size that can lead the population to extinction. It is based on the fact that in a population with a reduced census size, where genetic drift is the predominant factor of change, there is an increase in the probability of fixation of deleterious mutations, which implies a reduction in fitness and a consequent decrease in the census size.

Mutational variance (V_M). It is the increase in additive variance of a trait per generation as a consequence of mutation. It is usually scaled by the environmental variance (mutational heritability, h_M^2) or by the square of the trait mean (squared mutational coefficient of variation, CV_M).

N

Neutral mutation. It is the mutation without effect on the fitness of the individual.

Neutral mutation–drift balance. It is the equilibrium value reached by the allele frequencies and the genetic variance of quantitative traits as a consequence of the appearance of neutral mutations and their elimination by genetic drift.

Neutralism. Theory developed by M. Kimura that maintains that most of the variation observed at the molecular level is neutral, that is, it is due to mutations without effect on fitness. Neutralism does not negate selection and serves as a null hypothesis for its verification.

Number of lethal equivalents (B). It is the number of deleterious mutations that, combined in an additive way, would produce the same inbreeding depression as a single recessive lethal. It constitutes the so-called inbreeding load.

O

Outbreeding depression. It is the reduction in the average fitness of a hybrid formed from the cross of different populations if these are genetically adapted to different local conditions.

Overdominance. It is the type of gene action in which the heterozygote value for a character is higher than those of the homozygotes at the locus considered.

P

Panmixia. It is the random pairing of gametes approximated, in practice, by the random pairing between individuals.

Phenotype. It is the expression of a character in an individual.

Phenotypic correlation (r_P). It is the correlation between the individual phenotypic values for two quantitative traits.

Phenotypic plasticity. It is the ability of a particular genotype to produce different phenotypes in relation to environmental variation. Sometimes it occurs as an adaptive response.

Phenotypic value (P). It is the observable value of a character in an individual, which is almost always influenced by the environment in the case of quantitative traits.

Phenotypic variance (V_P). It is the variance of the phenotypic values of the individuals.

Physiological theory of dominance. It is the theory initially postulated by S. Wright that explains, in terms of the kinetics of metabolic pathways, the observation of the tendency for small-effect mutations to be additive and those with a large effect to be recessive.

Pleiotropy. It is the situation in which a locus has an effect on more than one quantitative trait.

Population bottleneck. It is the temporary reduction of the census size of a population, with consequences on the magnitude of genetic drift and inbreeding.

Positive selection. It is the kind of natural selection that acts by increasing the frequency of beneficial mutations.

Probability of fixation of a mutation. It is the probability with which a mutation will finally be fixed in a population of finite census size according to its selective coefficient, its coefficient of dominance and the effective population size.

Progeny test. It is a procedure to estimate the additive value of an individual. It is based on crossing this with others taken at random from the population and evaluating the character in a number of descendants from each cross. Twice the deviation of the mean of the character in the descendants with respect to the mean of the population constitutes an estimate of the additive value of the individual.

Purged inbreeding coefficient (g). It is the coefficient of inbreeding that results from the purging of deleterious mutations by natural selection. Its magnitude depends on the purging coefficient and is equal to or less than the genealogical coefficient of inbreeding.

Purging coefficient (d). It is the dominance effect that represents the part of the effect on fitness of a deleterious recessive or partially recessive allele that remains hidden in heterozygosis but is expressed in homozygosis. For deleterious mutations this value represents the effect on fitness attributable to a copy of the allele. For example, for a recessive lethal allele ($s = 1$ and $h = 0$), $d = 0.5$, so that in the homozygote, the two copies of the allele produce their effect in homozygosis ($s = 1$).

Purging selection. It is the type of selection that acts on the mutation load masked in heterozygotes (inbreeding load) which is expressed by inbreeding.

Q

QTL. It is a term widely used in the scientific literature that refers to the abbreviation of quantitative trait locus.

Quantitative traits. They are those traits characterized by a continuous variation, either directly observed or as an underlying character (called liability). They are also called continuous or metric traits. Because of their multifactorial nature, they are also called polygenic or multifactorial traits and, in the context of medicine, complex traits.

Quasi-neutral mutation. It is said of that non-neutral mutation whose destiny is determined fundamentally by genetic drift and, therefore, its probability of fixation is approximately that corresponding to a neutral one. *Quasi*-neutral mutations are those that hold the inequality $|N_e s| < 1$, where N_e is the effective population size and s the mutation effect on fitness.

R

Rate of allelism of lethals (I_a). It is the frequency of allelic lethal genes of the same locus. A method for estimating the effective population size is based on this frequency.

Rate of increase in inbreeding (ΔF). It is the increase in inbreeding coefficient from one generation to the next, relative to the precise change of that coefficient to take the maximum value of 1. In the ideal population of Wright–Fisher of census size N, inbreeding increases at a rate $\Delta F = 1/2N$ per generation.

Reaction norm. It is the representation of the change in the phenotypic value of the genotypes or individuals with respect to the environment.

Realized heritability. It is the heritability that is estimated from the result of selection as $h^2 = R/S$ or from the regression of the cumulative response R on the accumulated selection differential S if several generations of selection have been carried out.

Recombination. Consequence of the genetic exchange between homologous chromosomes during the formation of gametes. The typical recombination rate (c) of mammals is about 1 cM (1% recombination) per 1 Mb of sequence.

REML. Acronym for restricted maximum likelihood, which is the most commonly used procedure for the estimation of genetic variance components with the animal model.

Repeatability (r). It is the proportion of the phenotypic variance explained by the genotypic variance and the general environmental variance (see 'Environmental variance').

Response to selection. It is the change in the mean of the selected trait per generation by the effect of selection. It is predicted as $R = h^2S$, which is the so-called Breeder's equation, where h^2 is the heritability in the population considered and S is the selection differential.

Rule 50/500. Rule assumed in conservation programmes relative to the minimum effective size that a population must have in order to be viable in the short or long term, respectively.

Runs or regions of homozygosity (ROH). They are portions of the genome in which all or the vast majority of the bases are homozygous.

S

Selection coefficient (s). It is the fitness effect associated with an allele in homozygosis.

Selection differential (S). It is the difference between the average of the selected individuals and the average of the population for a character subject to selection. The selection differential scaled by the phenotypic standard deviation is the intensity of selection.

Selection gradients. They are the coefficients of partial regression of fitness on two or more quantitative traits, which estimate the intensity of natural selection on those traits.

Selection index. It is the combination of the phenotypic value of an individual and that of its relatives, or of the values of an individual for different traits, in order to be used as a selection criterion to increase the response of the trait, in the first case, or of the different traits, in the second.

Selection intensity (i). It is the selection differential, or mean of the trait for the selected individuals, deviated from the population mean, scaled by the phenotypic standard deviation.

Selective sweep. It is the effect that occurs when selection rapidly increases the frequency of a beneficial allele, dragging other linked neutral variants and drastically reducing genetic diversity in the surrounding regions. The alleles or neutral sequences that increase in frequency carried by the selected allele are called hitchhiking alleles.

Self-incompatibility systems. It is the existence of genetic systems of self-sterility, very common in plants, that prevent or reduce the likelihood of self-fertilization and mating between relatives in order to minimize the effects of inbreeding.

Self-sterility loci. Loci with multiple alleles that avoid self-fertilization and reduce the probability of mating between relatives in plants. It is a genetic system of self-incompatibility determined by the genotype of the pollen or the plant that produces it. The pollen grains carrying some of the alleles of the self-sterility locus present in the genotype of the recipient plant will not be able to form a pollen tube, or the fertilization will be aborted.

Sexual selection. It is the kind of natural selection that affects the characters of sexual nature, such as those that intervene in mate choice.

Soft selection. It is the type of selection that does not imply changes in the population census size but results in changes in allele frequencies and fitness, depending on the density and frequency of the individuals.

Stabilizing selection. It is the type of selection that applies to characters with intermediate optimum, so that the most extreme individuals of the phenotypic distribution are those at selective disadvantage. Its effect is the maintenance of the population mean at the optimum and a reduction of the genetic variance.

Synergistic or reinforcing epistasis and antagonistic or attenuating epistasis. In a multilocus fitness model, the combined effect of the mutations may be greater (synergistic or reinforcing epistasis) or lower (antagonistic or attenuating epistasis) than the product or the sum of the individual effects of the mutations.

Synthetic population. It is a population that derives from the random mating of individuals of two or more lines, sometimes inbred ones, in order to produce heterosis and an increase in genetic variation.

T

Threshold traits. They are the characters that are only expressed in two or three phenotypic classes, such as resistance or susceptibility to a disease. It is assumed that, although the expression is discrete, the character determination is due to an underlying continuous trait that is called liability, predisposition or propensity in the context of medicine.

Traits with intermediate optimum. They are the quantitative traits on which selection acts against the individuals with extreme phenotypes being, therefore, favoured those of intermediate phenotype. The type of selection that acts on them is the stabilizing selection.

Two-fold cost of sex. It occurs because, for the same number of descendants per individual, the census size of an asexual population doubles in each generation, whereas that of a sexual population remains constant.

U

Underdominance. It is the type of gene action in which the heterozygote value for a character is lower than those of the homozygotes in the same locus. If applied to fitness, it is also called heterozygote disadvantage.

V

Variance of parental offspring contributions (S_k^2). It is the variance of the number of descendants contributed by parents, on which depends the value of the effective population size.

W

Wahlund's effect. It is the property of the heterozygosity of a subdivided population being always lower than or equal to that of the same population without subdivision.

Within-family selection. It is the type of artificial selection in which the best individual of each sex is selected per family.

Wright's statistics (F_{IS}, F_{ST} and F_{IT}). They are the coefficients of inbreeding applicable to populations structured in sub-populations of common origin. The F_{IS} statistic indicates the amount of inbreeding due to non-random mating of individuals within sub-populations. F_{ST}, or fixation index, indicates the inbreeding due to population subdivision, and it is the parameter most frequently used to quantify genetic differentiation among sub-populations. The F_{IT} statistic denotes the total inbreeding in the population due to both non-random mating and population subdivision.

References

Agrawal, A. F. and M. C. Whitlock (2012). Mutation load: the fitness of individuals in populations where deleterious alleles are abundant. Annual Review of Ecology, Evolution, and Systematics, 43: 115–35.

Allendorf, F. W., G. Luikart and S. N. Aitken (2013). *Conservation and the Genetics of Populations*, 2nd edn. Wiley-Blackwell, Oxford.

Álvarez-Castro, J. M., A. Le Rouzic and O. Carlborg (2008). How to perform meaningful estimates of genetic effects. PLoS Genetics, 5: e1000062.

Amador, C., J. Fernández and T. H. E. Meuwissen (2013). Advantages of using molecular coancestry in the removal of introgressed genetic material. Genetics Selection Evolution, 45(1): 13.

Amador, C., A. García-Dorado, D. Bersabé and C. López-Fanjul (2010). Regeneration of the variance of metric traits by spontaneous mutation in a Drosophila population. Genetics Research, 92: 91–102.

Andersson, L. and L. Georges (2004). Domestic-animal genomics: deciphering the genetics of complex *traits*. Nature Reviews Genetics, 5: 202–12.

Armbruster, P. and P. H. Reed (2005). Inbreeding in benign and stressful environments. Heredity, 95: 235–42.

Arnold, S. J. and M. J. Wade (1984a). On the measurement of natural and sexual selection: theory. Evolution, 38: 709–19.

Arnold, S. J. and M. J. Wade (1984b). On the measurement of natural and sexual selection: applications. Evolution, 38: 720–34.

Auton, A. The 1000 Genomes Project Consortium (2015). A global reference for human genetic variation. Nature, 526: 68–74.

Ávila, V., A. Pérez-Figueroa, A. Caballero, W. G. Hill, A. García-Dorado and C. López-Fanjul (2014). The action of stabilizing selection, mutation, and drift on epistatic quantitative traits. Evolution, 68: 1974–87.

Ávila, V., A. Vilas, J. Fernández and A. Caballero (2013). An experimental assessment of artificial within-family selection for fitness in conservation programs. Conservation Genetics, 14: 1149–59.

Ballou, J. D. (1997). Ancestral inbreeding only minimally affects inbreeding depression in mammalian populations. Journal of Heredity, 88: 169–78.

Ballou, J. D. and R. C. Lacy (1995). Identifying genetically important individuals for management of genetic diversity in pedigreed populations. Pp. 76–111 in *Population Management for Survival and Recovery*. Ed. J. D. Ballou, M. Gilpin and T. J. Foose. Columbia University Press, New York.

Barton, N. H. (1990). Pleiotropic models of quantitative variation. Genetics, 124: 773–82.

Barton, N. H. and M. Turelli (1989). Evolutionary quantitative genetics: how little do we know? Annual Review of Genetics, 23: 337–70.

Barton, N. H. and M. Turelli (1991). Natural and sexual selection on many loci. Genetics, 127: 229–255.

Barton, N. H. and M. Turelli (2004). Effects of genetic drift on variance components under a general model of epistasis. Evolution, 58: 2111–32.

Bataillon, T. and S. F. Bailey (2014). Effects of new mutations on fitness: insights from models and data. Annals of the New York Academy of Sciences, 1320: 76–92.

Bateman, A. J. (1959). The viability of near-normal irradiated chromosomes. International Journal of Radiation Biology, 1: 170–80.

Beaumont, M. A. and R. A. Nichols (1996). Evaluating loci for use in the genetic analysis of population structure. Proceedings of the Royal Society of London, Series B, 263: 1619–26.

Bersabé, D. and A. García-Dorado (2013). On the genetic parameter determining the efficiency of purging: an estimate for *Drosophila* egg-to-pupae viability. Journal of Evolutionary Biology, 26: 375–85.

Bierne, N., J. Welch, E. Loire, F. Bonhomme and P. David (2011). The coupling hypothesis: why genome scans may fail to map local adaptation genes. Molecular Ecology, 20: 2044–72.

Bijma P., J. A. M. van Arendonk and J. A. Woolliams (2001). Predicting rates of inbreeding for livestock improvement schemes. Journal of Animal Science, 79: 840–53.

Blasco, A. and M. A. Toro (2014). A short critical history of the application of genomics to animal breeding. Livestock Science, 166: 4–9.

Boakes, E. H., J. Wang and W. Amos (2007). An investigation of inbreeding depression and purging in captive pedigreed populations. Heredity, 98: 172–82.

Bookstein, F. L. (1991). *Morphometric Tools for Landmark Data*. Cambridge University Press, New York.

Boyko, A. R., S. H. Williamson, A. R. Indap, J. D. Degenhardt, R. D. Hernandez, K. E. Lohmueller, M. D. Adams, S. Schmidt, J. J. Sninsky, S. R. Sunyaev et al. (2008). Assessing the evolutionary impact of amino acid mutations in the human genome. PLoS Genetics, 4: e1000083.

Brookes, J. I. and R. Rochette (2007). Mechanism of a plastic phenotypic response: predator-induced shell thickening in the intertidal gastropod *Littorina obtusata*. Journal of Evolutionary Biology, 20: 1015–27.

Bulmer, M. G. (1971). The effect of selection on genetic variability. American Naturalist, 105: 201–11.

Bulmer, M. G. (1985). *The Mathematical Theory of Quantitative Genetics*. Clarendon Press, Oxford.

Butlin, R. K., M. Saura, G. Charrier, B. Jackson, C. André, A. Caballero, J. A. Coyne, J. Galindo, J. W. Grahame, J. Hollander et al. (2014). Parallel evolution of local adaptation and reproductive isolation in the face of gene flow. Evolution, 68: 935–49.

Caballero, A. (1994). Developments in the prediction of effective population size. Heredity, 73: 657–79.

Caballero, A. (1995). On the effective size of populations with separate sexes, with particular reference to sex-linked genes. Genetics, 139: 1007–11.

Caballero, A. (2006). Analysis of the biases in the estimation of deleterious mutation parameters from natural populations at mutation–selection balance. Genetical Research, 88: 177–89.

Caballero, A. and A. García-Dorado (2013). Allelic diversity and its implications for the rate of adaptation. Genetics, 195: 1373–84.

Caballero, A. and W. G. Hill (1992a). Effective size of nonrandom mating populations. Genetics, 130: 909–16.

Caballero, A. and W. G. Hill (1992b). Effects of partial inbreeding on fixation rates and variation of mutant genes. Genetics, 131: 493–507.

Caballero, A. and P. D. Keightley (1994). A pleiotropic nonadditive model of variation in quantitative traits. Genetics, 138: 883–900.

Caballero, A., P. D. Keightley and M. Turelli (1997). Average dominance for polygenes: drawbacks of regression estimates. Genetics, 147: 1487–90.

Caballero, A., E. Santiago and M. A. Toro (1996). Systems of mating to reduce inbreeding in selected populations. Animal Science, 62: 431–42.

Caballero, A., A. Tenesa and P. D. Keightley (2015). The nature of genetic variation for complex traits revealed by GWAS and regional heritability mapping analyses. Genetics, 201: 1601–13.

Caballero, A. and M. A. Toro (2000). Interrelations between effective population size and other pedigree tools for the management of conserved populations. Genetical Research, 75: 331–43.

Caballero, A. and M. A. Toro (2002). Analysis of genetic diversity for the management of conserved subdivided populations. Conservation Genetics, 3: 289–99.

Caballero, A., M. A. Toro and C. López-Fanjul (1991). The response to artificial selection from new mutations in *Drosophila melanogaster*. Genetics, 127: 89–102.

Carvajal-Rodríguez, A., P. Conde-Padín and E. Rolán-Alvarez (2005). Decomposing shell form into size and shape by geometric morphometric methods in two sympatric ecotypes of *Littorina saxatilis*. Journal of Molluscan Studies, 71: 313–18.

Castle, W. E. (1921). An improved method of estimating the number of genetic factors concerned in cases of blending inheritance. Science, 54: 223.

Cavalli-Sforza, L. L. and W. F. Bodmer (1971). *The Genetics of Human Populations*. Freeman, San Francisco, CA.

Charlesworth, B. (2009). Fundamental concepts in genetics: effective population size and patterns of molecular evolution and variation. Nature Reviews Genetics, 10: 195–205.

Charlesworth, B. (2015). Causes of natural variation in fitness: evidence from studies of Drosophila populations. Proceedings of the National Academy of Sciences of the USA, 112: 1662–9.

Charlesworth, B. and D. Charlesworth (1999). The genetic basis of inbreeding depression. Genetical Research, 74: 329–40.

Charlesworth, B. and D. Charlesworth (2010). *Elements of Evolutionary Genetics*. Roberts and Company, Greenwood Village, CO.

Charlesworth, D. and B. Charlesworth (1979). A model for the evolution of distyly. American Naturalist, 114: 467–98.

Charlesworth, D. and J. H. Willis (2009). The genetics of inbreeding depression. Nature Reviews Genetics, 10: 783–96.

Clark, S. A. and J. van der Werf (2013). Genomic best linear unbiased prediction (gBLUP) for the estimation of genomic breeding values. Methods in Molecular Biology, 1019: 321–30.

Cockerham, C. C. (1969). Variance of gene frequencies. Evolution, 23: 72–84.

Collard, B. C. Y. and D. J. Mackill (2008). Marker-assisted selection: an approach for precision plant breeding in the twenty-first century. Philosophical Transactions of the Royal Society of London, Series B, 363: 557–72.

Colleau, J. J. and M. Sargolzaei (2008). A proximal decomposition of inbreeding, coancestry and contributions. Genetical Research, 96: 191–8.

Comstock, R. E., H. F. Robinson and P. H. Harvey (1949). A breeding procedure designed to make maximum use of both general and specific combining ability. Journal of the American Society of Agronomy, 41: 360–7.

Conde-Padín, P., A. Carvajal-Rodríguez, M. Carballo, A. Caballero and E. Rolán-Alvarez (2007). Genetic variation for shell traits in a direct developing marine snail involved in a putative sympatric ecological speciation process. Evolutionary Ecology, 21: 635–50.

Cooper, V. S. and R. E. Lenski (2000). The population genetics of ecological specialization in evolving *Escherichia coli* populations. Nature, 407: 736–9.

Corbett-Detig, R. B., D. L. Hartl and T. B. Sackton (2015). Natural selection constrains neutral diversity across a wide range of species. PLoS Biology, 13(4): e1002112.

Crnokrak, P. and S. C. H. Barrett (2002). Purging the genetic load: a review of the experimental evidence. Evolution, 56: 2347–58.

Crnokrak, P. and D. A. Roff (1995). Dominance variance: associations with selection and fitness. Heredity, 75: 530–40.

Crow, J. F. (1970). Genetic loads and the cost of natural selection. Pp. 128–77 in *Mathematical Topics in Population Genetics*. Ed. Ken-ichi Kojima. Springer, Heidelberg, Germany.

Crow, J. F. (1986). *Basic Concepts in Population, Quantitative, and Evolutionary Genetics*. W. H. Freeman, New York.

Crow, J. F. (1997). The high spontaneous mutation rate: is it a health risk? Proceedings of the National Academy of Sciences of the USA, 94: 8380–6.

Crow, J. F. (2000). The origins, patterns and implications of human spontaneous mutation. Nature Reviews Genetics, 1: 40–7.

Crow, J. F. and C. Denniston (1988). Inbreeding and variance effective population numbers. Evolution, 42: 482–95.

Crow, J. F. and M. Kimura (1965). Evolution in sexual and asexual populations. American Naturalist, 99: 439–50.

Crow, J. F. and M. Kimura (1970). *An Introduction to Population Genetics Theory*. Harper and Row, New York.

Crow, J. F. and N. F. Morton (1955). Measurement of gene frequency drift in small populations. Evolution, 9: 202–14.

Cummings, M. R. (2014). *Human Heredity: Principles and Issues*, 11th edn. Cengage Learning, Boston.

Darwin, C. R. (1859). *On the Origin of Species by Means of Natural Selection, or the Preservation of Favoured Races in the Struggle for Life*. John Murray, London.

Darwin, C. R. (1876). *The Effects of Cross- and Self-Fertilization in the Vegetable Kingdom*. John Murray, London.

Darwin, C. R. (1877). *The Different Forms of Flowers on Plants of the Same Species*. John Murray, London.

de Cara, M. A. R., B. Villanueva, M. A. Toro and J. Fernández (2013). Using genomic tools to maintain diversity and fitness in conservation programmes. Molecular Ecology, 22: 6091–9.

de los Campos, G., D. Gianola and D. B. Allison (2010). Predicting genetic predisposition in humans: the promise of whole-genome markers. Nature Reviews Genetics, 11: 880–6.

Dempfle, L. (1975). A note on increasing the limit of selection through selection within families. Genetical Research, 24: 127–35.

Deng, H.-W. (1998). Estimating within-locus non-additive coefficient and discriminating dominance versus overdominance as the genetic cause of heterosis. Genetics, 148: 2003–14.

Deng, H.-W. and M. Lynch (1996). Estimation of deleterious-mutation parameters in natural populations. Genetics, 144: 349–60.

Deng, H.-W. and M. Lynch (1997). Inbreeding depression and inferred deleterious mutation parameters in Daphnia. Genetics, 147: 147–55.

DeRose, M. A. and D. A. Roff (1999). A comparison of inbreeding depression in life-history and morphological traits in animals. Evolution, 53: 1288–92.

Desta, Z. A. and R. Ortiz (2014). Genomic selection: genome-wide prediction in plant improvement. Trends in Plant Science, 19: 592–601.

Dohm, M. R. (2002). Repeatability estimates do not always set an upper limit to heritability. Functional Ecology, 16: 273–80.

Domínguez-García, S., C. García, H. Quesada and A. Caballero (2019). Accelerated inbreeding depression suggests synergistic epistasis for deleterious mutations in *Drosophila melanogaster*. Heredity, 123: 709–22.

Drake, J. W., B. Charlesworth, D. Charlesworth and J. F. Crow (1998). Rates of spontaneous mutation. Genetics, 148: 1667–86.

Dudley, J. W. and R. J. Lambert (2004). 100 generations of selection for oil and protein in corn, 1. Plant Breeding Reviews, 24: 79–110.

Dumont, B. and B. Payseur (2008). Evolution of the genomic rate of recombination in mammals. Evolution, 62: 276–94.

Edmands, S. (2007). Between a rock and a hard place: evaluating the relative risks of inbreeding and outbreeding for conservation and management. Molecular Ecology, 16: 463–75.

Eichler, E. E., J. Flint, G. Gibson, A. Kong, S. M. Leal, J. H. Moore and J. H. Nadeau (2010). Missing heritability and strategies for finding the underlying causes of complex disease. Nature Reviews Genetics, 11: 446–50.

Elena, S. F. and R. E. Lenski (2003). Evolution experiments with microorganisms: the dynamics and genetic bases of adaptation. Nature Reviews, 4: 457–69.

Endler, J. A. (1986). *Natural Selection in the Wild*. Princeton University Press, Princeton, NJ.

Eyre-Walker, A. (2010). Genetic architecture of a complex trait and its implications for fitness and genome-wide association studies. Proceedings of the National Academy of Sciences of the USA, 107: 1752–6.

Eyre-Walker, A. and P. D. Keightley (2007). The distribution of fitness effects of new mutations. Nature Reviews Genetics, 8: 610–18.

Falconer, D. S. (1952). Asymmetrical responses in selection experiments. International Symposium on Genetics and Population Structure. International Union of Biological Sciences, Naples, Series B, 15: 16–41.

Falconer, D. S. and T. F. C. Mackay (1996). *Introduction to Quantitative Genetics*, 4th edn. Longman, London.

Felsenstein, J. (1974). The evolutionary advantage of recombination. Genetics, 78: 737–56.

Fernández, B., A. García-Dorado and A. Caballero (2004). Analysis of the estimators of the average coefficient of dominance of deleterious mutations. Genetics, 168: 1053–69.

Fernández, B., A. García-Dorado and A. Caballero (2005). The effect of antagonistic pleiotropy on the estimation of the average coefficient of dominance of deleterious mutations. Genetics, 171: 2097–112.

Fernández, J., J. Galindo, B. Fernández, A. Pérez-Figueroa, A. Caballero and E. Rolán-Alvarez (2005b). Genetic differentiation and estimation of effective population size and migration rates in two sympatric ecotypes of the marine snail *Littorina saxatilis*. Journal of Heredity, 96: 460–4.

Fernández, J., M. A. Toro and A. Caballero (2003). Fixed contributions designs versus minimization of global coancestry to control inbreeding in small populations. Genetics, 165: 885–94.

Fernández, J., M. A. Toro and A. Caballero (2008). Management of subdivided populations in conservation programs: development of a novel dynamic system. Genetics, 179: 683–92.

Fernando, R. L. and M. Grossman (1989). Marker assisted selection using best linear unbiased prediction. Genetics Selection Evolution, 21: 467–77.

Fisher, R. A. (1918). The correlation between relatives on the supposition of Mendelian inheritance. Transactions of the Royal Society of Edinburgh, 52: 399–433.

Fisher, R. A. (1921). On the mathematical foundations of theoretical statistics. Philosophical Transactions of the Royal Society of London, Series A, 222: 309–68.

Fisher, R. A. (1928). The possible modification of the response of the wild type to recurrent mutations. American Naturalist, 62: 115–26.

Fisher, R. A. (1930). *The Genetical Theory of Natural Selection*. Clarendon Press, Oxford.

Fisher, R. A. (1941). Average excess and average effect of a gene substitution. Annals of Eugenics, 11: 53–63.

Fisher, R. A. (1949). *The Theory of Inbreeding*. Oliver and Boyd, Edinburgh.

Flather, C. H., G. D. Hayward, S. R. Beissinger and P. A. Stephens (2011). Minimum viable populations: is there a 'magic number' for conservation practitioners? Trends in Ecology and Evolution, 26: 307–16.

Foll, M. and O. Gaggiotti (2008). A genome-scan method to identify selected loci appropriate for both dominant and codominant markers: a Bayesian perspective. Genetics, 180: 977–95.

Fox, C. W. and D. H. Reed (2011). Inbreeding depression increases with environmental stress: an experimental study and meta-analysis. Evolution, 65: 246–58.

Frankham, R. (1980). The founder effect and response to artificial selection in *Drosophila*. Pp. 87–90 in *Heritage from Mendel*. Ed. R. A. Brink. University of Wisconsin Press, Madison.

Frankham, R. (1990). Are responses to artificial selection for reproductive fitness characters consistently asymmetrical? Genetical Research, 53: 35–42.

Frankham, R. (1995). Effective population size/adult population size ratios in wildlife: a review. Genetical Research, 66: 95–107.

Frankham, R., J. D. Ballou and D. A. Briscoe (2010). *Introduction to Conservation Genetics*. Cambridge University Press, Cambridge.

Frankham, R., C. J. A. Bradshaw and B. W. Brook (2014). Genetics in conservation management: revised recommendations for the 50/500 rules, Red List criteria and population viability analyses. Biological Conservation, 170: 56–63.

Franklin, I. R. (1980). Evolutionary change in small populations. Pp. 135–49 in *Conservation Biology: An Evolutionary-Ecological Perspective*. Ed. M. E. Soulé and B. A. Wilcox. Sinauer, Sunderland, MA.

Galindo, J. (2007). Divergencia adaptativa en dos ecotipos de un caracol marino relacionados con un proceso incompleto de especiación simpátrica y ecológica. Tesis doctoral, Universidade de Vigo.

Gallego, A. and C. López-Fanjul (1983). The number of loci affecting a quantitative trait in *Drosophila melanogaster* revealed by artificial selection. Genetical Research, 42: 137–49.

Galton, F. (1889). *Natural Inheritance*. Macmillan, London.

García, C., V. Ávila, H. Quesada and A. Caballero (2012). Gene-expression changes caused by inbreeding protect against inbreeding depression in *Drosophila*. Genetics, 192: 161–72.

García de Leaniz, C., E. Verspoor and D. Hawkins (1989). Genetic determination of the contribution of stocked and wild Atlantic salmon, *Salmo salar* L., to the angling fisheries in two Spanish rivers. Journal of Fish Biology, 35(Suppl. A): 261–70.

García-Dorado, A. (1997). The rate and effects distribution of viability mutation in *Drosophila*: minimum distance estimation. Evolution, 51: 1130–9.

García-Dorado, A. (2012). Understanding and predicting the fitness decline of shrunk populations: inbreeding, purging, mutations, and standard selection. Genetics, 190: 1461–76.

García-Dorado, A., V. Ávila, E. Sánchez-Molano, A. Manrique and C. López-Fanjul (2007). The build up of mutation-selection-drift balance in laboratory *Drosophila* populations. Evolution, 61: 653–65.

García-Dorado, A. and A. Caballero (2000). On the average coefficient of dominance of deleterious spontaneous mutations. Genetics, 155: 1991–2001.

García-Dorado, A., A. Caballero and J. F. Crow (2003). On the persistence and pervasiveness of a new mutation. Evolution, 57: 2644–6.

García-Dorado, A., C. López-Fanjul and A. Caballero (1999). Properties of spontaneous mutations affecting quantitative traits. Genetical Research, 74: 341–50.

García-Dorado, A., C. López-Fanjul and A. Caballero (2004). Rates and effects of deleterious mutations and their evolutionary consequences. Pp. 20–32 in *Evolution: From Molecules to Ecosystems*. Ed. A. Moya and E. Font. Oxford University Press, Oxford.

Gardner, A., S. A. West and N. H. Barton (2007). The relation between multilocus population genetics and social evolution theory. American Naturalist, 169: 207–26.

George, A. W., P. M. Visscher and C. S. Haley (2000). Mapping quantitative trait loci in complex pedigrees: a two-step variance component approach. Genetics, 156: 2081–92.

Gianola, D. (2013). Priors in whole-genome regression: the Bayesian alphabet returns. Genetics, 194: 573–96.

Gianola, D. and G. J. M. Rosa (2015). One hundred years of statistical developments in animal breeding. Annual Review of Animal Biosciences, 3: 19–56.

Gillespie, J. H. (2000). Genetic drift in an infinite population: the pseudohitchhiking model. Genetics, 155: 909–19.

Goddard, M. (2008). Genomic selection: prediction of accuracy and maximization of long term response. Genetica, 136: 245–57.

Goddard, M. E., B. J. Hayes and T. H. E. Meuwissen (2010). Genomic selection in livestock populations. Genetical Research, 92: 413–21.

Goodale, H. D. (1938). A study of the inheritance of body weight in the albino mouse by selection. Journal of Heredity, 29: 101–12.

Goodwillie, C., S. Kalisz and C. G. Eckert (2005). The evolutionary enigma of mixed mating systems in plants: occurrence, theoretical explanations, and empirical evidence. Annual Review of Ecology, Evolution, and Systematics, 36: 47–79.

Goudet, J. and L. Büchi (2006). The effects of dominance, regular inbreeding and sampling design on Q_{ST}, an estimator of population differentiation for quantitative traits. Genetics, 172: 1337–47.

Gowe, R. S., A. Robertson and B. D. H. Latter (1959). Environment and poultry breeding problems. 5. The design of poultry control strains. Poultry Science, 38: 462–71.

Grant, B. R. and P. R. Grant (1989). Natural selection in a population of Darwin's finches. American Naturalist, 133: 377–93.

Groenen, M. A. M., A. L. Archibald, H. Uenishi, C. K. Tuggle, Y. Takeuchi, M. F. Rothschild, C. Rogel-Gaillard, C. Park, D. Milan, H.-J. Megens et al. (2012). Analyses of pig genomes provide insight into porcine demography and evolution. Nature, 491: 393–8.

Gutiérrez, J. P., I. Cervantes, A. Molina, M. Valera and F. Goyache. (2008). Individual increase in inbreeding allows estimating effective sizes from pedigrees. Genetics Selection Evolution, 40: 359–78.

Haag-Liautard, C., M. Dorris, X. Maside, S. Macaskill, D. L. Halligan, B. Charlesworth and P. D. Keightley (2007). Direct estimation of per nucleotide and genomic deleterious mutation rates in *Drosophila*. Nature, 445: 82–5.

Haigh, I. (1978). The accumulation of deleterious genes in a population – Muller's Ratchet. Theoretical Population Biology, 14: 251–7.

Haldane, J. B. S. (1927). A mathematical theory of natural and artificial selection. Part V: Selection and mutation. Proceedings of the Cambridge Philosophical Society, 23: 838–44.

Haldane, J. B. S. (1931). A mathematical theory of natural and artificial selection. Part VII. Selection intensity as a function of mortality rate. Mathematical Proceedings of the Cambridge Philosophical Society, 27: 131–6.

Haldane, J. B. S. (1932). *The Causes of Evolution*. Longmans, Green, London.

Haldane, J. B. S. (1937). The effect of variation on fitness. American Naturalist, 71: 337–49.

Haley, C. S. and S. A. Knott (1992). A simple regression method for mapping quantitative trait loci in line crosses using flanking markers. Heredity, 69: 315–24.

Halligan, D. L. and P. D. Keightley (2009). Spontaneous mutation accumulation studies in evolutionary genetics. Annual Review of Ecology, Evolution and Systematics, 40: 151–72.

Handley, L. J. L., A. Manica, J. Goudet and F. Balloux (2007). Going the distance: human population genetics in a clinal world. Trends in Genetics, 23: 432–9.

Hartfield, M., S. P. Otto and P. D. Keightley (2010). The role of advantageous mutations in enhancing the evolution of a recombination modifier. Genetics, 184: 1153–64.

Hartl, D. L. and A. G. Clark (2007). *Principles of Population Genetics*, 3rd edn. Sinauer, Sunderland, MA.

Hayes, B. J. and M. E. Goddard (2001). The distribution of the effects of genes affecting quantitative traits in livestock. Genetics Selection Evolution, 33: 209–29.

Hayes, B. J., H. A. Lewin and M. E. Goddard (2013). The future of livestock breeding: genomic selection for efficiency, reduced emissions intensity, and adaptation. Trends in Genetics, 29: 206–14.

Hayes, B. J., P. M. Visscher, H. C. McPartlan and M. E. Goddard (2003). Novel multilocus measure of linkage disequilibrium to estimate past effective population size. Genome Research, 13: 635–43.

Hazel, L. N. (1943). The genetic basis for constructing selection indexes. Genetics, 28: 476–90.

Hazel, L. N. and J. L. Lush (1943). The efficiency of three methods of selection. Journal of Heredity, 33: 393–9.

Hedrick, P. W. (2005). A standardized genetic differentiation measure. Evolution, 59: 1633–8.

Hedrick, P. W. (2012). What is the evidence for heterozygote advantage selection? Trends in Ecology and Evolution, 27: 698–704.

Hedrick, P. W. and C. C. Cockerham (1986). Partial inbreeding: equilibrium heterozygosity and the heterozygosity paradox. Evolution, 40: 856–61.

Hedrick, P. W. and A. García-Dorado (2016). Understanding inbreeding depression, purging, and genetic rescue. Trends in Ecology and Evolution, 31: 940–52.

Hemani, G., S. Knott and C. Haley (2013). An evolutionary perspective on epistasis and the missing heritability. PLoS Genetics, 9(2): e1003995.

Henderson, C. R. (1976). A simple method for computing the inverse of a numerator relationship matrix used in prediction of breeding values. Biometrics, 32: 69–83.

Henderson, C. R. (1984). *Applications of Linear Models in Animal Breeding*. University of Guelph, Guelph, ON.

Henderson, C. R. (1985). Best linear unbiased prediction of non-additive genetic merits in noninbred populations. Journal of Animal Science, 60: 111–17.

Hill, W. G. (1971). Design and efficiency of selection experiments for estimation of genetic parameters. Biometrics, 27: 293–311.

Hill, W. G. (1979). A note on effective population size with overlapping generations. Genetics, 92: 317–22.

Hill, W. G. (1980). Design of quantitative genetic selection experiments. Pp. 1–13 in *Selection Experiments in Laboratory and Domestic Animals*. Ed. A. Robertson. Commonwealth Agricultural Bureaux, Slough, UK.

Hill, W. G. (1981). Estimation of effective population size from data on linkage disequilibrium. Genetical Research, 38: 209–16.

Hill, W. G. (1982). Predictions of response to artificial selection from new mutations. Genetical Research, 40: 255–78.

Hill, W. G. (2012). Quantitative genetics in the genomics era. Current Genomics, 13: 196–206.

Hill, W. G. (2014). Applications of population genetics to animal breeding, from Wright, Fisher and Lush to genomic prediction. Genetics, 196: 1–16.

Hill, W. G. and L. Bunger (2004). Inferences on the genetics of quantitative traits from long-term selection in laboratory and farm animals, 2. Plant Breeding Reviews, 24: 169–210.

Hill, W. G. and A. Caballero (1992). Artificial selection experiments. Annual Review of Ecology and Systematics, 23: 287–310.

Hill, W. G., A. Caballero and L. Dempfle (1996). Prediction of response to selection within families. Genetics Selection Evolution, 28: 379–83.

Hill, W. G., M. E. Goddard and P. M. Visscher (2008). Data and theory point to mainly additive genetic variance for complex traits. PLoS Genetics, 4: e1000008.

Hill, W. G. and H. A. Mulder (2011). Genetic analysis of environmental variation. Genetics Research, 92: 381–95.

Hill, W. G. and J. Rasbash (1986). Models of long term artificial selection in finite populations. Genetical Research, 48: 41–50.

Hill, W. G. and A. Robertson (1966). The effect of linkage on limits to artificial selection. Genetical Research, 8: 269–94.

Hill, W. G. and A. Robertson (1968). Linkage disequilibrium in finite populations. Theoretical and Applied Genetics, 38: 226–31.

Hill, W. G. and B. S. Weir (2011). Variation in actual relationship as a consequence of Mendelian sampling and linkage. Genetics Research, 93: 47–64.

Hospital, F. (2009). Challenges for effective marker-assisted selection in plants. Genetica, 136: 303–10.

Houle, D. (1992). Comparing evolvability and variability of quantitative traits. Genetics, 130: 195–204.

Houle, D., B. Morikawa and M. Lynch (1996). Comparing mutational variabilities. Genetics, 143: 1467–83.

Ibanez-Escriche, N. and H. Simianer (2016). Animal breeding in the genomics era. Animal Frontiers, 6: 4–5.

James, J. W. (1970). The founder effect and response to artificial selection. Genetical Research, 16: 241–50.

Jódar, B. and C. López-Fanjul (1977). Optimum proportions selected with unequal sex numbers. Theoretical and Applied Genetics, 50: 57–61.

Johannsen, W. (1903). *Über Erblichkeit in Populationen und in Reinen Linien*. Gustav Fischer, Jena, Germany.

Jost, L. (2008). G(ST) and its relatives do not measure differentiation. Molecular Ecology, 17: 4015–26.

Jost, L., F. Archer, S. Flanagan, O. Gaggiotti, S. Hoban and E. Latch (2017). Differentiation measures for conservation genetics. Evolutionary Applications, 11: 1139–48.

Kacser, H. and J. A. Burns (1981). The molecular basis of dominance. Genetics, 97: 639–66.

Kardos, M., G. Luikart and F. W. Allendorf (2015). Measuring individual inbreeding in the age of genomics: marker-based measures are better than pedigrees. Heredity, 115: 63–72.

Kearsey, M. J. and B. W. Barnes (1970). Variation for metrical characters in *Drosophila populations*. II. Natural selection. Heredity, 25: 11–21.

Keightley, P. D. (1994). The distribution of mutation effects on viability in *Drosophila melanogaster*. Genetics, 138: 1315–22.

Keightley, P. D. (2012). Rates and fitness consequences of new mutations in humans. Genetics, 190: 295–304.

Keightley, P. D. and W. G. Hill (1990). Variation maintained in quantitative traits with mutation-selection balance: pleiotropic side effects on fitness traits. Proceedings of the Royal Society of London, Series B, 242: 95–100.

Keightley, P. D., T. F. C. Mackay and A. Caballero (1993). Accounting for bias in the estimation of the rate of polygenic mutation. Proceedings of the Royal Society of London, Series B, 253: 291–6.

Kempthorne, O. (1957). *An Introduction to Genetic Statistics*. John Wiley, New York.

Kevles, D. J. (1998). *In the Name of Eugenics: Genetics and the Uses of Human Heredity*. Harvard University Press, Cambridge, MA.

Kimura, M. (1957). Some problems of stochastic processes in genetics. Annals of Mathematical Statistics, 28: 882–901.

Kimura, M. (1965). A stochastic model concerning the maintenance of genetic variability in quantitative characters. Proceedings of the National Academy of Sciences of the USA, 54: 731–6.

Kimura, M. (1983). *The Neutral Theory of Molecular Evolution*. Cambridge University Press, Cambridge.

Kimura, M. and J. F. Crow (1963a). The measurement of effective population number. Evolution, 17: 279–88.

Kimura, M. and J. F. Crow (1963b). On the maximum avoidance of inbreeding. Genetical Research, 4: 399–415.

Kimura, M. and J. F. Crow (1964). The number of alleles that can be maintained in a finite population. Genetics, 49: 725–38.

Kimura, M. and T. Ohta (1969). The average number of generations until extinction of an individual mutant gene in a finite population. Genetics, 63: 701–9.

King, J. L. (1966). The gene interaction component of genetic load. Genetics, 53: 403–13.

Kingsolver, J. G., H. E. Hoekstra, J. M. Hoekstra, D. Berrigan, S. N. Vignieri, C. E. Hill, A. Hoang, P. Gibert and P. Beerli (2001). The strength of phenotypic selection in natural populations. American Naturalist, 157: 245–61.

Kirkpatrick, M. (2009). Patterns of quantitative genetic variation in multiple dimensions. Genetica, 136: 271–84.

Kirkpatrick, M., T. Johnson and N. H. Barton (2002). General models of multilocus evolution. Genetics, 161: 1727–50.

Knott, S. A., J. M. Elsen and C. S. Haley (1996). Methods for multiple-marker mapping of quantitative trait loci in half-sib populations. Theoretical and Applied Genetics, 93: 71–80.

Kojima, K.-I. (1959). Role of epistasis and overdominance in stability of equilibria with selection. Proceedings of the National Academy of Sciences of the USA, 97: 984–9.

Kondrashov, A. S. and J. F. Crow (1988). King's formula for the mutation load with epistasis. Genetics, 120: 853–6.

Kondrashov, A. S. and J. F. Crow (1993). A molecular approach to estimating the human deleterious mutation rate. Human Mutation, 2: 229–34.

Kondrashov, F. A. and A. S Kondrashov (2010). Measurements of spontaneous rates of mutations in the recent past and the near future. Philosophical Transactions of the Royal Society of London, Series B, 365: 1169–76.

Krimbas, C. B. and S. Tsakas (1971). The genetics of *Dacus oleae* V. Changes of esterase polymorphism in a natural population following insecticide control: selection or drift? Evolution, 25: 454–60.

Kristensen, T. N., P. Sørensen, M. Kruh, K. S. Pedersen and V. Loeschcke (2005). Genome-wide analysis on inbreeding effects on gene expression in *Drosophila melanogaster*. Genetics, 171: 157–67.

Kruuk, L. E. B. (2003). Estimating genetic parameters in natural populations using the 'animal model'. Philosophical Transactions of the Royal Society of London, Series B, 359: 873–90.

Kruuk, L. E. B., T. H. Clutton-Brock, J. Slate, J. M. Pemberton, S. Brotherstone and F. E. Guinness (2003). Heritability of fitness in a wild mammal population. Proceedings of the National Academy of Sciences of the USA, 97: 698–703.

Lande, R. (1975). The maintenance of genetic variability by mutation in a polygenic character with linked loci. Genetical Research, 26: 221–35.

Lande, R. (1979). Quantitative genetic of multivariate evolution, applied to brain: body allometry. Evolution, 33: 402–16.

Lande, R. (1992). Neutral theory of quantitative genetic variance in an island model with local extinction and colonization. Evolution, 46: 381–9.

Lande, R. and S. J. Arnold (1983). The measurement of selection on correlated characters. Evolution, 37: 1212–26.

Lande, R. and R. Thompson (1990). Efficiency of marker-assisted selection in the improvement of quantitative traits. Genetics, 124: 743–56.

Lander, E. S. and D. Botstein (1989). Mapping Mendelian factors underlying quantitative traits using RFLP linkage maps. Genetics, 121: 185–99.

Latter, B. D. H. (1965). The response of artificial selection due to autosomal genes of large effect. I. Changes in gene frequency at an additive locus. Australian Journal of Biological Sciences, 18: 585–98.

Latter, B. D. H. and A. Robertson (1962). The effects of inbreeding and artificial selection on reproductive fitness. Genetical Research, 3: 110–38.

Lawson, H. A., J. M. Cheverud and J. B. Wolf (2013). Genomic imprinting and parent-of-origin effects on complex traits. Nature Reviews Genetics, 14: 609–17.

Leberg, P. L. and B. D. Firmin (2008). Role of inbreeding depression and purging in captive breeding and restoration programmes. Molecular Ecology, 17: 334–43.

LeCorre, V. and A. Kremer (2003). Genetic variability at neutral markers, quantitative trait loci and trait in a subdivided population under selection. Genetics, 164: 1205–19.

Leffler, E. M., K. Bullaughey, D. R. Matute, W. K. Meyer, L Ségurel, A. Venkat, P. Andolfatto and M. Przeworski (2012). Revisiting an old riddle: what determines genetic diversity levels within species? PLoS Biology, 10(9): e1001388.

Legarra, A., I. Aguilar and I. Misztal (2009). A relationship matrix including full pedigree and genomic information. Journal of Dairy Science, 92: 4656–63.

Leinonen, T., R. J. S. McCairns, R. B. O'Hara and J. Merilä (2013). Q_{ST}–F_{ST} comparisons: evolutionary and ecological insights from genomic heterogeneity. Nature Reviews Genetics, 14: 179–90.

Lerner, I. M. (1950). *Population Genetics and Animal Improvement*. Cambridge University Press, Cambridge.

Lerner, I. M. (1954). *Genetic Homeostasis*. Oliver and Boyd, Edinburgh.

Leroy, G. (2014). Inbreeding depression in livestock species: review and meta-analysis. Animal Genetics, 45: 618–28.

Lesecque, Y., P. D. Keightley and A. Eyre-Walker (2012). A resolution of the mutation load paradox in humans. Genetics, 191: 1321–30.

Lewontin, R. C. (1964). The interaction of selection and linkage. I. General considerations; heterotic models. Genetics, 49: 49–67.

Lewontin, R. C. (1974). *The Genetic Basis of Evolutionary Change*. Columbia University Press, New York.

Lewontin, R. C. and K.-I. Kojima (1960). The evolutionary dynamics of complex polymorphisms. Evolution, 14: 458–72.

Lewontin, R. C. and J. Krakauer (1973). Distribution of gene frequency as a test of the theory of the selective neutrality of polymorphisms. Genetics, 74: 175–95.

Li, W. and L. Sadler (1991). Low nucleotide diversity in man. Genetics, 129: 513–23.

Lloyd, D. G. and C. J. Webb (1992). The evolution of heterostyly. Pp. 151–78 in *Evolution and Function of Heterostyly*. Ed. S. C. H. Barrett. Springer, Berlin.

López, M. A. and C. López-Fanjul (1993a). Spontaneous mutation for a quantitative trait in *Drosophila melanogaster*. I. Response to artificial selection. Genetical Research, 61: 107–16.

López, M. A. and C. López-Fanjul (1993b). Spontaneous mutation for a quantitative trait in *Drosophila melanogaster*. II. Distribution of mutant effects on the trait and fitness. Genetical Research, 61: 117–26.

López-Cortegano, E. and A. Caballero (2019). Inferring the nature of missing heritability in human traits using data from the GWAS Catalog. Genetics, 12: 891–904.

López-Cortegano, E., A. Vilas, A. Caballero and A. García-Dorado (2016). Estimation of genetic purging under competitive conditions. Evolution, 70: 1856–70.

López-Fanjul, C., A. Fernández and M. A. Toro (2000). Epistasis and the conversion of non-additive to additive genetic variance at population bottlenecks. Theoretical Population Biology, 58: 49–59.

López-Fanjul, C., A. Fernández and M. A. Toro (2003). The effect of neutral non-additive gene action on the quantitative index of population divergence. Genetics, 164: 1627–33.

López-Fanjul, C., A. Fernández and M. A. Toro (2006). The effect of genetic drift on the variance/covariance components generated by multilocus additive × additive epistatic systems. Journal of Theoretical Biology, 239: 161–71.

Luikart, G., N. Ryman, D. A. Tallmon, M. K. Schwartz and F. W. Allendorf (2010). Estimation of census and effective population sizes: the increasing usefulness of DNA-based approaches. Conservation Genetics, 11: 355–73.

Lush, J. L. (1945). *Animal Breeding Plans*, 3rd edn. Iowa State College Press, Ames.

Lush, J. L. (1947). Family merit and individual merit as bases for selection. American Naturalist, 81: 241–61.

Lynch, M. (2016). Mutation and human exceptionalism: our future genetic load. Genetics, 202: 869–75.

Lynch, M. and J. Conery (2003). The origins of genome complexity. Science, 302: 1402–4.

Lynch, M., J. Conery and R. Bürger (1995). Mutation accumulation and the extinction of small populations. American Naturalist, 146: 489–518.

Lynch, M. and W. Gabriel (1990). Mutation load and survival of small populations. Evolution, 44: 1725–37.

Lynch, M. and W. G. Hill (1986). Phenotypic evolution by neutral mutation. Evolution, 40: 915–35.

Lynch, M. and R. Lande (1998). The critical effective size for a genetically secure population. Animal Conservation, 1: 70–2.

Lynch, M. and B. Walsh (1998). *Genetics and Analysis of Quantitative Traits*. Sinauer, Sunderland, MA.

MacArthur, J., E. Bowler, M. Cerezo, L. Gil, P. Hall, E. Hastings, H. Junkins, A. McMahon, A. Milano, J. Morales et al. (2017). The new NHGRI-EBI Catalog of published genome-wide association studies (GWAS Catalog). Nucleic Acids Research, 45(D1): D896–901.

Mackay, T. F. C. (2001). The genetic architecture of quantitative traits. Annual Review of Genetics, 35: 303–39.

Mackay, T. F. C. (2009). Mutations and quantitative genetic variation: lessons from *Drosophila*. Philosophical Transactions of the Royal Society of London, Series B, 365: 1229–39.

Mackay, T. F. C. (2014). Epistasis and quantitative traits: using model organisms to study gene–gene interactions. Nature Reviews Genetics, 15: 22–33.

Mackay, T. F. C., S. Richards, E. A. Stone, A. Barbadilla, J. F. Ayroles, D. H. Zhu, S. Casillas, Y. Han, M. M. Magwire, J. M. Cridland et al. (2012). The *Drosophila melanogaster* genetic reference panel. Nature, 482: 173–8.

Mackay, T. F. C., E. A. Stone and J. F. Ayroles (2009). The genetics of quantitative traits: challenges and prospects. Nature Reviews Genetics, 10: 565–77.

Maher, B. (2008). Personal genomes: the case of the missing heritability. Nature, 456: 18–21.

Malécot, G. (1948). *Les Mathématiques de l'Hérédité*. Masson et Cie, Paris.

Malecót, G. (1952). Les processus stochastiques et la méthode des fonctions génératrices ou caracteréstiques. Publications de l'Institut de Statistique de l'Université de Paris, 1(3): 1–16.

Manolio, T., F. Collins, N. Cox, D. Goldstein, L. Hindorff, D. Hunter, M. McCarthy, E. Ramos, L. Cardon, A. Chakravarti et al. (2009). Finding the missing heritability of complex diseases. Nature, 461: 747–53.

Martin, G., C. Elodie and J. Goudet (2008). Multivariate $Q_{ST}–F_{ST}$ comparisons: a neutrality test for the evolution of the G Matrix in structured populations. Genetics, 180: 2135–49.

Martin, G. and T. Lenormand (2006). The fitness effects of mutations across environments: a survey in the light of fitness landscape models. Evolution, 60: 2413–27.

Mather, K. and J. L. Jinks (1982). *Biometrical Genetics*. 3rd edn. Chapman and Hall, London.

Maynard Smith, J. (1978). *The Evolution of Sex*. Cambridge University Press, Cambridge.

Maynard Smith, J. and J. Haigh (1974). The hitch-hiking effect of a favourable gene. Genetical Research, 23: 23–35.

McPhee, C. P. and A. Robertson (1970). The effect of suppressing crossing-over on the response to selection in *Drosophila melanogaster*. Genetical Research, 16: 1–16.

McQuillan, R., A. L. Leutenegger, R. Abdel-Rahman, C. S. Franklin, M. Pericic, L. Barac-Lauc, N. Smolej-Narancic, B. Janicijevic, O. Polasek, A. Tenesa et al. (2008). Runs of homozygosity in European populations. American Journal of Human Genetics, 83: 359–72.

Meuwissen, T. H. E. and M. E. Goddard (1996). The use of marker haplotypes in animal breeding schemes. Genetics Selection Evolution, 28: 161–76.

Meuwissen, T. H. E., B. J. Hayes and M. E. Goddard (2001). Prediction of total genetic value using genome-wide dense marker maps. Genetics, 157: 1819–29.

Meyer, K. (1989). Restricted maximum-likelihood to estimate variance components for animal models with several random effects using a derivative-free algorithm. Genetics Selection Evolution, 21: 317–40.

Mills, L. S. and F. W. Allendorf (1996). The one-migrant-per-generation rule in conservation and management. Conservation Biology, 10: 1509–18.

Misztal, I. (2006). Challenges of application of marker assisted selection – a review. Animal Science Papers and Reports, 24(1): 5–10.

Mittell, E. A., S. Nakagawa and J. D. Hadfield (2015). Are molecular markers useful predictors of adaptive potential? Ecology Letters, 18: 772–8.

Morton, N. E., J. F. Crow and H. J. Muller (1956). An estimate of the mutational damage in man from data on consanguineous marriages. Proceedings of the National Academy of Sciences of the USA, 42: 855–63.

Mousseau, A. and D. A. Roff (1987). Natural selection and the heritability of fitness components. Heredity, 59: 181–97.

Mukai, T. (1964). The genetic structure of natural populations of *Drosophila melanogaster*. I. Spontaneous mutation rate of polygenes controlling viability. Genetics, 50: 1–19.

Mukai, T. (1969a). The genetic structure of natural populations of *Drosophila melanogaster*. VII. Synergistic interaction of spontaneous mutant polygenes controlling viability. Genetics, 61: 749–61.

Mukai, T. (1969b). The genetic structure of natural populations of *Drosophila melanogaster*. VIII. Natural selection on the degree of dominance of viability polygenes. Genetics, 63: 467–78.

Mukai, T. (1985). Experimental verification of the neutral theory. Pp. 125–45 in *Population Genetics and Molecular Evolution*. Ed. T. Ohta and K. Aoki. Japan Scientific Society Press/Springer, Tokyo.

Mukai, T. (1988). Genotype–environment interaction in relation to the maintenance of genetic variability in populations of Drosophila melanogaster. Pp. 21–31 in *Proceedings of the Second International Conference on Quantitative Genetics*, chapter 3. Ed. B. S. Weir, E. J. Eisen, M. M. Goodman and G. Namkoong. Sinauer, Sunderland, MA.

Mukai, T., S. I. Chigusa, L. E. Mettler and J. F. Crow (1972). Mutation rate and dominance of genes affecting viability in *Drosophila melanogaster*. Genetics, 72: 335–55.

Mukai, T. and T. Yamazaki (1968). The genetic structure of natural populations of *Drosophila melanogaster*. V. Coupling-repulsion effects of spontaneous mutant polygenes controlling viability. Genetics, 59: 513–35.

Mulder, H. A., P. Bijma and W. G. Hill (2008). Selection for uniformity in livestock by exploiting genetic heterogeneity of residual variance. Genetics Selection Evolution, 40: 37–59.

Muller, H. J. (1928). The measurement of gene mutation rate in Drosophila, its high variability, and its dependence upon temperature. Genetics, 13: 279–357.

Muller, H. J. (1932). Some genetic aspects of sex. American Naturalist, 66: 118–38.

Muller, H. J. (1950). Our load of mutations. American Journal of Human Genetics, 2: 111–76.

Muller, H. J. (1964). The relation of recombination to mutational advance. Mutation Research, 1: 2–9.

Nagylaki, T. (1978). The correlation between relatives with assortative mating. Annals of Human Genetics, London, 42: 131–7.

Nei, M. (1968). The frequency distribution of lethal chromosomes in finite populations. Proceedings of the National Academy of Sciences of the USA, 60: 517–24.

Nei, M. (1973). Analysis of gene diversity in subdivided populations. Proceedings of the National Academy of Sciences of the USA, 70: 3321–3.

Nei, M. (1987). *Molecular Evolutionary Genetics*. Columbia University Press, New York.

Nei, M. and M. Murata (1966). Effective population size when fertility is inherited. Genetical Research, 8: 257–60.

Neimann-Sorensen, A. and A. Robertson (1961). The association between blood groups and several production characteristics in three Danish cattle breeds. Acta Agriculturale Scandinavica, 11: 163–96.

Nicholas, F. W. and A. Robertson (1980). The conflict between natural and artificial selection in finite populations. Theoretical and Applied Genetics, 56: 57–64.

Nilsson-Ehle, H. (1909). Kreuzung Untersuchungen an Hafer und Weizen. Lunds University, Arsskrift, n. s. series 2, 5(2): 1–122.

Nolte, I. M., P. J. van der Most, B. Z. Alizadeh, P. I. de Bakker, H. M. Boezen, M. Bruinenberg, L. Franke, P. van der Harst, G. Navis, D. Postma et al. (2017). Missing heritability: is the gap closing? An analysis of 32 complex traits in the Lifelines Cohort Study. European Journal of Human Genetics, 25: 877–85.

Nomura, T. (2000). Effective population size under marker assisted selection. Japanese Journal of Biometrics, 21: 1–12.

Nomura, T. (2002). Effective size of populations with unequal sex ratio and variation in mating success. Journal of Animal Breeding and Genetics, 118: 297–310.

Nomura, T. (2008). Estimation of effective number of breeders from molecular coancestry of single cohort sample. Evolutionary Applications, 1: 462–74.

Nosil, P. and J. L. Feder (2012). Genomic divergence during speciation: causes and consequences. Philosophical Transactions of the Royal Society of London, Series B, 367: 332–42.

Nussey, D. H., A. J. Wilson and J. E. Brommer (2007). The evolutionary ecology of individual phenotypic plasticity in wild populations. Journal of Evolutionary Biology, 20: 831–44.

O'Grady, J. J., B. W. Brook, D. H. Reed, J. D. Ballou, D. W. Tonkyn and R. Frankham (2006). Realistic levels of inbreeding depression strongly affect extinction risk in wild populations. Biological Conservation, 133: 42–51.

Palstra, F. P. and D. E. Ruzzante (2008). Genetic estimates of contemporary effective population size: what can they tell us about the importance of genetic stochasticity for wild population persistence? Molecular Ecology, 17: 3428–47.

Patterson, H. D. and R. Thompson (1971). Recovery of interblock information when block sizes are unequal. Biometrika, 58: 545–54.

Pérez-Enciso, M. (2014). Genomic interrelationships computed from either next-generation sequence or array SNP data. Journal of Animal Breeding and Genetics, 131: 85–96.

Pérez-Figueroa, A., A. Caballero, A. García-Dorado and C. López-Fanjul (2009b). The action of purifying selection, mutation and drift on fitness epistatic systems. Genetics, 183: 299–313.

Pérez-Figueroa, A., M. J. García-Pereira, M. Saura, E. Rolán-Alvarez and A. Caballero (2010). Comparing three different methods to detect selective loci using dominant markers. Journal of Evolutionary Biology, 23: 2267–76.

Pérez-Figueroa, A., M. Saura, J. Fernández, M. A. Toro and A. Caballero (2009a). METAPOP – a software for the management and analysis of subdivided populations in conservation programs. Conservation Genetics, 10: 1097–9.

Phillips, P. C. (2008). Epistasis – the essential role of gene interactions in the structure and evolution of genetic systems. Nature Review Genetics, 9: 855–67.

Pigliucci, M. (2005). Evolution of phenotypic plasticity: where are we going now? Trends in Ecology and Evolution, 20: 481–6.

Polderman, T. J. C., B. Benyamin, C. A. de Leeuw, P. F. Sullivan, P. M. Visscher and D. Posthuma (2015). Meta-analysis of the heritability of human traits based on fifty years of twin studies. Nature Genetics, 47: 702–9.

Poon, A. and L. Chao (2005). The rate of compensatory mutation in the DNA bacteriophage φX174. Genetics, 170: 989–99.

Powell, J. E., P. M. Visscher and M. E. Goddard (2010). Reconciling the analysis of IBD and IBS in complex trait studies. Nature Reviews Genetics, 11: 800–805.

Price, G. R. (1970). Selection and covariance. Nature, 227: 520–1.

Pudovkin, A. I., D. V. Zaykin and D. Hedgecock (1996). On the potential for estimating the effective number of breeders from heterozygote-excess in progeny. Genetics, 144: 383–7.

Qanbari, S., D. Gianola, B. Hayes, F. Schenkel, S. Miller, S. Moore, G. Thaller and H. Simianer (2011). Application of site and haplotype-frequency based approaches for detecting selection signatures in cattle. BMC Genomics, 12: 318.

Queller, D. C. (2017). Fundamental theorems of evolution. American Naturalist, 189: 345–53.

Quesada, H., D. Posada, A. Caballero, P. Morán and E. Rolán-Alvarez (2007). Phylogenetic evidence for multiple sympatric ecological diversification in a marine snail. Evolution, 61: 1600–12.

Rands, C. M., S. Meader, C. P. Ponting and G. Lunter (2014). 8.2% of the human genome is constrained: variation in rates of turnover across functional element classes in the human lineage. PLoS Genetics, 10(7): e1004525.

Reed, D. H., C. W. Fox, L. S. Enders and T. N. Kristensen (2012). Inbreeding-stress interactions: evolutionary and conservation consequences. Annals of the New York Academy of Sciences, 1256: 33–48.

Reed, D. H. and R. Frankham (2001). How closely related are molecular and quantitative measures of genetic variation? A meta-analysis. Evolution, 55: 1095–103.

Reed, D. H. and R. Frankham (2003). Correlation between fitness and genetic diversity. Conservation Biology, 17: 230–7.

Reeve, E. C. R. and F. W. Robertson (1954). Studies in quantitative inheritance VI. Sternite chaeta number in *Drosophila*: a metameric quantitative character. Zeitschrift für Induktive Abstammungs- und Vererbungslehre, 86: 269–88.

Reich, D. (2018). *Who We Are and How We Got Here*. Oxford University Press, Oxford.

Ritland, K. (1996). A marker-based method for inferences about quantitative inheritance in natural populations. Evolution, 50: 1062–73.

Robertson, A. (1952). The effect of inbreeding on the variation due to recessive genes. Genetics, 37: 189–207.

Robertson, A. (1959). The sampling variance of the genetic correlation coefficient. Biometrics, 15: 469–85.

Robertson, A. (1960). A theory of limits in artificial selection. Proceedings of the Royal Society of London, Series B, 153: 234–49.

Robertson, A. (1961). Inbreeding in artificial selection programmes. Genetical Research, 2: 189–94.

Robertson, A. (1964). The effect of nonrandom mating within inbred lines on the rate of inbreeding. Genetical Research, 5: 164–7.

Robertson, A. (1965). The interpretation of genotypic ratios in domestic animal populations. Animal Production, 7: 319–24.

Robertson, A. (1966). A mathematical model of the culling process in dairy cattle. Animal Production, 8: 95–108.

Robertson, F. W. (1957). Studies in quantitative inheritance X. Genetic variation of ovary size in Drosophila. Journal of Genetics, 55: 410–27.

Rodríguez-Ramilo, S. T., P. Morán and A. Caballero (2006). Relaxation of selection with equalization of parental contributions in conservation programs: an experimental test with *Drosophila melanogaster*. Genetics, 172: 1043–54.

Rodríguez-Ramilo, S. T., M. A. Toro, A. Caballero and J. Fernández (2007). The accuracy of a heritability estimator using molecular information. Conservation Genetics, 8: 1189–98.

Roff, D. A. (1996). The evolution of genetic correlations: an analysis of patterns. Evolution, 50: 1392–403.

Roff, D. A. (2001). *Life History Evolution*. Sinauer, Sunderland, MA.

Rogers, A. R. and H. C. Harpending (1983). Population structure and quantitative characters. Genetics, 105: 985–1002.

Rolán-Alvarez, E., C. J. Austin and E. G. Boulding (2015). The contribution of *Littorina* to the field of evolutionary ecology. Oceanography and Marine Biology: an Annual Review, 53: 157–214.

Romiguier, J., P. Gayral, M. Ballenghien, A. Bernard, V. Cahais, A. Chenuil, Y. Chiari, R. Dernat, L. Duret, N. Faivre et al. (2014). Comparative population genomics in animals uncovers the determinants of genetic diversity. Nature, 515: 261–3.

Rosenberg, N. A., J. K. Pritchard, J. L. Weber, H. M. Cann, K. K. Kidd, L. A. Zhivotovsky and M. W. Feldman (2002). Genetic structure of human populations. Science, 298: 2381–5.

Rutter, M. T., F. H. Shaw and C. B. Fenster (2010). Spontaneous mutation parameters for *Arabidopsis thaliana* measured in the wild. Evolution, 64: 1825–35.

Sabeti, P. C., D. E. Reich, J. M. Higgins, H. Z. Levine, D. J. Richter, S. F. Schaffner, S. B. Gabriel, J. V. Platko, N. J. Patterson, G. J. McDonald et al. (2002). Detecting recent positive selection in the human genome from haplotype structure. Nature, 419: 832–7.

Sabeti, P. C., P. Varilly, B. Fry, J. Lohmueller, E. Hostetter, C. Cotsapas, X. Xie, E. H. Byrne, S. A. McCarroll, R. Gaudet et al. (2007). Genome-wide detection and characterization of positive selection in human populations. Nature, 449: 913–18.

Sánchez, L., P. Bijma and J. A. Woolliams (2003). Minimizing inbreeding by managing genetic contributions across generations. Genetics, 164: 1589–95.

Sanjuán, R. and S. F. Elena (2006). Epistasis correlates to genomic complexity. Proceedings of the National Academy of Sciences of the USA, 103: 14402–5.

Santiago, E. and A. Caballero (1995). Effective size of populations under selection. Genetics, 139: 1013–30.

Santiago, E. and A. Caballero (1998). Effective size and polymorphism of linked neutral loci in populations under directional selection. Genetics, 149: 2105–17.

Santiago, E. and A. Caballero (2001). Application of reproductive technologies to the conservation of genetic resources. Conservation Biology, 14: 1831–6.

Santiago, E. and A. Caballero (2016). Joint prediction of the effective population size and the rate of fixation of deleterious mutations. Genetics, 204: 1267–79.

Santiago, E., I. Novo, A. F. Pardiñas, M. Saura, J. Wang and A. Caballero (2019). Recent demographic history inferred by high-resolution analysis of linkage disequilibrium. Unpublished manuscript.

Santos, J., M. Pascual, P. Simões, I. Fragata, M. Lima, B. Kellen, M. Santos, A. Marques, M. R. Rose and M. Matos (2012). From nature to the laboratory: the impact of founder effects on adaptation. Journal of Evolutionary Biology, 25: 2607–22.

Santure, A. W. and J. Wang (2009). The joint effects of selection and dominance on the Q_{ST} $–F_{ST}$ contrast. Genetics, 181: 259–76.

Sargolzaei, M., H. Iwaisaki and J.-J. Colleau (2005). A fast algorithm for computing inbreeding coefficients in large populations. Journal of Animal Breeding and Genetics, 122: 325–31.

Saura, M., A. Caballero, P. Caballero and P. Morán (2008). Impact of precocious male parr on the effective size of a wild population of Atlantic salmon. Freshwater Biology, 53: 2375–84.

Schluter, D. (2001). Ecology and the origin of species. Trends in Ecology and Evolution, 16: 372–80.

Sgrò, C. M. and A. A. Hoffmann (2004). Genetic correlations, tradeoffs and environmental variation. Heredity, 93: 241–8.

Shaw, F. H., C. J. Geyer and R. G. Shaw (2002). A comprehensive model of mutations affecting fitness and inferences for *Arabidopsis thaliana*. Evolution, 56: 453–63.

Shaw, R. G. and J. R. Etterson (2012). Rapid climate change and the rate of adaptation: insight from experimental quantitative genetics. New Phytologist, 195: 752–65.

Shrimpton, A. E. and A. Robertson (1988). The isolation of polygenic factors controlling bristle score in *Drosophila melanogaster*. II. Distribution of third chromosome bristle effects within chromosome sections. Genetics, 118: 445–59.

Simmons, M. J. and J. F. Crow (1977). Mutations affecting fitness in *Drosophila* populations. Annual Review of Genetics, 11: 49–78.

Slate, J. and J. M. Pemberton (2002). Comparing molecular measures for detecting inbreeding depression. Journal of Evolutionary Biology, 15: 20–31.

Smith, S. P. and A. Maki-Tanila (1990). Genotypic covariance matrices and their inverses for models allowing dominance and inbreeding. Genetics Selection Evolution, 23: 65–91.

Sniegowski, P. D. and P. J. Gerrish (2010). Beneficial mutations and the dynamics of adaptation in asexual populations. Philosophical Transactions of the Royal Society of London, Series B, 365: 1255–63.

Sorensen, D. and D. Gianola (2002). *Likelihood, Bayesian and MCMC Methods in Quantitative Genetics*. Springer, New York.

Sorensen, D. A. and B. W. Kennedy (1984). Estimation of genetic variances from unselected and selected populations. Journal of Animal Science, 58: 1097–106.

Soulé, M. E. (1980). Thresholds for survival: maintaining fitness and evolutionary potential. Pp. 151–69 in *Conservation Biology: An Evolutionary-Ecological Perspective*. Ed. M. E. Soulé and B. A. Wilcox. Sinauer, Sunderland, MA.

Spitze, K. (1993). Population structure in *Daphnia obtusa*: quantitative genetic and allozymic variation. Genetics, 135: 367–74.

Stern, C. (1973). *Principles of Human Genetics*. Freeman, San Francisco, CA.

Sved, J. A. (1971). Linkage disequilibrium and homozygosity of chromosome segments in finite populations. Theoretical Population Biology, 2: 125–41.

Taft, H. R. and D. A. Roff (2012). Do bottlenecks increase additive genetic variance? Conservation Genetics, 13: 333–42.

Tajima, N. (1983). Evolutionary relationship of DNA sequences in finite populations. Genetics, 105: 437–60.

Tajima, N. (1989). Statistical method for testing the neutral mutation hypothesis by DNA polymorphism. Genetics, 123: 585–95.

Takahata, N. (1983). Gene identity and genetic differentiation of populations in the finite island model. Genetics, 104: 497–512.

Tanksley, S. D. (1993). Mapping polygenes. Annual Review of Genetics, 27: 205–33.

Thornton, K., A. Foran and A. Long (2013). Properties and modeling of GWAS when complex disease risk is due to non-complementing, deleterious mutations in genes of large effect. PLoS Genetics, 9: e1003258.

Toro, M. A., C. Barragán, C. Óvilo, J. Rodrigañez, C. Rodríguez and L. Silió (2002). Estimation of coancestry in Iberian pigs using molecular markers. Conservation Genetics, 3: 309–20.

Toro, M. A., L. A. García-Cortés and A. Legarra (2011). A note on the rationale for estimating genealogical coancestry from molecular markers. Genetics Selection Evolution, 43: 27.

Toro, M. A. and M. Pérez-Enciso (1990). Optimization of selection response under restricted inbreeding. Genetics Selection Evolution, 22: 93–107.

Traill, L. W., C. J. A. Bradshaw and B. W. Brook (2007). Minimum viable population size: a meta-analysis of 30 years of published estimates. Biological Conservation, 139: 159–66.

Trussell, G. C. (1996). Phenotypic plasticity in an intertidal snail: the role of a common crab predator. Evolution, 50: 448–54.

Turelli, M. (1984). Heritable genetic variation via mutation-selection balance: Lerch's zeta meets the abdominal bristle. Theoretical Population Biology, 25: 138–93.

Uemoto, Y., R. Pong-Wong, P. Navarro, V. Vitart, C. Hayward, J. F. Wilson, I. Rudan, H. Campbell, N. D. Hastie, A. F. Wright and C. S. Haley (2013). The power of regional heritability analysis for rare and common variant detection: simulations and application to eye biometrical traits. Frontiers in Genetics, 4: 232.

UK10K Consortium (2015). The UK10K project identifies rare variants in health and disease. Nature, 526: 82–90.

Van Buskirk, J. and Y. Willi (2006). The change in quantitative genetic variation with inbreeding. Evolution, 60: 2428–34.

Van Raden, P. M. (2008). Efficient methods to compute genomic predictions. Journal of Dairy Science, 91: 4414–23.

Vencovsky, R., L. J. Chaves and J. Crossa (2012). Variance population size for dioecious species. Crop Science, 52: 79–90.

Vermeulen, C. J. and R. Bijlsma (2004). Characterization of conditionally expressed mutants affecting age-specific survival in inbred lines of *Drosophila melanogaster*: lethal conditions and temperature-sensitive periods. Genetics, 167: 1241–8.

Verrier, E., J. J. Colleau and J. L. Foulley (1991). Methods for predicting response to selection in small populations under additive genetic models: a review. Livestock Production Science, 29: 93–114.

Via, S. (2012). Divergence hitchhiking and the spread of genomic isolation during ecological speciation-with-gene-flow. Philosophical Transactions of the Royal Society of London, Series B, 367: 451–60.

Vilas, A. (2014). Caracterización y Gestión de la Diversidad Genética en Poblaciones Estructuradas. Tesis doctoral, Universidade de Vigo.

Vilas, A., A. Pérez-Figueroa and A. Caballero (2012). A simulation study on the performance of differentiation-based methods to detect selected loci using linked neutral markers. Journal of Evolutionary Biology, 25: 1364–76.

Vilas, A., A. Pérez-Figueroa, H. Quesada and A. Caballero (2015). Allelic diversity for neutral markers retains a higher adaptive potential for quantitative traits than expected heterozygosity. Molecular Ecology, 24: 4419–32.

Vinkhuyzen, A. A., N. R. Wray, J. Yang, M. E. Goddard and P. M. Visscher (2013). Estimation and partition of heritability in human populations using whole-genome analysis methods. Annual Review of Genetics, 47: 75–95.

Visscher, P. M., M. Brown, M. McCarthy and J. Yang (2012). Five years of GWAS discovery. American Journal of Human Genetics, 90: 7–24.

Visscher, P. M. and M. E. Goddard (2015). A general unified framework to assess the sampling variance of heritability estimates using pedigree of marker-based relationships. Genetics, 199: 223–32.

Visscher, P. M., N. R. Wray, Q. Zhang, P. Sklar, M. I. McCarthy, M. A. Brown and J. Yang (2017). 10 years of GWAS discovery: biology, function, and translation. American Journal of Human Genetics, 101: 5–22.

Vitezica, Z. G., I. Aguilar, I. Misztal and A. Legarra (2011). Bias in genomic predictions for populations under selection. Genetical Research, 93: 357–66.

Voight, B. F., S. Kudaravalli, X. Wen and J. K. Pritchard (2006). A map of recent positive selection in the human genome. PLoS Biology, 4: e72.

Wade, M. J. and C. J. Goodnight (1998). The theories of Fisher and Wright in the context of metapopulations: when nature does many small experiments. Evolution, 52: 1537–53.

Wagner, G. P. and J. Zhang (2011). The pleiotropic structure of the genotype–phenotype map: the evolvability of complex organisms. Nature Reviews Genetics, 12: 204–13.

Wahlund, S. (1928). Zusammensetzung von populationen und korrelationserscheiningen vom standpunkt der vererbungslehre aus betrachtet. Hereditas, 11: 65–106.

Wallace, B. (1975). Hard and soft selection revisited. Evolution, 29: 465–73.

Walsh, B. and M. Lynch (2019). *Evolution and Selection of Quantitative Traits*. Oxford University Press, Oxford.

Wang, J. (1997). More efficient breeding systems for controlling inbreeding and effective size in animal populations. Heredity, 79: 591–9.

Wang, J. (2004). Monitoring and managing genetic variation in group breeding populations without individual pedigrees. Conservation Genetics, 5: 813–25.

Wang, J. (2009). A new method for estimating effective population sizes from a single sample of multilocus genotypes. Molecular Ecology, 18: 2148–64.

Wang, J. and A. Caballero (1999). Developments in predicting the effective size of subdivided populations. Heredity, 82: 212–26.

Wang, J. and W. G. Hill (2000). Marker assisted selection to increase effective population size

by reducing Mendelian segregation variance. Genetics, 154: 475–89.

Wang, J., E. Santiago and A. Caballero (2016). Prediction and estimation of effective population size. Heredity, 117: 193–206.

Waples, R. S. (2005). Genetic estimates of contemporary effective population size: to what time periods do the estimates apply? Molecular Ecology, 14: 3335–52.

Watterson, G. A. (1975). On the number of segregation sites. Genetics, 38: 405–17.

Weber, K. E. (2004). Population size and long-term selection. Plant Breeding Reviews, 24(I): 249–68.

Weber, K. E. and L. T. Diggins (1990). Increased selection response in larger populations. II. Selection for ethanol vapor resistance in *Drosophila melanogaster* at two population sizes. Genetics, 125: 585–97.

Wei, M., A. Caballero and W. G. Hill (1996). Selection response in finite populations. Genetics, 144: 1961–74.

Weir, B. S. and C. C. Cockerham (1984). Estimating F-statistics for the analysis of population structure. Evolution, 38: 1358–70.

Weller, J. I. (2009). *Quantitative Trait Loci Analysis in Animals*, 2nd edn. CABI, Oxford.

Welter, D., J. MacArthur, J. Morales, T. Burdett, P. Hall, H. Junkins, A. Klemm, P. Flicek, T. Manolio, L. Hindorff and H. Parkinson (2014). The NHGRI GWAS catalog, a curated resource of SNP-trait associations. Nucleic Acids Research, 42 (Database issue): D1001–6. www.ebi.ac.uk/gwas/.

Whitlock, M. C. (1999). Neutral additive genetic variance in a metapopulation. Genetical Research, 74: 215–21.

Whitlock, M. C. (2008). Evolutionary inference from Q_{ST}. Molecular Ecology, 17: 1885–96.

Whitlock, M. C. (2011). G'_{ST} and D do not replace F_{ST}. Molecular Ecology, 20: 1083–91.

Whitlock, M. C. and N. H. Barton (1997). The effective size of a subdivided population. Genetics, 146: 427–41.

Whitlock, M. C. and D. Bourguet (2000). Factors affecting the genetic load in Drosophila: synergistic epistasis and correlations among fitness components. Evolution, 54: 1654–60.

Whitlock, M. C. and R. Bürger (2004). Fixation of new mutations in small populations. Pp. 155–70 in *Evolutionary Conservation Biology*. Ed. R. Ferrière, U. Dieckmann and D. Couvet. Cambridge University Press, Cambridge.

Williams, S. E. and E. A. Hoffman (2009). Minimizing genetic adaptation in captive breeding programs: a review. Biological Conservation, 142: 2388–400.

Wilson, S. P., H. D. Goodale, W. H. Kyle and E. F. Godfrey (1971). Long term selection for body weight in mice. Journal of Heredity, 62: 228–34.

Witzenberger, K. A. and A. Hochkirch (2011). *Ex situ* conservation genetics: a review of molecular studies on the genetic consequences of captive breeding programmes for endangered animal species. Biodiversity and Conservation, 20: 1843–61.

Wolf, J. B., E. D. Brodie III, A. J. Cheverud, A. J. Moore and M. J. Wade (1998). Evolutionary consequences of indirect genetic effects. Trends in Ecology and Evolution, 13: 64–9.

Wood, A., T. Esko, J. Yang, S. Vedantam, T. Pers, S. Gustafsson, A. Chun, K. Estrada, J. Luan, Z. Kutalik et al. (2014). Defining the role of common variation in the genomic and biological architecture of adult human height. Nature Genetics, 46: 1173–86.

Woolliams, J. A. and R. Thompson (1994). A theory of genetic contributions. Proceedings of the 5th World Congress on Genetics Applied to Livestock Production, 19: 127–34.

Wray, N. R., M. E. Goddard and P. M. Visscher (2007). Prediction of individual genetic risk to disease from genome-wide association studies. Genome Research, 17: 1520–8.

Wray, N. R., J. A. Woolliams and R. Thompson (1994). Predicting of rates of inbreeding in populations undergoing index selection. Theoretical and Applied Genetics, 87: 878–92.

Wright, S. (1921). Systems of mating. Genetics, 6: 111–78.

Wright, S. (1922). Coefficients of inbreeding and relationship. American Naturalist, 56: 330–8.

Wright, S. (1931). Evolution in Mendelian populations. Genetics, 16: 97–159.

Wright, S. (1933). Inbreeding and homozygosis. Proceedings of the National Academy of Sciences of the USA, 19: 411–20.

Wright, S. (1934). Physiological and evolutionary theories of dominance. American Naturalist, 67: 24–53.

Wright, S. (1938). Size of population and breeding structure in relation to evolution. Science, 87: 430–1.

Wright, S. (1943). Isolation by distance. Genetics, 28: 114–38.

Wright, S. (1951). The genetical structure of populations. Annals of Eugenics, 15: 323–54.

Wright, S. (1952). The genetics of quantitative variability. Pp. 5–41 in *Quantitative Inheritance*. Agricultural Research Council, London.

Wright, S. (1969). *Evolution and the Genetics of Populations. Vol. 2. The Theory of Gene Frequencies*. University of Chicago Press, Chicago, IL.

Wright, S. (1977). *Evolution and the Genetics of Populations. Vol. 3. Experimental Results and Evolutionary Deductions*. University of Chicago Press, Chicago, IL.

Wu, M. C., S. Lee, T. Cai, Y. Li, M. Boehnke and X. Lin (2011). Rare-variant association testing for sequencing data with the sequence kernel association test. American Journal of Human Genetics, 89: 82–93.

Yang, J., A. Bakshi, Z. Zhu, G. Hemani, A. E. Vinkhuyzen, S. H. Lee, M. R. Robinson, J. R. B. Perry, I. M. Nolte, J. V van Vliet-Ostaptchouk et al. (2015). Genetic variance estimation with imputed variants finds negligible missing heritability for human height and body mass index. Nature Genetics, 47: 1114–20.

Yang, J., B. Benyamin, B. McEvoy, S. Gordon, A. Henders, D. Nyholt, P. Madden, A. Heath, N. Martin, G. W. Montgomery et al. (2010). Common SNPs explain a large proportion of the heritability for human height. Nature Genetics, 42: 565–9.

Yonezawa, K. (1997). Effective population size of plant species propagating with mixed sexual and asexual reproduction system. Genetical Research, 70: 251–8.

Yoo, B. H. (1980). Long-term selection for a quantitative trait in large replicate populations of *Drosophila melanogaster*. I. Response to selection. Genetical Research, 35: 1–17.

Yule, G. U. (1902). Mendel's laws and their probable relation to intra-racial heredity. New Phytologist, 1: 193–207, 222–38.

Zaitlen, N., P. Kraft, N. Patterson, B. Pasaniuc, G. Bhatia, S. Pollack and A. L. Price (2013). Using extended genealogy to estimate components of heritability for 23 quantitative and dichotomous traits. PLoS Genetics, 9: e1003520.

Zeng, Z.-B., D. Houle and C. C. Cockerham (1990). How informative is Wright's estimator of the number of genes affecting a quantitative character. Genetics, 126: 235–47.

Zhang, X.-S. and W. G. Hill (2005). Predictions of patterns of response to artificial selection in lines derived from natural populations. Genetics, 169: 411–25.

Zuk, O., E. Hechter, S. R. Sunyaev and E. S. Lander (2012). The mistery of missing heritability: genetic interactions create phantom heritability. Proceedings of the National Academy of Sciences of the USA, 109: 1193–8.

Index